Primates in Nutritional Research

ACADEMIC PRESS RAPID MANUSCRIPT REPRODUCTION

Proceedings from a workshop on primate nutrition held
October 23–24, 1978, in Beaverton, Oregon.

Primates in Nutritional Research

edited by

K. C. Hayes

Department of Nutrition
Harvard School of Public Health
Boston, Massachusetts
and
New England Regional Primate
Research Center
Southborough, Massachusetts

Academic Press 1979

A Subsidiary of Harcourt Brace Jovanovich, Publishers

New York London Toronto Sydney San Francisco

ACADEMIC PRESS, INC.
111 Fifth Avenue, New York, New York 10003

United Kingdom Edition published by
ACADEMIC PRESS, INC. (LONDON) LTD.
24/28 Oval Road, London NW1 7DX

Library of Congress Cataloging in Publication Data

Main entry under title:

Primates in nutritional research.

Papers presented at a workshop held Oct 23–24, 1978 in Beaverton, Oregon and sponsored by the Sudcommittee on Primate Research Centers of the Animal Resources Review Committee and by the Nutrition Study Section of the Division of Research Grants, National Institutes of Health.

1 Primates–Physiology–Congressess 2. Primates–Food–Congresses. 3 Primates–Diseases–Congresses. 4. Nutrition–Congresses. I. Hayes, Kenneth Cronise, Date. II. United States. National Institutes of Health. Animal Resources Review Committee. Subcommittee on Primate Research Centers. III. United States. National Institutes of Health. Nutrition Study Section.
QL737. P9P678 559: 8 '0413 79-20836
ISBN 0–12–334250–3

PRINTED IN THE UNITED STATES OF AMERICA

79 80 81 82 9 8 7 6 5 4 3 2 1

Contents

Contributors

Numbers in parentheses indicate the pages on which authors' contributions begin.

MANFRED ALEXANDER (139), *Department of Nutrition and Metabolic Diseases, Oregon Regional Primate Research Center, Beaverton, Oregon*

LYNNE M. AUSMAN (39, 225), *Department of Nutrition, Harvard School of Public Health, Boston, Massachusetts*

WILLIAM R. BEISEL (315), *United States Army Medical Research, Institute of Infectious Diseases, Fort Detrick, Frederick, Maryland*

JAMES G. FOX (249), *Department of Nutrition and Food Science, Massachusetts Institute of Technology, Cambridge, Massachusetts*

DANIEL L. GALLINA (39, 225), *Department of Nutrition, Harvard School of Public Health, Boston, Massachusetts*

LEONARD S. GOTTLIEB (249), *Department of Pathology, Boston University School of Medicine and Mallory Institute of Pathology, Boston City Hospital, Boston, Massachusetts*

CLAYTON L. HADICK, JR. (315), *United States Army Medical Research, Institute of Infectious Diseases, Fort Detrick, Frederick, Maryland*

BARBARA C. HANSEN (59, 291), *University of Michigan, Ann Arbor, Michigan*

K. C. HAYES (99, 181), *Department of Nutrition, Harvard School of Public Health, Boston, Massachusetts and New England Regional Primate Research Center, Southborough, Massachusetts*

CHARLES F. HOWARD, JR. (215), *Department of Nutrition and Metabolic Diseases, Oregon Regional Primate Research Center, Beaverton, Oregon*

RONALD D. HUNT (11), *New England Regional Primate Research Center, Southborough, Massachusetts*

KAI-LIN C. JEN (59), *University of Michigan, Ann Arbor, Michigan*

JOSEPH J. KNAPKA (121), *Division of Research Services, Veterinary Resources, National Institutes of Health, Bethesda, Maryland*

GREGORY LENHART (249), *Department of Nutrition and Food Science, Massachusetts Institute of Technology, Cambridge, Massachusetts*

CHARLES W. McPHERSON (1), *Animal Resources Program Branch, Division of Research Resources, National Institutes of Health, Bethesda, Maryland*

M. R. MALINOW (199), *Oregon Regional Primate Research Center, Beaverton, Oregon, and University of Oregon Health Sciences Center, Portland, Oregon*

MARTIN L. MORIN (121), *Division of Research Services, Veterinary Resources, National Institutes of Health, Bethesda, Maryland*

PAUL M. NEWBERNE (99), *Department of Nutrition and Food Science, Massachusetts Institute of Technology, Cambridge, Massachusetts*

ROBERT J. NICOLOSI (11), *New England Regional Primate Research Center, Southborough, Massachusetts*

TOSHIAKI OSUGA (139), *Department of Medicine, University of Tsukuba School of Medicine, Sakamura, Niihari-gun, Ibaraki-ken, Japan, and First Department of Internal Medicine, University of Tokyo, Tokyo 111, Japan*

OSCAR W. PORTMAN, (139), *Department of Nutrition and Metabolic Diseases, Oregon Regional Primate Research Center, Beaverton, Oregon, and Department of Biochemistry, University of Oregon Health Sciences Center, Portland, Oregon*

KATHLEEN MAHER RASMUSSEN (73), *Division of Nutritional Sciences, Cornell University, Ithaca, New York*

ARTHUR J. RIOPELLE (341), *Department of Psychology, Louisiana State University, Baton Rouge, Louisiana*

ADRIANNE E. ROGERS (249), *Department of Nutrition and Food Science, Massachusetts Institute of Technology, Cambridge, Massachusetts*

NAOMI TANAKA (139), *Department of Medicine, University of Tsukuba School of Medicine, Sakamura, Niihari-gun, Ibaraki-ken, Japan, and First Department of Internal Medicine, University of Tokyo, Tokyo 111, Japan*

ALBERT WALLSTROM (249), *Mallory Institute of Pathology, Boston City Hospital, Boston, Massachusetts*

ROBERT W. WANNEMACHER, JR. (315), *United States Army Medical Research, Institute of Infectious Diseases, Fort Detrick, Frederick, Maryland*

KATHERINE WHITNEY (249), *Department of Nutrition and Food Science, Massachusetts Institute of Technology, Cambridge, Massachusetts*

Preface

The impetus for this publication originated with the concern for basic nutritional husbandry of nonhuman primates voiced by the Subcommittee on Primate Research Centers of the Animal Resources Review Committee (AARC) through the auspices of the Animal Resources Branch of the Division of Research Resources (DRR) and with the concern regarding the lack of adequate information related to nutritional methodology applicable to these invaluable animal models voiced by the Nutrition Study Section of the Division of Research Grants (DRG), National Institutes of Health.

Most species of nonhuman primates are threatened or endangered, and it is increasingly apparent that their successful maintenance and breeding in captivity will depend on adequate nutrition. Unfortunately, there is still much to learn concerning both general and specific nutrient requirements for most of the nonhuman primate species involved. Thus, a two-day workshop on the subject of primate nutrition was held on October 23–24, 1978. This workshop was sponsored by the above-mentioned review groups and was hosted by the Oregon Regional Primate Research Center, Beaverton, Oregon. The format of the worshop is best summarized by the table of contents of this book. Its objective was to present information on the nutrient requirements and various techniques of feeding monkeys (day 1) and to provide in-depth reviews of currently investigated aspects of nutritional research in which nonhuman primates are the preferred model of human disease (day 2). This volume includes the information presented at the workshop. The first seven chapters deal with practical matters of nutritional husbandry, and the following nine chapters describe various models under investigation today. In addition, authors have identified topics of particular need or interest that require further attention and research.

Individuals who contributed extensively to the planning of the workshop include Dr. Robert Prichard and Dr. Robert Wilson of the Subcommittee on Primate Research Centers, ARRC; Dr. Paul Newberne, Dr. Peter Dallman, and Dr. K. C. Hayes of the Nutritional Study Section, DRG; and Dr. Noel Lehner of Wake Forest University. The organizational encouragement and support of Dr. Leo Whitehair and Dr. Dennis Johnsen of the Animal Resources Branch and Dr. John R. Schubert, Executive Secretary of the Nutrition Study Section, are acknowledged and appreciated. Our most sincere thanks are extended to the host institution

(Oregon Regional Primate Research Center), its Director, Dr. William Montagna, and Center staff who rendered the workshop both effective and enjoyable. Our thanks and appreciation are given to Dr. Robert Prichard for providing special support in the programming and for chairing one of the sessions, and to Rita Kenneally and Ann Blanchard for their editiorial assistance in preparing the manuscripts that were dutifully and promptly submitted by the contributing authors.

K. C. Hayes

INDICATION/CONSIDERATION FOR NONHUMAN PRIMATES AS RESEARCH MODELS

Charles W. McPherson

Animal Resources Program Branch
Division of Research Resources
National Institutes of Health
Bethesda, Maryland

I. SELECTION OF ANIMAL MODEL

In selecting an animal model for use in any area of research, the primary consideration is whether the animal possesses those characteristics which will allow the investigator to obtain information which is important to the particular research question being studied. The second section of this Workship addresses a number of nutritional research areas where nonhuman primates possess characteristics which make them particularly valuable.

The selection of an animal model must also involve certain other (practical) considerations. Namely, can the animal be kept in an environment suitable for its research use; is the cost of its acquisition and care within affordable limits; and can it be obtained in sufficient numbers. All three of these considerations represent problems to a certain degree when it comes to using nonhuman primates in research. Let us look at them one at a time.

ISBN 0-12-334250-3

••••Can the animal be kept in an environment suitable for its research use? Keeping primates in the laboratory is not nearly as difficult as it was 20 years ago. This problem was then one of the forces that resulted in the establishment of the Primate Research Centers. The Centers in turn, together with a number of other laboratories, developed effective procedures for the care and management of nonhuman primates in the laboratory. Nonetheless, primates are still more difficult to care for and manage than any other common laboratory species. We still need research on primate care, management, and diseases. One aspect of this area will be dealt with in more detail in the first section of this workshop; namely, the nutritional requirements of nonhuman primates.

••••Is the cost of the animal's acquisition and care within affordable limits? Investigators using nonhuman primates know that the cost of these animals has markedly increased in the past few years. Nonetheless, in most projects the cost of the animal is still a small fraction of the total project cost. If the nonhuman primate is truly a superior model for a research project, then use of a cheaper but inferior model is false economy.

••••Can the animal be obtained in sufficient numbers? The availability of nonhuman primates has been a focus of major concern for the past several years. The population of many species has been declining. This decline is due largely to habitat destruction because of increased agriculture, logging and exploration for minerals in countries with indigenous primate populations. This population decline, coupled with a nationalistic concern for conservation of natural resources, has led a number of countries to ban or drastically restrict the export of primates. A further indication of this problem is that the Convention on International Trade in Endangered Species of Wild Fauna and Flora lists all nonhuman primates in either its Appendix I (endangered) or Appendix II (threatened with endangerment). The 1977 use of nonhuman primates in the United States is summarized in Table I, the present availability of wild caught primates is summarized in Table II.

II. SUPPLY AND USE OF NONHUMAN PRIMATES

In response to problems of nonhuman primate supply, the Federal government has established an Interagency Primate Steering Committee. This Committee was formed to coordinate

TABLE I United States Estimated 1977 Use of Nonhuman Primates by National Health Need and Species

National Health Need	*Required By Law Or Regulation For Biologics Testing*	*Production Of Biologics*	*Other Testing*	*Research*	*Total*
Rhesus	5300	200	1160	7355	14015
Cynomolgus	900	-0-	100	5005	6005
Other Macaque	-0-	-0-	-0-	995	995
Saguinus Marmoset	-0-	-0-	800	1390	2190
Common Marmoset	-0-	-0-	-0-	170	170
Owl Monkey	-0-	-0-	250	1215	1465
Chimpanzee	-0-	-0-	40	140	180
Gibbon	-0-	-0-	-0-	100	100
Baboon	-0-	-0-	-0-	1282	1282
African Green Monkey	750	500	15	810	2075
Squirrel Monkey	-0-	300	1100	3045	4445
Other	-0-	-0-	-0-	990	990
Total	6590	1000	3465	22497	33912

[a]*Adapted from Draft National Primate Plan, Interagency Primate Sterring Committee, Department of Health, Education, and Welfare; National Institutes of Health, November 1977.*

TABLE II Present Availability of Wild-Caught Primates 1978

SPECIES	*LOCATION OF INDIGENOUS POPULATIONS*	*AVAILABILITY ASSESSMENT*
Macaca Mulatta (rhesus monkey)	*India and surrounding countries*	*Not available from India. Quantities available from other countries have not been established.*
M. fascicularis (cynomolgus)	*Southeast Asia, Indonesia, Philippines*	*Commercially available in sufficient numbers to satisfy current demand.*
Other macaques	*Asia*	*Supplies extremely limiited to nonexistent*
Saguinus spp. (tamarins)	*South America*	*Commercial trade prohibited by most South American countries; small numbers available through special international arrangements.*
Callithrix jacchus (common marmoset)	*Brazil*	*Export prohibited by Brazil*
Aotus trivirgatus (owl monkey)	*Central and South America*	*Commercially available subspecies in greatest demand not commercially available*

TABLE II (cont'd)

SPECIES	LOCATION OF INDIGENOUS POPULATIONS	AVAILABILITY ASSESSMENT
Pan troglodytes (common chimp)	West Africa	Commercial trade prohibited; endangered species
Hylobates spp. (gibbons)	Southeast Asia	Commercial trade prohibited; endangered species
Papio spp. (baboons)	Sub-Sahara Africa	Commercially available in sufficient numbers to satisfy current demand
Cercopithecus aethiops (African green monkey)	Africa	Commercially available in sufficient numbers to satisfy current demand
Saimiri sciureus (squirrel monkey)	Central and South America	Commercially available; subspecies in greatest demand not commercially available
Other	--	Generally supplies are limited; varies according to species

[a]Adapted from Draft National Primate Plan, Interagency Primate Steering Committee, Department of Health, Education, and Welfare; National Institutes of Health, November 1977.

the activities of the various Federal agencies to develop a unified approach to assure both short and long-term supplies of nonhuman primates for biomedical activities. This Committee has developed a National Primate Plan (1). This plan recommends the following actions to assure an adequate supply of primates.

••••action to assure the most effective use of available primates

••••development of international programs for the production and conservation of primates

••••expansion of domestic production of primates

Programs are underway to accomplish each of these recommendations. An activity to promote the effective use of available primates is the Primate Supply Information Clearinghouse. The Clearinghouse is an activity of the Primate Information Center of the Regional Primate Research Center at the University of Washington. It functions to match available primates with investigators who require animals, tissues, or specimens. Emphasis is placed on reuse of animals that may no longer be needed in the research for which they were originally obtained. A computer based system is used to match listings of available animals and services with requests for the same. New listings of animals available or wanted are published weekly and widely distributed. (2) There is no charge for Clearinghouse services, and offerors and requestors of animals negotiate their own terms for transfer of animals.

Another activity to promote effective use of nonhuman primates is the Steering Committee's recommendation that the following criteria be adhered to in using primates in biomedical activities. These criteria are:

••••That the research proposed can be done best with primates; i.e., that no other known system or other kind of animals could produce comparable results.

••••That the species of primate proposed is the most appropriate and that some other more plentiful species would not be adequate.

••••That the number of primates proposed is the minimum that will produce acceptable scientific results.

••••That the primates will not be sacrificed during or at the end of the study except in those cases requiring termination as part of the investigation.

••••That, if sacrifice is deemed necessary, positive action will be taken to share body material when feasible.

International programs for the production and conservation of primates are also underway. The Interagency Primate Steering Committee has worked with the Pan American Health Organization to establish primate programs in South America.

Squirrel monkeys *(Saimiri sciureus)* and mystax marmosets *(sagunius mystax)* are currently being produced by this project.

III. BREEDING PROGRAMS

A number of domestic breeding programs are being established. We currently estimate that by 1980 over 9,000 animals will be born in these breeding colonies. The anticipated production of these colonies is summarized in Table III. The production of most of the colonies is targeted to meet the needs of the specific sponsoring organization. The Primate Research Centers Program supported by the Division of Research Resources (DRR), has an overall goal of breeding a sufficient number of animals to be self-sustaining for their own research programs. The Centers currently produce about 1,300 of the 2,000 animals they need annually. In addition, the Animal Resources Program Branch of the DRR, supports several colonies from which investigators with extramural grants and contracts from NIH may obtain animals. The availability of these animals is periodically announced through a supplement to the NIH Guide for Grants and Contracts. The anticipated production from these colonies is summarized in Table IV.

The use of domestically produced nonhuman primates has several advantages over the use of wild-caught animals. These include the ability to provide animals of known age and parentage and the potential to produce animals with specific genetic characteristics. Another advantage is that generally healthier animals and ones free from certain diseases can be produced. For example, tuberculosis testing and control programs are in effect in all breeding colonies; and one commercial producer has had apparent success in maintaining a colony free from Herpesvirus simiae (B-virus).

A disadvantage of domestic production is the cost. Although when one considers the uncertain quality of wild-caught animals and the long quarantine period required for them, the cost of domestically produced animals is competitive.

IV. CONCLUSION

Selection of an animal model should be based primarily on how well its biological characteristics meet the requirements of a research project, but to be useful animals

TABLE III U.S. Domestic Production of Nonhuman Primates 1980 Projected Levels

Animal	Estimated Production
Rhesus	6,485
Cynomolgus	280
Other Macaques	470
Saguinus Marmosets	460
Common Marmosets	120
Squirrel Monkey	800
Owl Monkey	105
Chimpanzee	90
Gibbon	10
Baboon	60
African Green Monkey	200
Other	460
Total	9,540

TABLE IV Domestic Production of Nonhuman Primates Animal Resources Branch, DRR 1980 Targeted Production for Distribution to NIH Extramural Projects

Animal	Production Sites	Targeted 1980 Production
Rhesus	Yemassee, SC; Alice, Texas	1,200
Cynomolgus	Alice, Texas	150
Squirrel monkeys	Winston-Salem, NC and Sabana Seca, Puerto Rico	400

must also be feasible to acquire and maintain in the laboratory. Nonhuman primates are well suited for a number of nutritional research areas, but they are becoming increasingly difficult to obtain. Declining populations due to habitat destruction have caused many countries to reduce export of primates. Domestic breeding programs are, in part, going to be able to supply primates previously imported and some species are still in good supply from wild-caught sources. If primates are, in fact, the best animal for a research project, their use is justified; however, they should not be used if other animals or systems will effectively serve the project needs.

REFERENCES

1. Primate Supply Information Clearinghouse, Primeat Information Center, Regional Primate Research Center, University of Washington, Seattle, Washington 98195.

2. Adapted from Draft National Primate Plan, Interagency Primate Steering Committee, Department of Health, Education, and Welfare, National Institutes of Health, November 1977.

DIETARY ALLOWANCES FOR NUTRIENTS IN NONHUMAN PRIMATES

Robert J. Nicolosi
Ronald D. Hunt

New England Regional Primate Research Center
Southborough, Massachusetts

I. INTRODUCTION

The increased use of nonhuman primates in biomedical research emphasizes the importance of expanding existing knowledge of the factors that influence the response of these animal models to experimental treatment. While much of the available information on nonhuman primates derives from studies of the disease state, it is obvious that an understanding of the normal is a necessary corollary to meaningful results in biomedical research. Diet is one of the more important factors that can have a profound effect on an animal's response to experimental manipulation. A marginal nutrient deficiency introduced by purified diets or commercial rations may go undetected by the investigator and thereby compromise the results of an otherwise well-

ISBN 0-12-334250-3

designed and well-executed experiment. To reduce the occurrence of such a possibility and better insure the well-being of the monkey species to be studied, the investigator should have at his disposal all available information pertaining to the nutrient requirements of the particular experimental animal.

While far from adequate, a considerable amount of data has accumulated on the nutrient requirements of the Old World rhesus monkey largely due to its predominant use in biomedical research. But with increased utilization of New World monkeys as models of human disease, more investigative effort is needed to establish nutrient requirements of these nonhuman primates. While the assessment of nutrient requirements has great practical application, the task is far from easy, at least from a methodological standpoint. Such things as "nonphysiological adaptive responses" to alterations in nutrient intake, the effect that limiting one essential nutrient may have on the metabolism of another, period of life cycle, and what constitutes the actual index or criteria of nutrient well-being of a monkey all confound interpretation of data. The last point is a critical issue since the current use of equating nutrient adequacy with growth rate, especially in immature animals during short term experiments, is likely inadequate and an oversimplication, and more consideration should be given to establishing other indices such as life span, resistance to diseases, such as atherosclerosis, reproductive performance and social behavior. The major problem with utilizing these criteria is the necessary long term studies associated with establishing them; so that while idealistic in principle, some happy medium is the objective we must pursue if we are to continue to utilize nonhuman primates in the medical sciences.

Since, by definition, the "precise" nutrient requirements of nonhuman primates can never be known, this report will summarize the current data on nonhuman primates drawing from information made available in the most recent NRC publication on the subject (1) supplemented with additional comments where further information exists. Its objective will be to provide the investigator with a workable outline of nutrient requirements of nonhuman primates under "normal" captive conditions and thereby contribute to controlled experimental conditions. It will also consider areas where future research is needed to gather data on specific nutrients.

II. ENERGY ALLOWANCES

The major foodstuffs in the form of protein, carbohydrate and fat not only provide an individual or animal species with specific nutrients but also the necessary energy during periods of maintenance, growth, reproduction, lactation, etc. During fasting or malnutrition, stores of carbohydrates as glycogen, fat as adipose tissue triglyceride, and tissue protein are called upon to satisfy maintenance energy requirements. In describing factors that can influence the determination of nutrient requirements, age of the animal is an important consideration since the high energy or food consumption in immature animals that results in rapid growth would be expected to induce obesity in mature animals. This emphasizes the need for better criteria of nutritional adequacy than growth rate since this logic would suggest that conditions which lead to excessive energy intake and greater weight gain may be superior or healthier than the "lean" state, a situation not necessarily true especially when one utilizes humans as an example. Table 1 indicates that species size has an important influence on nutrient requirements which, when considered together with age, suggests that the smaller, younger species have greater energy requirements than their larger counterparts (2). Based on ad libitum intake, infants of smaller New World monkeys appear to require 300-500 kcal energy/kg/day as compared to the 200-300 kcal figure of larger Old World monkeys. Both Old and New World monkeys reduce their energy requirements by 30-50% as adults when growth ceases. To what extent periods of gestation and lactation increase these requirements remains to be determined.

III. PROTEIN ALLOWANCES

As in other animal species, nonhuman primates require protein for growth and anabolic processes. While excess dietary protein has no apparent physiological advantage, and may in fact compromise renal function, deficiency in quantity or quality of protein places the animal not only at risk of limited growth, but also introduces possible abnormalities associated with protein deficiency syndromes such as kwashiorkor.

The conventional methods of measurement of adequacy of dietary protein are based upon determinations of growth

TABLE I *Recommended and Minimum Daily Allowances for Energy and Protein in Primates*

Species and Age	Energy	Protein[a]	
	kcal/kg body wt	% calories	g/kg body wt
Human			
Infant	120 - 60	10 (6.0)[b]	2.4 (2.0)
Adult	30 - 40	8 (5.0)	0.8 (0.5)
Baboon			
Infant	<290	12	5.0
Adult	53 - 72	12	2.0
Chimpanzee			
Infant	120 - 100	12	5.0
Adult	50 - 60	12	2.0
Rhesus and Cynomolgus			
Infant	270 - 190	12 (6.6)	8.0 (4.0)
Adult	75 - 120	12 (6.6)	3.0
Cebus			
Infant	400 - 250	13 (7.0)	10.0 (5.25)
Adult	100 - 150	13 (7.0)	5.0 (2-3)

Species and Age	*Energy*	*Protein*[a]	
	kcal/kg body wt	*% calories*	*g/kg body wt*
Squirrel			
Infant	*600 - 300*	*16 (13.0)*	*18.0 (18.0)*
Adult	*100 - 300*	*13 (6.0)*	

[a] *High quality protein*
[b] *Values in parentheses are minimal requirements; not determined for the baboon and chimpanzee.*

rates, balance between dietary intake and urinary and fecal output of nitrogen, and plasma amino acid patterns. What is often forgotten in these measurements is that both the quantity of protein (nitrogen content) and quality of protein (specific ratios of constituent amino acids) contribute or influence the total protein requirement of a given species. For example, in discussing the quality of protein, it is important that the distinction be made between animal protein, which is usually of "high quality", and plant protein which can be low in one or more essential amino acids and must be either supplemented with the absent amino acid or with animal protein. The importance of maintaining an adequate quality of ingested protein is exemplified by the studies of Follis et al. (3) who demonstrated signs of protein deficiency in grivet monkeys fed solely on maize and by Kerr et al. (4) who produced protein deficiency in infant rhesus monkeys fed a diet limiting in phenylalanine. The importance of amino acid composition of protein diets is also demonstrated in the observations of Samonds and Hegsted (5) in which the amount of protein (g /kg body wt/day) needed to meet maintenence requirements in 2-6 mo. old cebus monkeys was 2.0 (lactalbumin), 8.7 (wheat gluten) and 2.8 (wheat gluten and lysine). In similar manner, more recent data indicate that infant squirrel monkeys require 3.43, 3.46, and 7.96 g protein/kg/day as lactalbumin, casein, and soy protein, respectively, for weight maintenance.

As with energy requirements, protein requirements of smaller species are greater on a per unit body weight basis than larger species, and although poorly defined, it is expected that this requirement increases during pregnancy and lactation. While few in number, studies of protein requirements in young adult rhesus monkeys revealed that intakes of 0.5, 2.0 g protein/kg body wt/day caused weight loss in all animals (6,7), the lower figure also resulting increased fetal stillbirths and perinatal death in pregnant rhesus.

Table I briefly outlines the available data on protein requirements of growing and adult monkeys. Generally speaking, the larger rhesus and chimpanzee monkeys require 3-5 gr/kg/day of high quality protein while the infant squirrel and cebus monkeys need between 5-10 g /kg/day of protein during active growth. Both adult Old and New World monkeys reduced their requirements to approximately 4 g /kg/day.

IV. FAT AND FATTY ACID ALLOWANCES

The consequences of diets which induce essential fatty acid (EFA) deficiency have only been investigated in rhesus (8,9) and cebus (10,11) monkeys and appear to be similar to their human counterpart, i.e., loss of hair, dry scaly skin and an increase in plasma and RBC 20:3 ω9 with a reduction in 18:2 ω6 and 20:4 ω6.

While the EFA requirements have not been established for most nonhuman primates, it does appear that like humans, 1-2% of the total caloric requirement supplied as linoleic acid will prevent the obvious signs of EFA deficiency. There is one report by Fiennes et al. (12) suggesting a specific requirement for linolenic (18:3 ω3) acid in cebus monkeys but the data are inconclusive and are not supported by the experience of other investigators who have failed to note this requirement (13). Several investigators have shown that most nonhuman primates readily adjust to 10% total dietary fat (8,9,11,13) and to diets much higher in fat with or without cholesterol in atherosclerosis studies (14-16), so long as the requirement for EFAs is met. This requirement for growth can apparently be satisfied for cebus, squirrel, cynomolgus, and rhesus with as little as 0.4% of calories supplied as linoleic acid in coconut oil (13,17-19).

V. FAT SOLUBLE VITAMIN ALLOWANCES

Vitamin A. Both cebus and rhesus monkeys (20,21) have been used to investigate the consequences of vitamin A deficiency. The early stages of the syndrome involve a decline in growth, appetite and resistance to infection and in its later, more severe stages, xeropthalmia, night blindness, corneal destruction and retinal degeneration. While the requirements of vitamin A are not well-established in nonhuman primates, a considerable amount of data has accumulated to permit estimates of such requirements (Table II). In both cebus (20) and rhesus monkeys (21), the vitamin A deficiency syndrome could be produced after a year of a deficient diet fed to young monkeys and was associated with reductions in plasma vitamin A from 26 μg/100 ml to 10 μg/100 ml in rhesus and from 15-20 μg/100 ml to <5 μg/100 ml in cebus monkeys. When pregnant rhesus monkeys were maintained on vitamin A intakes of 400 IU, twice weekly, the plasma levels dropped below 10 μg/100 ml and resulting infants

TABLE II *Recommended Daily Allowances for Fat Soluble Vitamins in Nonhuman Primates*

Vitamin	*per kg diet*
Vitamin A	*12,500 IU*
Vitamin D	*1,250 IU*
Vitamin E	*20-100 IU*[a]
Vitamin K	*1 mg*

[a] *Requirement increases directly with dietary polyunsaturated fat.*

developed varying degrees of xeropthalmia (21). While these data indicate that 400 IU given twice weekly does not prevent deficiency in pregnant female rhesus monkeys, subsequent studies in monkeys have demonstrated that 1800-3000 IU of vitamin A daily appears sufficient to sustain normal growth and reproductive efficiencies (9,22,23).

Vitamin D. It has been demonstrated in several species of nonhuman primates that symptoms of rickets, such as softening and demineralization of bone can be reversed by supplements of vitamin D although this apparent simple cause/effect relationship is more complex and involves factors which influence both dietary calcium and phosphorous absorption. Effective prevention of rickets in New World monkeys has been demonstrated with vitamin D_3 (cholecalciferol), but not with vitamin D_2 (ergocalciferol) (24,25) which can be utilized by Old World monkeys (26). This difference may contribute to the apparent, greater resistance of Old World monkeys to rickets.

As little as 100-125 IU/100 g diet of cholecalciferol (D_3) has been shown to prevent rickets in cebus, cynomolgus (13), and squirrel (25) monkeys while as much as 400 IU/day was tolerated in rhesus monkeys with no signs of apparent vitamin D toxicity (22,27). The recommended allowance is listed in Table 2. While the essentiality of vitamin D in normal bone metabolism and in the prevention of rickets has been demonstrated, high levels of D_3 (50,000-100,000 IU) have been associated with renal failure, hypercalcemia and bone rarefication in squirrel and cebus monkeys which Hunt et al. (28) related to the greater ability of D_3 to enhance intestinal absorption of Ca^{++} as compared to D_2

Vitamin E. A deficiency of vitamin E is often associated with anemia, muscular dystrophy, and reproductive failure, which appear to be reversible with tocopherol supplementation. Vitamin E deficiency has been studied in rhesus (29, 30), cynomolgus and cebus monkeys (13,31), all of which developed the above clinical signs when deprived of tocopherol for periods in excess of a year. Development of the deficiency seems to be closely linked to the dietary content of unsaturated fat, with daily requirements ranging from 0.7 to 2.9 mg/kg body wt, the higher value derived from monkeys fed 8% lard and 3% cod liver oil and the lower value from rhesus monkeys in which starch replaced most of the fat (27). Values of 0.36-0.72 mg of d-α-tocopheryl acetate per gram of dietary linoleic acid have been found adequate in rhesus monkeys (32). Ausman and Hayes (13) reported a minimal requirement of 0.16 mg/kg body wt/day for the New World cebus monkey and 0.10 mg/kg/day in a single cynomolgus monkey (see Table II).

Vitamin K. Vitamin K plays a major role in the clotting process by maintaining normal levels of prothrombin and at least 3 other clotting factors via its ability to allow carboxylation and calcium binding by these factors. While little is known of the vitamin K requirement for nonhuman primates, the increased prothrombin time in rhesus monkeys fed vitamin K-deficient diets and its return to normal upon supplementation with 0.1 μg of the vitamin/kg body wt/day indicates a small but possible requirement (33). It has been suggested that in infant monkeys, supplements of 100-500 μg of vitamin K daily may be necessary (34), but seems unlikely (Table II).

V. WATER SOLUBLE VITAMIN ALLOWANCES

Thiamin. The role of thiamin as a coenzyme in the pathways of the Krebs cycle and carbohydrate metabolism has been demonstrated, and while not established, it is conceivable that the clinical disorders associated with the deficiency syndrome are linked to abnormalities in these metabolic pathways. The ramifications of thiamin deficiency in monkeys similar to man in many respects, include weight loss, apathy, weakness, paralysis, loss of reflexes, convulsions and death. Heart failure which is often associated with necrosis of myocardial fibers and degeneration of both peripheral and certain nervous system function has also been reported in the rhesus, all of which can be prevented by administration of 15 30 µg thiamin/kg body wt/day (35). (Table III)

Riboflavin. The essentiality of riboflavin in the synthesis of flavoprotein coenzymes and subsequent enzyme reactions prompted several investigations of riboflavin deficiency. Studies by Day et al. (36, Waisman, 37, Cooperman et al. (38), and Greenberg and Moon (39) on the deficiency state in rhesus monkeys, indicated that poor growth dermatitis, impaired vision, incoordination, scanty hair coat, hypochromic normocytic anemia and death were characteristics of the deficiency syndrome. Administration of 50-100 µg of riboflavin per day prevented the syndrome. Mann et al. (40,41) described the development of similar clinical signs in riboflavin deficient cebus with the exception of lack of any demonstrable anemia. These studies in rhesus and cebus suggested a daily requirement of riboflavin of 25-30 µg/kg body wt (Table III).

Niacin. A deficiency of niacin (nicotinic acid) has been reported in rhesus monkeys to result in loss of weight, skin pigmentation, anemia and diarrhea (42-45). These conditions, brought on by feeding a purified diet deficient only in niacin, can be prevented by administering 10-35 mg/week or 1-2 gr tryptophan (a niacin precursor) weekly (43). Belavady et al. (44) successfully treated the reduction in albumin and blood nucleotides and the atrophic gastritis and enterocolitis associated with niacin deficiency by administering 10-25 mg niacin per day to rhesus monkeys. Similar reversibility of the deficiency syndrome with 40 mg niacin was also demonstrable in rhesus monkeys in which the niacin deficiency was induced by supplementing the diet with 1.5 g L-leucine/day (a niacin antagonist) (45). (Table III)

Pyridoxine. The primary importance of vitamin B_6 as a coenzyme in amino acid metabolism, was initially studied by McCall et al. (46) who reported that the deficiency syndrome in rhesus included weight loss, anemia, ataxia, apathy and microcytic anemia. Supplementation of 1 mg of pyridoxine daily for 2 weeks to 2 kg rhesus monkeys improved the clinical symptoms. On the other hand, Emerson (47) suggested a minimum requirement of 500 µg/kg/day for 3-4 kg rhesus. Later studies (48-57) added fatty liver, hepatic necrosis and cirrhosis, dental caries and cerebral cortex degeneration to the list of clinical findings, all of which could be prevented by supplementing the diet of infant rhesus monkeys with 50 µg/kg body wt/day of pyridoxine. A requirement of 50-100 µg of pyridoxine/kg body wt/day was suggested

TABLE III *Recommended Daily Allowances for Water Soluble Vitamins in Nonhuman Primates*

Vitamin	*mg per kg diet*
Ascorbic Acid	*500-600*
Niacin	*40-50*
Thiamin	*4-6*
Riboflavin	*8-10*
Pyridoxine	*4-6*
Folacin	*4-6*
Biotin	*0.2*

Requirements for pantothenic acid, choline, inositol, and cyanocobalamin have not been determined but are generally supplemented in the diet.

by Mann (58) as a result of studies of pyridoxine deficiency in cebus monkeys. Obviously, the exact requirements have not been established in monkeys but 50-500 μg/kg/day is the requirement felt necessary for infant rhesus monkeys (Table 3) (1,47). Of possible significance is the induction of B_6 deficiency associated with the administration of isoniazid for prophylaxis or treatment of tuberculosis (59). Routine use of this drug in nonhuman primate laboratories should be accompanied by supplementation of vitamin B_6 to the diet.

Pantothenic Acid. Investigations of this vitamin in nonhuman primates are too few to permit assessment of its requirement. While McCall (46) reported that a deficiency with symptoms of anemia, hair loss, ataxia and growth retardation could be partially prevented upon addition of 1-2 mg of pantothenic acid/kg body wt/day, Greenberg (14) reported more dramatic recovery with 3 mg/kg/day.

Biotin. Biotin forms the prosthetic group of several enzymes, particularly fatty acid synthetic enzymes, in intermediary metabolism. The studies of Waisman et al. (53) in adult rhesus monkeys are the only reports in nonhuman primates. Biotin deficiency was induced by feeding avidin as raw egg whites, resulting in scaly dermatitis, alopecia and loss of hair color. These conditions were corrected by the addition of 20 μg biotin/day to the diet. (Table III)

Choline and Inositol. Neither the role nor the dietary requirement of choline or inositol have been established in nonhuman primates. Both nutrients are assumed to be synthesized to some extent by all primates. Although choline deficiency has been reported to result in fatty livers and hepatic fibrosis in cebus (61), rhesus (62-64) and baboons (65), this diet was also low in protein and high in fat, preventing a direct cause/effect relationship due to choline deficiency from being established. Inositol studies in primates have not been reported. Nonetheless, it is considered prudent to supplement diets with 0.3% choline and 0.1% inositol, especially when high levels of fat are fed.

Folic Acid. A deficiency of folacin was produced in rhesus monkeys (fed a purified diet) in the 1930's (66,67). Signs included gingivitis, stomatitis, leukopenia, macrocytic anemia, weight loss and diarrhea, which were corrected by a factor present in liver or yeast identified at the time as vitamin M and later as folic acid. Subsequent studies by Cooperman et al. (68) and Day and Trotter (69) confirmed these early findings and suggested a daily requirement of 40 μg/kg body wt/day for rhesus monkeys. While Huser and Beard (70) suggested that rhesus monkeys are a good species for investigating folic acid deficiency, similar deficiency syndromes have also been reported in the patas monkey *(Erythrocebus patas)* (71) and the marmoset (72), the latter responding to a supplement of 100 μg/day. More recent data, (see page 73), indicated that cebus and squirrel monkeys may require in excess of 100 μg/day for adequate growth (Table 3) and that the cebus may be the most appropriate primate model for studies of folacin deficiency in man.

Cynocobalamin. Vitamin B_{12} plays a role in folic acid and nucleic acid metabolism and is involved in the transfer of methyl groups during the synthesis of methionine from homocysteine. While a deficiency syndrome has been reported in man which results in pernicious anemia, investigators, in general, have been unable to demonstrate a B_{12} deficiency anemia in monkeys despite reductions in serum vitamin B_{12} levels. There have been reports (73-80) of peripheral and central nervous system dysfunction in several species of monkeys ascribed to vitamin B_{12} deficiency resulting from a vegetarian diet, but these data are inconclusive. Rhesus monkeys fed a vitamin B_{12} deficient diet for 4 years, developed both muscular and nervous system disorders that involved demyelination and axon loss in the optic nerve and papillomacular bundle, degenerative features of the optic chiasm, and enucleation of several layers of the lateral geniculate body (81). Earlier investigations of these monkeys at 31 months revealed reductions in vitamin B_{12} of serum, RBC, and liver, but no overt signs of a deficiency syndrome (82). Using weight gain as a criterion, depleted rhesus monkeys appear to respond to the addition of 500 µg of vitamin B_{12}/week to the diet (83).

Ascorbic Acid. Vitamin C has an essential role in steroid hydroxylations and collagen metabolism and is an essential nutrient for most nonhuman primates with the exception of the prosimian tree shrew and slow loris which reportedly synthesize vitamin C (84). Deficiency of this vitamin is common in monkeys, producing scurvy with lesions that include gingival hemorrhage, normocytic normochromic anemia, loose teeth, epiphyseal fractures, subperiosteal hemorrhage and exophthalmus (85,86). While the dietary requirements for most nonhuman primates remain to be established, requirements of 0.5-1.0 mg/kg/day have been reported for rhesus monkeys (87,88) and 7.5-10 mg/kg/day for squirrel monkeys (89-91). (Table III).

VII. MINERAL ALLOWANCES

Despite the important and varied role of minerals in such functions as coagulation, acid-base balance, membrane transport and enzyme catalysts, there have been relatively few studies that have been concerned with the mineral requirements of nonhuman primates. While the requirements for sodium, potassium, chlorine, calcium, phosphorous, magnesium and sulphur are appreciable, trace elements thought to function in intermediary metabolism include zinc, copper, cobalt, iron, iodine, manganese, chromium, selenium and molybdenum (Table 4). Aluminum, tin, nickel, arsenic, vanadium and fluorine are also thought to be trace minerals, although their functions are less understood.

Sodium, Potassium and Chlorine. As the major extra-and intracellular cation and anion, one of the functions of these elements is to regulate cell membrane potentials and transport processes. Excessive losses as a result of vomiting or diarrhea can lead to growth failure and death. While their exact requirements have not been established, diets containing sodium at 0.22-0.64% potassium at 0.24-1.09% and

TABLE IV Recommended Daily Allowances for Minerals in Nonhuman Primates

	As percent of Diet
Sodium	*0.22 - 0.44*
Potassium	*0.24 - 1.09*
Chloride	*0.27 - 0.62*
Calcium	*0.60 - 0.80*
Phosphorus	*0.30 - 0.40*
Magnesium	*0.10 - 0.15*
	As ppm in Diet
Iron	*100*
Zinc	*20*
Iodine	*2*
Chromium	*0.5*

Requirements for manganese, copper, selenium and fluoride not determined for nonhuman primates, but they can be assumed to be required at approximate levels of 20, 2, 0.1 and 0.5 ppm, respectively.

chlorine in the range of 0.27-0.62% (92) have been fed to monkeys with no apparent problem.

Calcium. Many of the functions attributed to calcium, such as membrane transport and skeletal muscle metabolism, have been investigated in nonhuman primates. One of the first studies in monkeys that defined the metabolism of calcium were those of Harris et al. (93) in adult rhesus monkeys using radioactive ^{45}CA. Use of Hegsted IV (94) salt mixture in these studies established consumption and accretion rates of 500 and 243 mg Ca/day, respectively, and suggested a calcium requirement for growing monkeys of approximately 150 mg Ca/kg body wt. Rhesus monkeys receiving a diet containing 0.15% calcium or approximately 150 mg/Ca/kg body wt/day at 2 years of age and 60-75 mg Ca/kg body wt/day at 7 years of age developed osteoporosis (95). Other studies of osteoporosis in rhesus monkeys (96,97), suggest that dietary fluoride sequesters calcium and protects against osteoporosis when accompanied by the appropriate dietary levels of calcium, protein and vitamin C. The effect of various hormones such as thyrocalcitonin and parathyroid hormone on calcium metabolism has been investigated in rhesus monkeys (98,99) and has been shown to be similar to man and other laboratory animals. Interestingly, in studies on the effects of 36-60 days of immobilization on calcium metabolism, the loss of bone density and strength and the resultant osteoporosis could not be reversed by increasing dietary calcium (100,101). These various studies indicate that 0.5% by weight of dietary calcium satisfy the requirements of both Old and New World monkeys, while levels of 0.15% causes deficiency syndromes.

Phosphorous. While the specific requirements of phosphorous for nonhuman primates is not known, diets containing mineral mixes supplying phosphorous at 0.2% of the diet, by weight, have been found to be adequate. Studies in cebus monkeys (102) fed diets in which the calcium:phosphorous ratios ranged from 0.5 to 0.23 were unable to demonstrate any of the deleterious effects attributed to these ratios in studies with other laboratory animals such as the dog and cat (103-106).

Magnesium. Magnesium functions as a cofactor in many enzyme systems. A deficiency syndrome induced in cebus fed with 6 mg magnesium/100 gr diet included weight loss, irritability, convulsions and reduced serum potassium levels, and was prevented by feeding 40 mg magnesium/kg body wt/day (107). A possible role of magnesium on plasma and aortic cholesterol homeostasis was suggested from these studies since an increase in plasma cholesterol concentration and aortic sudanophilia coincided with the deficiency state. Magnesium deficiency has also been reported in rhesus monkeys fed diets containing 3 mg magnesium/100 gr diet (108).

Manganese. There have been no investigations of manganese requirements or induced deficiencies in nonhuman primates.

One study of manganese toxicity has been reported in adult rhesus monkeys to result in marked central nervous system dysfunction (109).

Zinc. Zinc deficiency has been produced in bonnet monkeys *(Macaca radiata)* (110) and squirrel monkeys (111,112) fed diets containing less than 0.5 ppm zinc. The deficiency syndrome was characterized by growth failure, hair loss, parakeratosis of the tongue, severe dermal lesions, testicular atrophy and cessation of menstrual cycles. When compared to normal squirrel monkeys fed a diet containing 15 ppm zinc, serum zinc levels of the deficient animals were lower (75.3 μg/dl) than the normal group (104.8 μg/dl). These studies suggest that a zinc requirement of 1 mg/kg body wt/day for squirrel monkeys (or approximately 20 ppm in the diet) would be adequate (Table IV).

Chromium. Chromium deficiency, which has been reported in squirrel monkeys fed commercial monkeys diets, has been associated with the development of glucose intolerance and atheroma (113,114). An improved glucose intolerance was noted with dietary supplements of trivalent chromium acetate (10 ppm in drinking water), but not with the divalent ion. Subsequent studies in squirrel monkeys fed a chromium deficient diet for 9-10 weeks, with the total intake of chromium being approximately 4 μg/day, resulted in corneal opacities that could not be reversed by chromium supplementation (115). While these studies did not establish a chromium requirement, a daily chromium intake of 150 μg/day as part of a commercial monkey diet did not result in corneal lesions.

Selenium. The only report of a selenium deficiency in nonhuman primates was by Muth et al. (116) in squirrel monkeys fed a semipurified seleniumdeficient diet for 6-9 months. The symptoms of weight loss, alopecia and cardiac, skeletal muscle, and hepatic necrosis were prevented by a single injection of 0.04 mg sodium selenite.

Iron. While the association of dietary iron intake and anemia has been known for many years, Fitch et al. (22) and Sood et al. (117) first reported that low-protein diets based on soy protein result in a microcytic hypochromic anemia in rhesus monkeys which was characteristic of iron deficiency. The anemia associated with the soybean protein was thought to be due to reduced gastrointestinal absorption of iron. Subsequent studies in squirrel monkeys also demonstrated that an iron deficiency anemia could also be produced in monkeys fed a modified cow's milk diet (118) and in scorbutic rhesus monkeys (119) as well. While the specific iron requirements have not been determined in nonhuman primates, diets supplemented with .018% iron (Hegsted IV salts) have been used for growing and adult monkeys. This level is probably substantially more than is needed (Table IV).

Iodine. Except for studies by Pickering (120) and Schultz et al. (121) which showed that fetal thyroid glands incorporated more radioiodide than the maternal organs, no studies of iodine requirements in nonhuman primates have been reported. Nevertheless, levels of 2 ppm iodine appear to be adequate for most growing and adult monkeys.

Fluorine. There have been no reports of fluorine deficiency in nonhuman primates. However, fluorosis has been described in rhesus monkeys given an oral dose of 10 mg NaF/kg body wt/day (96,97). The characteristic abnormal bone deposits and increased calcium accretion rates could be accentuated by low dietary levels of calcium and vitamin C. In the presence of a low-protein (2.5% casein) diet, fluoride appeared to increase bone resorption. In subsequent studies, rhesus monkeys fed 5-10 mg NaF/kg body wt/day revealed no signs of fluorosis regardless of high (1%) or low (0.15%) calcium levels, but did resist the development of calcium deficiency osteoporosis (95).

VIII. WATER REQUIREMENT

In their native habitat most nonhuman primates ingest a vegetablebased diet and thereby satisfy their water requirements. However, animals fed more concentrated commercial diets require a supplemental source of water. Consideration of energy expenditure can provide a rough estimation of water requirements as long as there are no unexpected water losses due to abnormal kidney function or as a result of a particular disease state. Pace et al. (122) first reported a water requirement for pigtail macaques consuming 70 kcal of gross energy/kg/day at 1 ml of water per kcal of gross energy/day, a figure subsequently verified by Kerr et al. (34) for most primates. The water requirement for rhesus monkeys fed a commercial diet was reported to be 80 ml/kg/day (123). Water requirements increase with smaller, younger animals (infant squirrel (124) and rhesus monkeys (125) consume 500 and 420 ml/kg/day respectively during the 2nd month of life) and also with febrile states, lactation, increased environmental temperature and increased fluid loss that may accompany vomiting and diarrhea.

REFERENCES

1. National Research Council (NRC), *Nutrient Requirements of Nonhuman Primates*, National Academy of Sciences, Washington, D. C., 1978.
2. Stahl, W.R., and Malinow, M.R. (1967) A survey of physiological measurements in *M. mulatta*. *Folia Primatol* 7: 12-33.
3. Follis, R.H. (1957) A kwashiorkor-like syndrome observed in monkeys fed maize. *Proc. Soc. Exp. Biol. Med.* 96: 523-528.
4. Kerr, G.R., Chamove, A.S., Harlow, H.F., and Waisman, H. A. (1969a) The development of infant monkeys fed low phenylalanine diets. *Pediatr. Res.* 3: 305-312.
5. Samonds, K.W., and Hegsted, D.M. (1973) Protein Requirements of young cebus monkeys *(Cebus albifrons and apella)*. *Am. J. Clin. Nutr.* 26: 30-40.
6. Kohrs, M.B. Harper, A.E., and Kerr, G.R. (1976) Effects of a low-protein diet during pregnancy of the rhesus monkey. I. Reproductive effeciency. *Am. J. Clin. Nutr.* 29: 136-145.
7. Riopelle, A.J., Hill, C.W., Li, S.C., Wolf, R.H., Seibold, H.R., and Smith, J. L. (1974) Protein deprivation in primates. I. Nonpregnant adult rhesus monkeys. *Am. J. Clin. Nutr.* 27: 13-21.
8. Fitch, C.D., Dinning, J.S., Witting, L.A., and Horwitt, M.K. (1961) Influence of dietary fat on the fatty acid composition of monkey erythrocytes. *J. Nutr.* 75: 409-413.
9. Greenberg, L.D., and Moon, H.D. (1961) Alterations in the blood fatty acids in single and combined deficiencies of essential fatty acids and vitamin B_6 in monkeys. *Arch. Biochem, Biophys.* 94: 405-416.
10. Portman, O.W., Pinter, K. and Hayashida, T. (1959) Dietary fat and hypercholesteremia in the cebus monkey. III. Serum polyunsaturated fatty acids. *Am. J. Clin. Nutr.* 7: 63-69.
11. Portman, O.W., Andrus, S.B., Pollard, D. and Bruno, D. (1961) Effects of longterm feeding of fat-free diets to cebus monkeys. *J. Nutr.* 74: 429-440.
12. Fiennes, R.N.T.-W., Sinclair, A.J., and Crawford, M.A. (1973) Essential fatty acid studies in primates. Linolenic acid requirements of capuchins. *J. Med. Primatol.* 2: 155.
13. Ausman, L.M., and Hayes, K.C. (1974) Vitamin E. deficiency anemia in Old and New World monkeys. *Am. J. Clin. Nutr.* 27: 1141-1151.

14. Greenberg, L.D. (1970a) Plasma cholesterol levels of monkeys maintained on various controlled and deficient diets. *Am. J. Clin. Nutr.* 23: 1101-1104.
15. Portman, O.W. (1970) Nutritional requirements of non-human primates. pp. 87-115 in R.S. Harris, ed. Feeding and Nutrition of Nonhuman Primates. Academic Press, New York.
16. Eggen, D.A. (1974) Cholesterol metabolism in rhesus monkey, squirrel monkey, and baboon. *J. Lipid Res.* 15: 139-145.
17. Corey, J.E., Hayes, K.C., Dorr, B., and Hegsted, D.M. (1974) Comparative lipid response of four primate species to dietary changes in fat and carbohydrate. *Atherosclerosis.* 19: 119-134.
18. Corey, J.E., Nicolosi, R.J. and Hayes, K.C. (1976) Effect of Dietary fat on cholesterol turnover in Old and New World monkeys. *Exp. Molec. Pathol.* 25: 311-321.
19. Nicholosi, R.J., Hojnacki, J.L., Llansa, Norma, and Hayes K.C. (1977) Diet and lipoprotein influence on primate atherosclerosis. *Proc. Soc. Exp. Biol. Med.* 156: 1-7.
20. Hayes, K.C. (1974) Retinal degeneration in monkeys induced by deficiencies of vitamin E or A. *Invest. Ophthalmol.* 13: 499-510.
21. O'Toole, B.A., Fradkin, R., Warkany, J., Wilson, J.G., and Mann, G.V. (1974) Vitamin A deficiency and reproduction in rhesus monkeys. *J. Nutr.* 104: 1513-1524.
22. Fitch, C.D., Harville, W.E., Dinning, J.S., and Porter, F.S. (1964) Iron deficiency in monkeys fed diets containing soybean protein. *Proc. Soc. Exp. Biol. Med.* 116: 130-133.
23. Fitch, C.D., Dinning, J.S., Porter, F.S., Folkers, K., Moore, H.W., and Smith, J.S. (1965) Vitamin E deficiency in the monkey. VII. Role of vitamin E and vitamin-like activity of coenzyme Q. *Arch. Biochem. Biophys.* 112: 488-493.
24. Hunt, R.D., Garcia, F.G., and Hegsted, D.M. (1967) A comparison of vitamin D_2 and D_3 in the New World primates. I. Production and regression of osteodystrophia fibrosa. *Lab. Anim. Care* 17: 222-234.
25. Legner, N.D.M., Bullock, B.C., Clarkson, T.B. and Lofland, H.B. (1967) Biological activities of vitamins D_2 and D_3 for growing squirrel monkeys. *Lab. Anim. Care* 17: 483-493.

26. Hunt, R.D., Garcia, F.G., and Walsh, R.J. (1972) A comparison of the toxicity of ergocalciferol and cholecalciferol in rhesus monkeys *(Macaca mulatta)*. *J. Nutr.* 102: 975-986.
27. Fitch, C.D., and Dinning, J.S. (1963) Vitamin E deficiency and antioxidants on the syndrome. *J. Nutr.* 79: 69-78.
28. Hunt, R.D., Garcia, F.G., and Hegsted, D.M. (1969) Hypervitaminosis D in New World monkeys. *Am. J. Clin. Nutr.* 22: 358-366.
29. Mason, K.E., and Telford, I.R. (1947) Some manifestations of vitamin E deficiency in the monkey. *Arch. Pathol.* 43: 363-373.
30. Dinning, J.S., and Day, P.L. (1957) Vitamin E deficiency in the monkey. I. Muscular dystrophy, hematologic changes, and the excretion of urinary nitrogenous constituents. *J. Exp. Med.* 105: 395-402.
31. Hayes, K.C. (1974b) Pathophysiology of vitamin E deficiency in monkeys. *Am. J. Clin. Nutr.* 27: 1130-1140.
32. Bieri, J.G., and Poukka-Evarts, R.H. (1972) Vitamin E nutrition in the rhesus monkey. *Proc. Soc. Exp. Biol. Med.* 140: 1162-1165.
33. Hill, R.B., Schendel, H.E., Rama Rao, P.B., and Johnson, B.C. (1964) Effect of dietary vitamin K on the nutritional status of the rhesus monkey. *J. Nutr.* 84: 259-264.
34. Kerr, G.R. (1972) Nutritional requirements of subhuman primates. *Physiol. Rev.* 52: 415-467.
35. Rinehart, J.F., Greenberg, L.D., and Ginzton, L.L. (1948) Thiamine deficiency in the rhesus monkey. Clinical, metabolic and hematologic observations. *Blood* 3:1453-1459.
36. Day, P.L., Langston, W.C., and Shukers, C.F. (1935) Leukopenia and anemia in the monkey resulting from vitamin deficiency. *J. Nutr.* 9: 637-644.
37. Waisman, H.A., and McCall, K.B. (1944) A study of thiamine deficiency in the monkey *(M. mulatta)*. *Arch. Biochem.* 4: 265-279.
38. Cooperman, J.M., Waisman, H.A., McCall, K.B. and Elvehjem, (1945) Studies on the requirements of the monkey for riboflavin and new factor found in liver. *J. Nutr.* 30: 45-57.
39. Greenberg, L.D., and Moon, H.D. (1963) Riboflavin deficiency in the monkey. *Fed. Proc.* 22: 592.
40. Mann, G.V., Watson, P.L., McNally, A., and Goddard, J. (1952a) Primate nutrition. II. Riboflavin deficiency in the cebus monkey and its diagnosis. *J. Nutr.* 47: 225-241.

41. Mann, G.V., Watson, P.L., and Adams, L. (1952b) Primate Nutrition I. The cebus monkey; Normal Values. *J. Nutr.* 47: 213-224.
42. Cooperman, J.M., McCall, K.B., Ruegamer, W.R., and Elvehjem, C.A. (1946b) Attempts to produce a niacin deficiency in the rhesus monkey. *J. Nutr.* 32: 37-45.
43. Tappan, D.V., Lewis, U.J., Register, U.D., and Elvehjem, C.A. (1952) Niacin deficiency in the rhesus monkey. *J. Nutr.* 46: 75-85.
44. Belavady, B., Madhavan, T.V., and Gopalan, C. (1968) Experimental production of niacin deficiency in adult monkeys by feeding jowar diets. *Lab. Invest.* 18: 94-99.
45. Belavady, B., and Rao, P.U. (1973) Production of nicotinic acid deficiency in monkeys fed leucine supplemented diets. *Int. J. Vitam. Nutr. Res.* 43: 454-460.
46. McCall, K.B., Waisman, H.A., Elvehjem, C.A., and Jones, E.S. (1946) A study of pyridoxine and pantothenic acid deficiencies in the monkey *(M. mulatta)*. *J. Nutr.* 31: 685-697.
47. Emerson, G.A., Walker, J.B., and Ganapathy, S.N. (1960) Vitamin B_6 and lipid metabolism in monkeys. *Am. J. Clin. Nutr.* 8: 424-433.
48. Rinehart, J.F., and Greenberg, L.D. (1949b). Arteriosclerotic lesions in pyridoxine-deficient monkeys. *Am. J. Pathol.* 25: 481-491.
49. Rinehart, J.F., and Greenberg, L.D. (1951) Pathogenesis of experimental arteriosclerosis in pyridoxine deficiency. *Arch. Pathol.* 51: 12-18.
50. Rinehart, J.F., and Greenberg, L.D. (1956) Vitamin B_6 deficiency in the rhesus monkey with particular reference to the occurrence of atherosclerosis, dental caries, and hepatic cirrhosis. *Am. J. Clin. Nutr.* 4: 318-328.
51. Poppen, K.J., Greenberg, L.D., and Rinehart, J.F. (1952) The blood picture of pyridoxine deficiency in the monkey. *Blood.* 7: 436-744.
52. Mushett, C.W., and Emerson, G.A. (1956) Arteriosclerosis in pyridoxine-deficient monkeys and dogs. *Fed. Proc.* 15: 526.
53. Victor, M., and Adams, R.D. (1956) The neuropathology of experimental vitamin B_6 deficiency in monkeys. *Am. J. Clin. Nutr.* 4: 346-353.
54. Greenberg, L.D., McIvor, B., and Moon, H.D. (1958) Experimental arteriosclerosis in pyridoxine-deficient rhesus monkeys. *Am. J. Clin. Nutr.* 6: 635-637.

55. Gershoff, S.N. (1964) *Vitamin B_6 and oxalate metabolism. Vitam. Horm.* 22: 581-589.
56. Greenberg, L.D. (1964) Arteriosclerotic, dental, and hepatic lesions in pyridoxine-deficient monkeys. *Vitam. Horm.* 22: 677-694.
57. Wizgird, J.P., Greenberg, L.D., and Moon, H.D. (1965) Hepatic lesions in pyridoxine-deficient monkeys. *Arch. Pathol.* 79: 317-323.
58. Mann, G.V. (1968) Blood changes in experimental primates fed purified diets: Pyridoxine and riboflavin deficiency. *Vitam. Horm.* 26: 465-485.
59. Manning, P.J., and Clarkson, T.B. (1971) Serum levels of vitamin B_6 in hypercholesterolemic rhesus monkeys on prophylactic isoniazid. *Lab. Anim. Sci.* 21: 898-900.
60. Waisman, H.A., McCall, K.B., and Elvehjem, C.A. (1945) Acute and chronic biotin deficiencies in the monkey *(M. mulatta)*. *J. Nutr.* 29: 1-11.
61. Wilgram, G.F., Lucas, C.C., and Best, C.H. (1958) Kwashiorkor type of fatty liver in primates. *J. Exp. Med.* 108: 361-371.
62. Cueto, J., Tajen, N., Gilbert, E., and Currie, R.A. (1967) Experimental liver injury in the rhesus monkey. I. Effects of cirrhogenic diet and ethanol. *Ann. Surg.* 166: 19-28.
63. Ruebner, B.H., Moore, J., Rutherford, R.B., Seligman, A.M., and Zuidema, G.D. (1969) Nutritional cirrhosis in rhesus monkeys: Electron microscopy and histochemistry. *Exp. Mol. Pathol.* 11: 53-70.
64. Rutherford, R.B., Boitnott, J.K., Donchoo, J.S., Margolis, S., Sebor, J., and Zuidema, G.D. (1969) Production of nutritional cirrhosis in *(M. Mulatta)* monkeys. *Arch. Surg.* 98: 720-730.
65. Hoffbauer, F.W., and Zaki, F.G. (1965) Choline deficiency in baboon and rat compared. *Arch. Pathol.* 79: 364-369.
66. Day, P.L., Langston, W.C., and Shukers, C.F. (1935) Leukopenia and anemia in the monkey resulting from vitamin deficiency. *J. Nutr.* 9: 637-644.
67. Langston, W.C., Darby, W.J., Shukers, C.F., and Day, P.L. (1938) Nutritional cytopenia (vitamin M deficiency) in the monkey. *J. Exp. Med.* 68: 923-940.
68. Cooperman, J.M., Elvehjem, C.A., McCall, K.B., and Ruegamer, W.R. (1946a) "Folic acid" active compounds in the nutrition of the monkey. *Proc. Soc. Exp. Biol. Med.* 61: 92-97.

69. Day, P.L., and Totter, J.R. (1947) The bioassay of the vitamin M group. *Biol. Symp.* 12: 313-338.
70. Huser, H.J., and Beard, M.E.J. (1969) Studies on folate and vitamin B_{12} metabolism in primates. I. Blood and bone marrow morphology, folate and vitamin B_{12} levels. *Folia Primatol.* 10: 172-180.
71. Gyr, K., Wolf, R.H., and Felsenfeld, O. (1974) Intestinal absorption of D-xylose and folic acid in protein-deficient Patas monkeys *(Erythrocebus patas)*. *Am. J. Clin. Nutr.* 27: 350-354.
72. Dreizen, S., Levy, B.M., and Bernick, S. (1970) Studies on the biology of the periodontium of marmosets. VIII. The effect of folic acid deficiency on the marmoset oral mucosa. *J. Dent. Res.* 49: 616-620.
73. Oxnard, C.E. (1966) Vitamin B_{12} nutrition in some primates in captivity. *Folia Primatol.* 4: 424-431.
74. Oxnard, C.E. (1969) A note on the ruminant-like digestion of langurs. *Lab. Primate Newslett.* April 8: 24-25.
75. Oxnard, C.E., and Smith, W.T. (1966) Neurological degeneration and reduced serum vitamin B_{12}-levels in captive monkeys. *Nature* 210: 507-508.
76. Oxnard, C.E., Smith, W.T., and Torres, I. (1967) Observaciones recientes acerca de la deficiencia de vitamina B_{12} y degeneracion neurologica en primates cautivos. *Acta. Neurol. Lat. Amer.* 13:1-10.
77. Oxnard, C.E., Smith, W.T., and Torres, I. (1968) I. Peripheral neuropathy and hypovitaminosis B_{12} in captive monkeys. pp. 162-168 in H. O. Hoker, ed. Proceedings of the Second International Congress of Primatology. S. Karger, Basel, Switzerland.
78. Oxnard, C.E., Smith, W.T., Torres, I. (1970) Vitamin B_{12} deficiency in captive monkeys and its effect on the nervous system and the blood. *Lab. Anim.* 4: 1-12.
79. Torres, I., Smith, W.T., and Oxnard, C.E. (1971) Peripheral neuropathy associated with vitamin-B_{12} deficiency in captive monkeys. *J. Pathol.* 105: 125-146.
80. Hind, V.M.D. (1970) Degeneration in the peripheral visual pathway of vitamin B_{12}-deficient monkeys. Trans. *Ophthalmol. Soc. U.K.* 90: 839-846.
81. Agamanolis, D.P., Chester, E.M., Victor, M., Kark, J.A. Hines, J.S., and Harris, J.W. (1976) Neuropathology of experimental vitamin B_{12} deficiency in monkeys. *Neurology.* 26: 905-914.

82. Kark, J.A., Victor, M., and Hines, J.D. (1974) Nutritional vitamin B_{12} deficiency in rhesus monkeys. *Am. J. Clin. Nutr.* 27: 470-478.
83. Flinn, R.M., and Oxnard, C.E. (1966) The relationship between growth and the administration of cyanocobalamin in the rhesus monkey. *Folia Primatol.* 4: 432-437.
84. Elliott, O., Yess, N.J., and Hegsted, D.M. (1963) Biosynthesis of ascorbic acid in the tree shrew and slow loris. *Nature.* 212: 739-740.
85. Banerjee, S., Bal, H. (1959) Studies on changes in bone marrow and blood picture in monkeys during progressive stages of scurvy. *Indian J. Med. Res.* 47: 646-651.
86. Demaray, S.Y., Altman, N.H., and Ferrell, T.L. (1978) Suspected ascorbic acid deficiency in a colony of squirrel monkeys *(Saimiri sciureus)*. *Lab. An. Sci.* 28: 457-460.
87. Day, P.L. (1944) The nutritional requirements of primates other than man. *Vitam. Horm.* 2: 71-105.
88. Soloveva, G.A., Alekseeva, L.V., and Trufanov, A.V. (1966) Vitamin C deficiency and its effect on the secretion of 17-ketosteroids and dehydroepiandrosterone in monkeys. *Biol. Abstr.* 47 (46127).
89. Shaw, J.H., Phillips, P.H., and Elvehjem, C.A. (1945) Acute and chronic ascorbic acid deficiencies in the rhesus monkey. *J. Nutr.* 29: 365-372.
90. Lehner, N.D.M., Bullock, B.C., and Clarkson, T.B. (1968) Ascorbic acid deficiency in the squirrel monkey. *Proc. Soc. Exp. Biol. Med.* 128: 512-514.
91. DeKlerk, W.A., Plessis, D.P., van der Watt, J.J., de Jager, A., and Laubscher, N.F. (1973) Vitamin C requirements of the vervet monkey *(Cercopithecus aethiops)* under experimental conditions. *S. Afr. Med. J.* 47: 705-708.
92. Hummer, R.L. (1970) Observations of the feeding of baboons. In Feeding and nutrition of nonhuman primates. pp. 183-203. R.S. Harns, ed., Academic Press, N. Y.
93. Harris, R.S., Moor, J.R., and Wanner, R.L. (1961) Calcium metabolism of the normal rhesus monkey. *J. Clin. Invest.* 40: 1766-1774.
94. Hebsted, D.M., Mills, R.C., Elvehjem, C.A., and Hart, E.B. (1941) Choline in the nutrition of chicks. *J. Biol. Chem.* 138: 459-466.

95. Griffiths, H.J., Hunt, R.D., Zimmerman, R.E., Finberg, H., and Cuttino, J. (1975) The role of calcium and fluoride in osteoporosis in rhesus monkeys. *Invest. Radiol.* 10: 263-268.
96. Reddy, G.S., and Narasinga Rao, B.S. (1971) Effect of dietary calcium, vitamin C and protein in development of experimental skeletal fluorosis. II. Calcium turnover with Ca; Calcium and phosphorus balances. *Metabolism* 20: 650-656.
97. Reddy, G.S., and Srikantia, S.R. (1971) Effect of dietary calcium, vitamin C and protein in development of experimental skeletal fluorosis. I. Growth, serum chemistry, and changes in composition, and radiological appearance of bones. *Metabolism* 20: 642-649.
98. Bell, N.H., Barrett, R.J., and Patterson, R. (1966) Effects of porcine thyrocalcitonin on serum calcium, phosphorus, and magnesium in the monkey and in man. *Proc. Soc. Exp. Biol. Med.* 123: 114-118.
99. Borle, A.B. (1968) Effects of purified parathyroid hormone on the calcium metabolism of monkey kidney cells. *Endocrinology* 83: 1316-1322.
100. Pyke, R.E., Mack, P.B., Hoffman, R.A., Gilchrist, W.W., Hood, W.N., and George, G.P. (1968) Physiologic and metabolic changes in *Macaca nemestrina* on two types of diets during restraint and nonrestraint. III. Excretion of calcium and phosphorus. *Aerosp. Med.* 39: 704-708.
101. Kazarian L.E., and VonGierke, H.E. (1969) Bone loss as a result of immobilization and chelation: Preliminary results in *Macaca mulatta*. *Clin. Orthop.* 65: 67-75.
102. Anderson, M.P., Hunt, R.D., Griffiths, H.J., McIntyre, K.W., and Zimmerman, R.E. (1977) Long-term effect of low dietary calcium: phosphate ratio on the skeleton of *Cebus albifrons* monkeys. *J. Nutr.* 107: 834-839.
103. Krook, L., Barrett, R.B., Usui, K., and Wolke, R.E. (1963) Nutritional secondary hyperparathyroidism in the cat. *Cornell Vet.* 53: 224-240.
104. Krook, L., and Lowe, J.E. (1964) Nutritional secondary hyperparathyroidism in the horse: with a description of the normal equine parathyroid gland. *Pathol. Vet.* 1(suppl.), 1-92.
105. Rowland, G.N., Capen, C.C., and Nagode, L.A. (1968) Experimental hyperparathyroidism in young cats. *Pathol. Vet.* 5: 505-519.

106. Jowsey, J., Reiss, E., and Canterbury, J.M. (1974) Long-term effects of high phosphate intake on parathyroid hormone levels and bone metabolism. *Acta.Orthop Scand*. 45: 801-808.

107. Vitale, J.J., Velez, H., Guzman, C., and Correa, P. (1963) Magnesium deficiency in the cebus monkey. *Cir. Res*. 12: 642-650.

108. Dunn, M.J. (1971) Magnesium depletion in the rhesus monkey: Induction of magnesium-dependent hypocalcemia. *Clin. Sci*. 41: 333-344.

109. Pentschew, A., Ebner, F.F., and Kovatch, R.M. (1963) Experimental manganese encepnalopathy in monkeys. *J. Neuropathol. Exp. Neurol*. 22: 488-499.

110. Swenerton, H., and Hurley, L.S. (1972) Zinc deficiency in the bonnet monkey *(Macaca radiata)*. *Fed. Proc*. 31: 667.

111. Barney, G.H., Macapinlac, M.P., Pearson, W.N., and Darby, W.J. (1967) Parakeratosis of the tongue: A unique histopathologic lesion in the zinc-deficient squirrel monkey. *J. Nutr*. 93: 511-517.

112. Macapinlac, M.P., Barney, G.H., Pearson, W.N., and Darby, W.J. (1967) Production of zinc-deficiency in the squirrel monkey *(Saimiri sciureus)*. *J. Nutr*. 93: 499-510.

113. Davidson, I.W.F., and Blackwell, W.L. (1968) Changes in carbohydrate metabolism of squirrel monkeys with chromium dietary supplementation. *Proc. Soc. Exp. Biol. Med*. 127: 66-70.

114. Davidson, I.W.F., Lang, C.M., and Blackwell, W.L. (1967) Impairment of carbohydrate metabolism of the squirrel monkey. *Diabetes*. 16: 395-401.

115. Martin, G.D., Stanley, J.A., and Davidson, I.W.F. (1972) Corneal lesions in squirrel monkeys maintained on a low-chromium diet. *Invest. Ophthalmol*. 11: 153-158.

116. Muth, O.H., Weswig, P.H., Whanger, P.D., and Oldfield, J.E. (1971) Effect of feeding selenium-deficient ration to the subhuman primate *(Saimiri sciureus)* *Am. J. Vet. Res*. 32: 1603-1605.

117. Sood, S.K., Deo, M.G., and Ramailingaswami, V. (1965) Anemia in experimental protein deficiency in the rhesus monkey with special reference to iron metabolism. *Blood* 26: 421-432.

118. Amine, E.E., Ausman, L.M., Hegsted, D.M., and Gershoff, S.N. (1972) Iron deficiency and absorption in infant squirrel monkeys. *Fed. Proc*. 31: 711. (Abstr.)

119. Banerjee, S., and Chakrabarty, A.S. (1965) Anemia and its relation with iron metabolism in scorbutic monkeys. *Indian J. Med. Res.* 53: 835-848.

120 Pickering, D.E. (1968) Thyroid physiology in the developing monkey fetus *(M. mulatta)*. *Gen. Comp. Endocrinol.* 10: 182-190.

121. Schultz, M.A., Forsander, J.B., Chez, R.A., and Hutchinson, D.L. (1965) The bidirectional placental transfer of I^{131}3:5:3'-triiodothyronine in the rhesus monkey. *Pediatrics* 35: 743-752.

122. Pace, N., Hansen, J.T., Rahlmann, D.F., Barnstein, N.J., and Cannon, M.D. (1964) Preliminary observations of some physiological characteristics of the pig-tailed monkey, *Macaca nemestrina*. *Aerosp. Med.* 35: 118-121.

123. Feldmahn, A.L., Smith, W.K., and Leventhal, C.M. (1960) Diurnal patterns of micturition and drinking in rhesus monkeys. *Ann. N.Y. Acad. Sci.* 85: 828-841.

124. Hinkle, D.K., and Session, H.L. (1972) A method for hand rearing of *Saimiri sciureus*. *Lab. Anim. Sci.* 22: 207-209.

125. Kerr, G.R., Scheffler, G., and Waisman, H.A. (1969b) Growth and development of infant *M. mulatta* fed a standardized diet. *Growth.* 33: 185-199.

LIQUID FORMULAS AND PROTEIN REQUIREMENTS OF NONHUMAN PRIMATES

Lynne M. Ausman
Daniel L. Gallina

Department of Nutrition
Harvard School of Public Health
Boston, Massachusetts

I. INTRODUCTION

Liquid semi-purified diets have been used for feeding adult animals of many species when drugs or other experimental compounds are being tested and when accurate food intake records must be kept. As early as 1958, liquid diets were used in programs of food reinforcement for squirrel monkeys (1). A modification of this diet (2), as well as several others, have been reported (3-8) or are now available from commercial suppliers. These diets vary considerably in dietary components and amounts, but each is reported to be an emulsion providing approximately 1 kcal/ml (range 1-3 kcal/ml) which is sufficient for weight maintenance and adequate health of adult monkeys over a period of several months. However, when using liquid semi-purified diets for rearing infant monkeys,

ISBN 0-12-334250-3

certain fixed guidelines for the formulation of these diets must be followed to complement the physiology of the neonate and the infant.

Commercial liquid formulas intended for human infants are often adequate for a portion of all of the experimental period (5,6,9-20) in infant monkeys. Indeed, some investigators have arranged for these commercial companies to formulate experimental diets using the same techniques available for commercial liquid formulas now on the market (16,21). The ease of storage and mixing, consistency between batches, and minimal technological problems are apparent advantages. However, the unsuitability of the commercial formulas in some instances as well as the special requirements of some nutritional studies may require that the investigator formulate and mix his own liquid diets (5-8,22). For example, when a commercial liquid formula was fed to infant squirrel monkeys for a period exceeding 16 weeks or was fed to adults as a solid diet for 2 months, the resulting hair growth was an abnormal bright yellow (10). Although the cause of the hair discoloration was never identified, the experience does serve as a reminder that human infant formulas may not be adequate for extended periods for all species. In a second example, it was found that feeding cow's milk to infant squirrel monkeys caused gastrointestinal bleeding comparable to that found in human infants (10). This review, therefore, addresses itself to some of the considerations necessary when formulating liquid diets for infant primates. General recommendations on formulation of liquid diets for human infants were recently reviewed (23). While not applicable in all instances, the issues considered are pertinent to the formulation of diets for infant monkeys and will be discussed here in conjunction with the "trial and error" experiences of the authors with two species of New World monkeys, *Saimiri sciureus* and *Cebus albifrons* and *apella*, and lesser experience with two Old World species, *Macaca mulatta* and *Macaca fascicularis*.

II. GENERAL CONSIDERATIONS

Infant primates may be removed from their dams immediately after birth and raised under nursery conditions if desired. Nursery care is described elsewhere (9,13,14,18-20). In general, liquid diets are prepared 3-4 times weekly from stock supplies of premixed dry ingredients and other dietary components. The liquid diets are aliquoted into feeding bottles (34,65, and 100 ml capacity) which are fitted with one of 2 sizes of nipples. Food is available *ad libitum* from 8 a.m. to

11 p.m. Young animals are hand fed frequently as needed until they begin to self-feed, a process that begins as early as a few days in some species and weeks in others.

Until the kidney has matured and until the animals are capable of vigorous self-feeding, infants are dependent on the water provided in their formula to maintain proper hydration. Hypertonic dehydration occurs when the osmolality of the diet, influenced mainly by the mineral and carbohydrate concentration exceeds 300 mOsm/kg water (24-26). The dilution of the diet and reduction in the salt concentration help to reduce osmolality. The substitution of fat (which contains no osmotic load) for carbohydrate further lowers the osmolality. Accordingly, and in keeping with the composition of human and other primate milks (27-30), the lipid content of formulas for newborn and infant animals commonly comprises between 45-55% of the dietary calories, limiting the contribution from lactose to less than 40% of the calories. In trials with newborn squirrel and cebus monkeys (see diet #74, Table 1) we have found that a dilution of 0.7 kcal/ml, 0.68 g salt mix/100 kcal and 38.5% of calories from lactose produces apparently normal growth and development when compared to infants reared with their dams (11,31) and is associated with insignificant morbidity and mortality. An inability to handle an increased renal solute load, contributed to by urea as well as salts, and an inability to handle certain excess amino acids (particularly sulfur containing amino acids) is also characteristic of the premature and newborn human infant (23,24) and would suggest placing an upper limit on the protein content of the diet. No data is available on the performance of infant primates when fed amounts of protein substantially greater than required.

As the animal matures, it is able to compensate for a wider range of dietary formulations, one example of which is a liquid diet for older animals in Table 2 (Diet #78). This diet contains 1 kcal/ml, has 68% more salt/kcal than the infant diet, and derives 51% of the calories from purified carbohydrate. It must be emphasized that when these diets were fed prematurely, they were associated with a significantly increased diarrhea, dehydration and mortality. On the other hand, if the diluted infant diet was continued too long as the sole diet (greater than 12 weeks of age), the growth rate, particularly in the squirrel monkey, was lower than that of monkeys receiving the "adult" type diet.

TABLE I Composition of a Semi-purified Liquid Diet for Infant Monkeys[a] (Diet #74)

Ingredient	Concentration g/100 g	kcal/100 g
Casein[b]	18.76	75.04
Lactose[c]	48.76	195.04
Corn oil[d]	25.98	233.82
Salt mix[e]	3.49	
Vitamin mix[f]	0.58	2.32
Choline	0.35	
Inositol	0.12	
Tween 80[g]	1.47	
Suspending agent[h]	0.49	
	100.00	506.22

a. Modified from Ausman et al. (6).
In preparing the diet, a premix is made of selected dry ingredients (casein, lactose, salt mix, vitamin mix, choline and inositol). To prepare the liquid diet, 1286 ml water, 3 ml of Tween 80, 1 g suspending agent, 59 ml of corn oil, and one ml banana flavoring are blended at high speed for 90 seconds. Then 147 g premix is added and the mixture blended for an additional 90 seconds. The resultant diet contains approximately 0.7 kcal/ml. The blending time, concentration of Tween 80 and amount of suspending agent will vary with the dietary protein and fat used. If coconut oil is used, it must first be melted and then incorporated into the diet with relatively longer blending times. Storage of saturated fat diets in the refrigerator necessitates rewarming and/or reblending on a daily basis.

b. Casein (vitamin free, fine ground, 100-200 mesh), Bioserv. Frenchtown, New Jersey. It contains 87.7% protein by Kjeldahl analysis (N x 6.25). Diet contains 13.0% of the calories from the casein protein.

c. D-lactose hydrate, Grand Island Biological Co., Grand Island, New York.

d. Mazola, Best Foods, Englewood Cliffs, New Jersey.

e. A modification of Hegsted IV Salt Mix containing the following in g/kg: $CaCO_3$, 290.4849; K_2HPO_4, 312.1673; $CaHPO_4 \cdot 2H_2O$, 72.597; $MgSO_4 \cdot 7H_2O$, 98.732; NaCl, 162.3664; MgO, 32.0395; $FeC_6H_6O_7 \cdot 5H_2O$, 26.5949; KI, 0.7737; $MnSO_4 \cdot H_2O$, 3.6638; $ZnCl_2$, 0.2418; $CuSO_4 \cdot 5H_2O$, 0.2901; $CH(C_2H_3O_2)_3$, 0.0443; Na_2SeO_3, 0.0043.

III. CARBOHYDRATE

Lactose is the main natural carbohydrate in mammalian milk and in the absence of specific mitigating factors such as allergy or a special experimental protocol, is probably the most suitable carbohydrate for infant primates for at least the first 2-3 months of life (Table 1). As the digestive system matures, other mono and disaccharides are introduced and/or substituted. A mixture of sucrose and dextrin has been well tolerated by both young and adult squirrel, cebus, rhesus and cynomolgus monkeys and has been incorporated into liquid as well as solid diets (Table 2). Other mixtures of carbohydrate are undoubtedly appropriate, but as they are discussed elsewhere in this workshop, will not be reviewed further. One note of caution is that a two year trial with dextrin as the sole carbohydrate source (70% of calories) in both liquid and solid diets was associated with depressed food intake and growth and clinical evidence of allergic conjunctivitis.

IV. FAT

A variety of fats, from solid, saturated coconut oil to safflower oil, have been successfully incorporated into the liquid diets in amounts varying from 10-47% of the calories. The choice of type and amount depends on the experimental objective and will not be reviewed further here. For routine purposes, we have found that soybean-cottonseed and/or corn oil produce apparently normal growth and development. Adequate essential fatty acids must be supplied, but as little as 0.4% of calories as linoleic acid provided as coconut oil has proved sufficient for normal growth of infant squirrel, cebus, rhesus, and cynomolgus (32, unpublished data).

---

f. *Vitamin mix contains in g/kg: Thiamin, 0.8; riboflavin, 1.6; pyridoxine, 0.8; CaPantothenate, 5.0; niacin, 8.0; folate, 0.8; biotin, 0.04;* B_{12}*, 0.03; ascorbic acid, 100; dl-a-tocopherol acetate (500 IU/g), 20; vitamin A acetate (500,000 IU/g), 5.0; vitamin* D_3 *(200,000 IU/g), 1.25; menadione, 1.00; dextrin, 855.68.*

g. *Polyoxyethylene sorbitan monooleate, ICN Pharmaceuticals, Cleveland, Ohio.*

h. *Bio-Serv, Inc. Frenchtown, New Jersey.*

TABLE II Composition of a Semi-purified Liquid Diet for Young Monkeys[a] (Diet #78)

Ingredient	Concentration g/100 grams	kcal/100 grams
Lactalbumin[b]	17.96	71.84
Dextrin[c]	27.68	110.72
Sucrose	27.68	110.72
Corn oil[d]	15.07	135.63
Salt mix[e]	5.02	
Vitamin mix[f]	0.50	2.00
Inositol	0.10	
Choline	0.30	
Cellulose[g]	4.93	
Tween 80[h]	0.36	
Suspending agent[i]	0.40	
	100.00	430.91

[a] In preparing the diet, a premix is made of the lactalbumin, sucrose, dextrin, salt and vitamin mix, inositol, choline and cellulose. To prepare the liquid diet, 838 ml water, 0.8 ml Tween 80, 0.9 grams suspending agent, 39 ml corn oil and 1 ml banana flavoring are blended at high speed for 90 seconds. The premix, 196 g, is then added and the diet beaten for an additional 90 seconds (blending times vary from batch to batch and depend largely on the type of fat and protein incorporated). The resultant diet will have a concentration of about 1 kcal/ml.

[b] Teklad, Inc., Madison Wisconsin. Lactalbumin contains 77.96% protein (N x 6.25) by Kjeldahl analysis. This diet contains 13.0% of the calories derived from lactalbumin protein.

[c] Bio-Serv, Frenchtown, New Jersey.

[d] Mazola

[e] See salt mix for Diet 74, Table I

[f] See vitamin mix for Diet 74, Table I

[g] Non-nutritive fiber, Teklad, Inc., Madison, Wisconsin.

[h] Polyoxyethylene sorbitan monooleate, ICN pharmaceuticals, Inc. Cleveland, Ohio.

[i] Bio-Serv, Frenchtown, New Jersey

V. PROTEIN

The selection of the type and amount of protein to be incorporated into the diets will be dealt with more extensively since the protein requirements of infant and young monkeys are most important to their growth. In general, when dietary protein is not an experimental variable, lactalbumin and/or casein serve as well-tolerated high biological value proteins that may be used in moderate amounts to promote adequate growth and development.

The quality of the protein source must be taken into account. As protein quality declines (33-34), the percent protein in the diet must increase accordingly to achieve the desired nutritional status. The physical characteristics of the protein such as granule size, charge and solubility are of utmost importance to liquid diet formulation. Most proteins are not soluble and must be stabilized. In our hands, very fine ground casein has been the easiest to suspend while lactalbumin, in coarser particles, has been more difficult. It has been difficult to prepare properly stabilized diets containing the large amount of unsupplemented soy or gluten required for maximum growth of infant primates.

In order to make comparisons between diets more uniform, the contribution of the protein to the diet should be calculated as the percent of calories as well as percent by weight. This avoids difficulties arising when diets vary in dilution, fat content, or contain non-nutritive bulk. As most dietary protein sources are not pure and contain small quantities of ash, carbohydrate, fat and moisture, the exact nitrogen content must be determined by Kjeldahl analysis. Ordinarily, to readily compare nitrogen contents of various diets, the protein content is routinely assumed to be the nitrogen content times 6.25 (N x 6.25).

Infant and young primates of various species have widely varying protein requirements. Since purified protein is an expensive dietary ingredient and because feeding too much protein may be detrimental, it is important to feed an amount of protein that will assure satisfactory growth and development without supplying an undue excess. On the other hand, when formulating diets containing limited amounts of protein for deprivation studies, it is particularly important to ascertain the approximate protein requirements of these infants in order to obtain the desired nutritional status. Both of these goals may be accomplished by predetermining the protein requirements of these infant monkeys for both body weight maintenance and for growth. Nitrogen balance techniques have been traditionally used for the determination of protein requirements in adult humans and animals although these have been difficult to accomplish with infant monkeys.

A dose-response, slope-ratio assay developed in the young growing rat has been adapted for use in monkeys (8). This assay allows the determination of both the protein requirements for weight maintenance and the efficiency of utilization of protein for new growth. Coupled with a knowledge of normal caloric intake and growth rate for age, it is possible to formulate diets to provide exactly enough protein for any desired rate of growth.

Examples of this type of experiment have recently been carried out for the infant squirrel monkeys using 3 proteins (35). Briefly, young squirrel monkeys in their rapid growth phase are fed, *ad libitum*, diets containing a single nitrogen source at several, growth restrictive levels. The regression of daily body weight change in a given period on the protein intake is calculated for animals of similar body size. The intersection of the regression line at the zero weight change axis yields an estimate of the protein requirement for weight maintenance and the weight gained per day (slope of the line) multiplied by the protein content of squirrel monkey carcass yields an estimate of the efficiency with which squirrel monkeys utilize the protein. In the case of casein the average maintenance requirement from monkeys of 3 sizes was 3.44 ± 0.19 g/kg/day, and the mean slope of $1.81 \pm 0.33 \times$ 20.6% protein (unpublished data) yields an efficiency of utilization of 37.3%.

A more convenient and statistically correct method which compensates for variations in body weight is the multiple regression analysis (8). Using the formula $Y = a + \beta I + \gamma W$ where Y = daily weight change (in grams), I = daily protein intake (in grams) and W = mean body weight (in kilograms), a general multiple regression can be calculated. The equation may be solved for protein requirements at any desired growth rate and body size by substituting values for Y and W. Efficiency of protein utilization is again derived from the slope, β, and the protein content of 20.6% protein. The multiple regression equation for casein-fed squirrel monkeys is: $Y = 0.13 + 1.56\ I - 5.48\ W$ $\quad s_b = 0.19 \quad r = 0.79$ $n = 43$ and the resultant maintenance requirement is 3.63 g casein protein/kg/day. Growth is accomplished with an efficiency of 32.1%. Similar data are obtained using lactalbumin which is utilized with an efficiency of 27.8% (35). Protein requirements for squirrel monkeys of varying body weight and growth rates calculated from the equation for casein are given in Table 3. Whereas protein requirements

TABLE III Protein and Calorie Requirements of Squirrel Monkeys Fed Casein Based Diets

BW (g)	Growth[a] Rate (g/day)	Kcal[b] Intake (kcal/day)	Protein requirements for maintenance (g/day)	(g/kg)	(% kcal)	Protein requirements for maintenance + growth (g/day)	(% kcal)
150	3.4	72	0.48	3.2	2.7	2.7	14.8
200	2.5	90	0.66	3.3	2.9	2.3	10.1
300	1.8	124	1.04	3.7	3.4	2.2	7.1
400	1.3	137	1.41	3.5	4.1	2.2	6.6
500	0.6	150	1.79	3.6	4.8	2.2	5.8

a and b derived from 42 male and female control monkeys fed 13% protein diets from birth (unpublished data).

for weight maintenance are more or less constant/kg body weight, protein needs for growth expressed as percent of calories are initially quite high in this species and decrease markedly with increasing body size.

Similar experiments with infant cynomolgus yield maintenance requirements of 1.04 g lactalbumin/kg and efficiencies of 65% (35). The values compare favorably with maintenance requirements 1.96 g lactalbumin/kg/day and an efficiency of protein utilization for growth of 74% in young cebus (8). If the data from these two species are compared to that of the squirrel, it seems obvious that the squirrel monkey is an inefficient utilizer of protein and that this is reflected by abnormally high protein requirements for growth as compared to the other two species. The requirement of 10-14% of the calories as protein for 150-200 gram squirrel monkeys cannot be met by most commercial formulas for human infants (36). Indeed, the protein content of squirrel monkey milk is 16% of calories (27).

In order to investigate the cause of the poor protein utilization in the squirrel monkey, factors such as malabsorption, nitrogen loss through hair, nails, and sweat, and poor calorie intake were ruled out (35). It was considered that the routine, obligatory 8-10 hour fasting period imposed on these infants may have been unphysiological and placed an undue drain upon nitrogen reserves for gluconeogenesis. However, provision of the diet on an *ad libitum* round-the-clock basis failed to alter the efficiency of protein utilization. Similarly, the casein diet was supplemented with sulfur amino acids--the most limiting amino acid in casein for rats; the efficiency, however, did not improve. It is possible to speculate, therefore, that the inefficient protein utilization may be due to inefficient post-prandial handling of the amino acids in this species.

Few studies are available in which the protein requirements, per se, of infant or adult monkeys, other than cebus, squirrel and cynomolgus discussed here, have been determined. Although a wealth of data exists in which infant and adult monkeys have been fed a variety of diets containing minimal levels of protein to produce protein and/or calorie deficiency syndromes (16-17, 37-49), it was difficult to use the majority of those data to determine minimal protein requirements for the following reasons: the exact nitrogen (or protein) content of the diet was not given; diets fed contained excess protein or protein of an unknown biological value; no food intake records were kept; and calories available to the animal were deliberately specified or restricted. However, data from one laboratory working with protein deprivation of infant rhesus monkeys for periods of 6 months (16) have been particularly useful. Infant rhesus showed slow growth (1-2 g/day) when fed diets containing only 1.8-2.0 g milk protein/kg/day (2.7% of calories as protein) and maximum growth at less than 4 g milk protein/kg/day (6.5% protein); these data agree well with estimates from young cynomolgus showing weight maintenance at 1-1.5 g/kg/day and maximum growth at 4 g/kg/day (6.6% protein for a 500 gram animal).

Available data from adult monkeys is based on long term body weight maintenance as well as some biochemical estimates of nutritional status such as total plasma protein and albumin concentrations or amino acid ratios. Adult cebus monkeys (2-3.5 kg) have been fed graded levels of lactalbumin protein for periods as long as one year. Maintenance of body weight and plasma albumin levels occurred in monkeys fed diets containing at least 8% of the calories as lactalbumin protein. Food intakes were 95 kcal/kg/day, thereby providing the protein maintenance requirement of 2 g/kg/day found for infant cebus earlier (8).

In another laboratory (49), adult rhesus monkeys were fed 0.5, 1.0, 2.0 and 4.0 g casein protein/kg/day for a 160-day period. Animals in the 1 and 2 g/kg group appeared physically healthy and normal but showed modest weight decreases and decreased biochemical indicies such as depressed albumin and a higher nonessential: essential amino acid ratio. The interpretation of the data is made difficult by a temporary inanition (and prolonged weight depression) in the 2 g/kg group. Taking these data into account, it seems that the minimal protein requirement for the adult rhesus monkey is probably somewhat greater than 2 g/kg/day. The figure is supported by data from adult rhesus which were in nitrogen balance consuming about 2.5 g of mixed protein/kg/day (45).

It is recognized that weight maintenance of an infant is, by definition, a highly unphysiological situation and that in both the infant and adult, weight maintenance alone may be accompanied by some degree of deterioration in nutritional status. The ultimate test of the minimal protein requirement is through its utilization in long term studies of both growing infants and maintaining adults. Nevertheless, these minimal protein requirements place a lower limit on the protein requirements and, coupled with an estimate of calorie needs, digestibility, and biological variation, can help in the planning of adequate protein supplies for animals of varying ages and growth rates.

VI. MINERALS

Little specific information is available concerning the mineral requirements of infant primates and, to date, more attention has been paid to avoiding excess mineral concentrations since they influence osmolality. The mineral content of Diet #74 is compared in Table IV to: a) a commercial human formula; b) recommended intakes for human infants (23); c) the mineral content of Diet #78 and d) the recommended intakes of adult monkeys recently published by the National Research Council (50). Such comparisons that span age and species can obviously be misleading although they can help to identify possible trouble areas. A case in point is the zinc content of Diets 74 and 78 which are probably low by any standard. Recommended intake of zinc for squirrel monkeys is 15-20 mg/kg diet (51) which corresponds to a dietary concentration of approximately 0.34-0.45 mg/100 kcal. Caloric intakes of infant squirrel monkeys consuming Diet 74 vary from 400-600 kcal/kg so that their total intake in mg/kg body weight is similar to that of adults. Nevertheless, the zinc content of our salt mix has recently been increased to provide 0.3 mg/100 kcal (15.2 mg/kg diet) in Diet 74 and 0.511 mg/100 kcal (22.4 mg/kg diet) in Diet 78.

VII. VITAMINS

Table V lists the vitamin content of Diet 74 and Diet 78 and compares them to a commercial infant formula and recommended allowances for humans. The data are expressed per 100 kcal of diet. Species peculiarities must, of course, be considered. Vitamin D_3, for example, is essential for New World primates and is also adequate for all Old World primates tested (52-53). Ascorbic acid deficiency in young, New World primates develops within a few weeks of deprivation (34) and has been a recurrent problem. The utilization of coated ascorbic acid beadlets[1] in the vitamin mixture has reduced oxidative loss, prevented scorbutic clinical signs, and circumvents oral dosing of every monkey on a routine basis. The concentration of ascorbate may seem excessive, but has been left at this level to compensate for unknown losses during preparation and storage. Choline and inositol, because

[1]*Ascorbic acid (USP-FCC) Type 5, Hoffman LaRoche, Inc.*

TABLE IV Comparison of Mineral Composition of Diets and Recommended Intakes for Monkeys and Humans (per 100 kcal)

Mineral	Semi-purified Dietary Formula		Commercial Infant Formula[2]	Recommended Daily Intakes	
	#74	#78		Human[3]	Monkey[4]
Potassium, mg	96.7	163.5	163.1	-----[1]	59-235
Calcium, mg.	91.8	155.2	76.2	60.0	141-212
Chloride, mg	69.0	114.9	73.8	-----[1]	59-118
Phosphorus, mg	47.3	79.9	58.6	30-50	70-140
Sodium, mg	44.0	74.4	46.7	-----[1]	59-118
Magnesium, mg	20.0	33.9	6.0	6.0	24-35
Sulfur, mg	9.4	15.8	-----	----	----
Iron, mg	3.0	5.16	1.78	1.0	2.4
Manganese, µg	821.0	1387.4	5.08	5.0	470
Iodine, µg	407.7	688.9	14.8	5.0	47
Zinc, µg	80.0	135.1	742.0	300.0	470
Copper, µg	50.9	86.0	60.5	30.0	47
Chromium, µg	6.9	11.7	-----	----	11.8
Selenium, µg	1.4	2.3	1.0	----	2.3

1 Recommended intake of potassium, chloride and sodium together are not to exceed 7.3 mEq/l.
2 Similac with Iron, Ross Laboratories, Columbus, Ohio.
3 Derived from Acta Paediatricà Scandinavica (23).
4 Derived from Nicolosi and Hunt (p.11) and NRC, 1978, (50).

TABLE V Comparison of Vitamin Compositions of Diets and Recommended Intakes for Monkeys and Humans (per 100 kcal)

Vitamin	Semi-purified Dietary Formula #74 and #78	Commercial Infant Formula[1]	Recommended Daily Intake Human[2]
Thiamin, μg	91.7	96.1	40.0
Riboflavin, μg	183.4	147.9	60.00
Vitamin B_6, μg	91.7	59.2	35.0
Calcium pantothenate, μg	573.1	443.8	300.0
Niacin, μg	917.0	1035.0	250.0
Folate, μg	91.7	7.4	4.0
Biotin, μg	4.6	----	1.5
Vitamin B_{12}, μg	3.44	0.22	0.15
Ascorbic acid[3], mg.	11.46	8.14	8.00
dl-α-tocopheryl acetate, IU	1.15	2.22	0.70
Vitamin A, IU	286.5	370.0	250.0
Vitamin D, IU	28.6	59.2	40.0
Vitamin K, μg	115.0	4.0	4.0

[1] Similac with Iron, Ross Laboratories, Columbus, Ohio.
[2] Derived from Acta Paediatrica Scandinavica (23).
[3] Ascorbic Acid (USP-FCC) Type 5, Hoffman-LaRoche, Inc.

of their hygroscopic properties, are not included in the vitamin mixture and are incorporated directly into the diet. The dietary folate supplement was recently quadrupled in light of evidence that prior levels were insufficient (see review by Rasmussen in this symposium).

VIII. OTHER CONSIDERATIONS

No investigation of the proper amount of bulk needed in these semi-purified liquid diets has been reported. Although a number of formulas have been utilized over time, the current infant diets contain no bulk and those for older animals contain approximately 5 grams/100 grams nutritive ingredients.

A recent concern has been the finding that carrageenan, commonly used as a stabilizing agent in human commercial formulas, is associated with gastrointestinal lesions in at least 5 species (21). Most of the lesions are attributable to degraded, hydrolyzed carrageenan. The effect of feeding native, undergraded carrageenan to infant baboons was studied over a period of 112 days (21) without deleterious effect. The suspending agents utilized in the diets described in the present report were derived from a microcrystalline cellulose. No data are available on its possible deleterious effects.

REFERENCES

1. Herndon, J.F., Greenburg, S.M., VanLoon, E.J., Kelleher, R.T., Cook, L., and Davidson, A. (1958) A liquid diet for animals in behavioral studies. *J. Exp. Anal. Behav. 6: 249-252.*
2. Ellison, T., and Riddle, W.C. (1961) Commercial liquid diet for animals in behavioral studies. *J. Exp. Anal. Behav. 4: 370.*
3. Jones, D.C., Lofland, H.S., Clarkson, T.B., and St.Clair, R.W. (1975) Plasma cholesterol concentrations in squirrel monkeys as influenced by diet and phenotype. *J. Food Science 40: 2-7.*
4. Lofland, H.B., Clarkson, T.B., St.Clair, R.W., and Lehner, H.D.M. (1972) Studies on the regulation of plasma cholesterol levels in squirrel monkeys of two genotypes. *J. Lipid Research 13: 39-47.*
5. Ausman, L.M., Hayes, K.C., and Hegsted, D.M. (1972) Protein deficiency and carbohydrate tolerance of the infant squirrel monkey (*Saimiri sciureus*) *J. Nutr. 102: 1519-1528.*

6. Ausman, L.M., Gallina, D.L., Camitta, B.M., Flath, L.C., and Hegsted, D.M. (1977) Acute erythroid hypoplasia in malnourished infant squirrel monkeys fed isolated soy protein. *Am. J. Clin. Nutr. 30: 1713-1720.*
7. Samonds, K.W., and Hegsted, D.M. (1978) Protein deficiency and energy restriction in young cebus monkeys. *Proc. Natl. Acad. Sci. 75: 1600-1604.*
8. Samonds, K.W., and Hegsted, D.M. (1973) Protein requirements of young cebus monkeys (*Cebus albifrons and apella*). *Am. J. Clin. Nutr. 26: 30-40.*
9. Ausman, L.M., Hayes, K.C., Lage, A., and Hegsted, D.M. (1970) Nursery care and growth of old and new world infant monkeys. *Lab. An. Care. 20: 907-913.*
10. Ausman, L.M. (1973) Harvard School of Public Health, Nutrition and Development of the Infant Squirrel Monkey. *(Saimiri sciureus).* Doctoral Dissertation.
11. Ausman, L.M., Gallina, D.L., Hayes, K.C., and Hegsted, D.M. (1976) Hematological development of the infant squirrel monkey. *Folia Primatologica 26: 292-300.*
12. Samonds, K.W., Ausman, L.M., and Hegsted, D.M. (1974) Hematological development of the cebus monkey *(Cebus albifrons and apella). Folia Primatologica 22: 72-79.*
13. Kaplan, J. (1974) Growth and behavior of surrogate-reared squirrel monkeys. *Developmental Psychobiology. 7: 7-13.*
14. Kaplan, J.N. (1977) Breeding and rearing squirrel monkeys *(Saimiri sciureus)* in captivity. *Laboratory Animal Science 27: 557-567.*
15. Kerr, G.R., Scheffler, G., and Waisman, H.A. (1969) Growth and development of infant *(Macaca mulatta)* fed a standardized diet. *Growth 33: 185-199.*
16. Kerr, G.R., Allen, J.R., Scheffler, G., and Waisman, H. A. (1970) Malnutrition studies in the rhesus monkey. I. Effect on physical growth. *Am. J. Clin. Nutr. 23: 739-748.*
17. Kerr, G.R., Chamove, A.S., Harlow, H.F., and Waisman, H. A. (1969) The development of infant monkeys fed low phenylalanine diets. *Pediat. Res. 3: 305-312.*
18. Fleischman, R.W. (1963) The care of infant rhesus monkeys (*Macaca mulatta*) *Lab. An. Care. 13: 703-710.*
19. Kaye, H., Povar, M.L., and Schrier, A.M. (1966) Rearing of *Macaca mulatta* from birth. *Lab. An. Care 16: 476-486.*
20. Hampton, S.H., and Hampton, J.K. (1967) Rearing marmosets from birth by artificial laboratory techniques. *Lab. An. Care. 17: 1-17.*

21. McGill, H.C., McMahan, C.A., Wigodsky, H.S., and Sprinz, H. (1977) Carrageenan in formula and infant baboon development. *Gastroenterology. 73: 512-517.*
22. Ordy, J.M., Samorajski, T., and Hershberger, T.J. (1970) Brain vulnerability to postnatal protein calorie deficiency in infant rhesus monkeys. *Proc. Soc. Exp. Biol. Med. 135: 680-684.*
23. Guideline on infant nutrition. I. Recommendations for the compostion of an adapted formula. *Acta Paediatrica Scandinavica.* Supplement 262. *European Society for Paediatric Gastroenterology and Committee on Nutrition.*
24. Tomarelli, R.M. (1976) Osmolality, osmolarity, and renal solute load of infant formulas. *J. Pediatrics. 88: 454-456.*
25. Segar, W.E., and Chesney, R.W. (1977) On certain physical-chemical properties of infant formulas. *Am. J. Dis. Child. 131: 137-138.*
26. Paxson, C.L., Adcock, E.W., and Morriss, F.H., Jr. (1977) Osmolalities of infant formulas. *Am. J. Dis. Child. 131: 139-141.*
27. Buss, D.H., and Cooper, R.W. (1972) Composition of squirrel monkey milk. *Folia. Primat. 17: 285-291.*
28. Evans, D.E. (1959) Milk composition of mammals whose milk is not normally used for human consumption. *Dairy Sci. Abs. 21: 277-278, 286.*
29. Lackey, T.D. (1961) General chemical components of milk: Mammals. In: Blood and Other Body Fluids. ed. P. L. Altman and D.S. Dittmer. Federation of American Societies for Experimental Biology. Bethesda, Maryland, 465-466.
30. Buss, D.H. (1969) Gross composition and variation of the components of baboon milk during natural lactation. *J. Nutr. 96: 421-426.*
31. Long, J.O., and Cooper, R.W. (1968) Physical growth and dental eruption in captive-bred squirrel monkeys, *Saimiri sciureus* (Leticia, Colombia) In: The Squirrel Monkey. Ed. L.A. Rosenblum and R.W. Cooper. *Academic Press, New York, 193-205.*
32. Ausman, L.M., and Hayes, K.C. (1974) Vitamin E deficiency anemia in Old and New World monkeys. *Am. J. Clin. Nutr. 27: 1141-1151.*
33. Hegsted, D.M., Worcester, J. (1966) Assessment of protein quality with young rats. *Proc. Seventh International Congress of Nutrition 5: 1-8.*

34. Hegsted, D.M. (1970) Nutritional research on the value of amino acid fortification---experimental studies in animals. In: Amino Acid Fortification of Protein Foods. Ed. N.S. Scrimshaw and A.M. Altshul. 157-178.

35. Ausman, L.M., Gallina, D.L., Samonds, K.W., and Hegsted, D.M. (1979 In press) Assessment of the efficiency of protein utilization in young squirrel and macaque monkeys. *Am. J. Clin. Nutr.*

36. Hambraeus, L. (1977) Proprietary milk versus human breast milk in infant feeding. A critical appraisal from the nutritional point of view. *Ped. Clin. North America 24: 17-47.*

37. Coward, D.G., and Whitehead, R.G. (1972) Experimental protein-energy malnutrition in baby baboons. *Br. J. Nutr. 28: 223-237.*

38. Deo, M.G., Sood, S.K., and Ramalingaswami, V. (1965) Experimental protein deficiency. *Arch. Path. 80: 14-23.*

39. Racela, A.S., Grady, H.J., Higginson, J., and Svoboda, D.J. (1966) Protein deficiency in rhesus monkeys. *Am. J. Pathol. 44: 419-443.*

40. Enwonwu, C.O., Stambaugh, R.V., and Jacobson, K.L. (1973) Protein-energy deficiency in nonhuman primates: biochemical and morphological alterations. *Am. J. Clin. Nutr. 26: 1287-1302.*

41. Kumar, V., and Chase, H.P. (1972) Progressive protein undernutrition and intestinal enzyme activities in monkeys. *Am. J. Clin. Nutr. 25: 485-489.*

42. Ordy, J.M., Samorajski, T., Zimmerman, R.R., and Rady, P.M. (1966) Effects of postnatal protein deficiency on weight gain, serum proteins, enzymes, cholesterol, and liver ultrastructure in a subhuman primate *(Macaca mulatta) Amer. J. Path. 48: 769-791.*

43. Kerr, G.R., Wolf, R.C., Waisman, H.A. (1966) A disorder of lipid metabolism associated with experimental hyperhistidinemia in *Macaca mulatta*. In: Some Recent Developments in Comparative Medicine. Symposia of the Zoological Society of London, No. 17. Academic Press, New York.

44. Geist, C.R., Zimmermann, R.R., and Strobel, D.A. (1972) Effect of protein-calorie malnutrition on food consumption, weight gain, serum proteins, and activity in the developing rhesus monkey *(Macaca mulatta) Lab. An. Sci. 22: 369-377.*

45. Robbins, R.C., and Gavan, J.A. (1966) Utilization of energy and protein of a commercial diet by rhesus monkeys *(Macaca mulatta) Lab. An. Care 16: 286-291.*

46. Wilgram, G.F., Lucas, C.C., and Best, C.H. (1958) Kwashiorkor type of fatty liver in primates. *J. Exp. Med. 108: 361-370.*
47. Follis, R.H. (1957) A kwashiorkor-like syndrome observed in monkeys fed maize. *Proc. Soc. Exp. Biol. Med. 96: 523-528.*
48. Hodson, H.H., Mesa, V.L., and VanRiper, D.C. (1967) Protein requirements of the young growing chimpanzee. *Lab. An. Care 17: 551-562.*
49. Riopelle, A.J., Hill, C.W., Su-Chen Li, Wolf, R.H., Weibold, H.R., and Smith, J.L. (1974) Protein deprivation in primates. I. Nonpregnant adult rhesus monkeys. *Am. J. Clin. Nutr. 27: 13-21.*
50. Nutrient Requirements of Nonhuman Primates (1978) No. 14 in a series of Nutrient Requirements of Domestic Animals. National Academy of Sciences, Washington, D.C.
51. Macapinlac, M.P., Barney, G.H., Pearson, W.N., and Darby, W.J. (1967) Production of zinc-deficiency in the squirrel monkey *(Saimiri sciureus)*. *J. Nutr. 93: 499-510.*
52. Hunt, R.C., Garcia, F.G., and Hegsted, D.M. (1967) A comparison of vitamin D_2 and D_3 in the New World primates. I. Production and regression of Osteodystrophia fibrosa. *Lab. An. Care 17: 222-234.*
53. Lehner, N.D.M., Bullock, B.C., Clarkson, T.B., and Lofland, H.B. Biological activities of vitamins D_2 and D_3 for growing squirrel monkeys. *Lab. An. Care 17: 483-493.*
54. Lehner, N.D.M., Bullock, B.C., and Clarkson, T.B. (1968) Ascorbic acid deficiency in the squirrel monkey. *Proc. Soc. Exp. Biol. Med. 128: 512-514.*

CALORIC INTAKE AND WEIGHT CHANGES IN ADULT RHESUS MONKEYS

Barbara C. Hansen
Kai-lin C. Jen

University of Michigan
Ann Arbor, Michigan

I. INTRODUCTION

Energy requirements and growth curves have been carefully studied in infant and very young rhesus monkeys (18,12,13,28, 4) and in infants of other primate species (1,24). Adult monkeys, on the other hand, have seldom been examined for nutrient requirements, and rarely has consideration been given to the optimal body weight of the adult rhesus monkey. Extrapolations for caloric requirements based on studies of monkeys up to 3 years of age have generally exaggerated the needs of the adult.

It is, therefore, important to analyze the nature of current information concerning the feeding and weight changes of adolescent and adult rhesus monkeys (limited here to monkeys weighing 4 kg or more). First consideration is given to the nature and variability of the diets used in the feeding of adult monkeys. Within the constraints of those diets

ISBN 0-12-334250-3

which have been adequately reported, various estimates and determinations of caloric requirements of adult rhesus monkeys will be compared. Finally, discussion will focus on expected body weight change in adult monkeys and the variability of estimated caloric needs based upon body weight criteria.

II. DIET AND ENERGY INTAKE OF ADULT RHESUS

In a review of the nutritional requirements of nonhuman primates, Kerr (10) concluded that adult macaques maintain "good health when fed a balanced diet providing 50-150 kcal/kg per day." Indeed, this wide range does cover the values reported, but provides only a gross index for assessing the adequacy of a monkey's intake. Clearly, intake is determined not only by the energy content of a diet, but also by minimal levels of essential nutrients. For purposes of comparison one can make tentative (although as yet not well tested) assumption that each of the diets reported in use for primates contains minimally acceptable levels of the essential nutrients, since each has described "adequate" maintenance of monkeys.

Table I presents the estimated percent of calories represented by protein, carbohydrate, and fat in a series of experiments using either adult pigtailed or rhesus macaques. The first three diets were specially formulated for experimental purposes with ingredients only partially specified in the publications. The fourth and fifth diets are commercially formulated, pelleted natural ingredients. The first five diets all have 13 to 17% of calories as protein, with a range of 4.5 to 20.5% as fat, and were formulated to provide approximately 4 kcal/gm. The sixth and seventh diets are commercial liquid formulas, originally developed for infant (#6) and for adult (#7) human use, with a caloric density of 1 kcal/ml. Unfortunately most of these studies were not designed to determine accurate spontaneous or *ad lib* caloric intakes, and none assessed both voluntary energy consumption and energy output.

Studies using solid diets (Diets 1,2 and 4) provided food according to fixed formulas. Robbins and Gavan (22,23) restricted intake to the amount calculated by the formula given by Crampton and Lloyd (3) to provide maintenance requirements (cal = 93 $(\text{wt.kg})^{0.75}$). Thus, their caloric levels of less than 400 kcal per day (57 kcal/kg/day) clearly were minimal or inadequate for "optimal" nutrition. Hoffman *et al.* (9) limited food available to 70 kcal/kg/day. Riopelle's (21) control diet (#3) was formulated to provide 4 g protein/kg/day if it was taken at a rate of 120 kcal/kg/day. Interestingly, they found (by their estimate) that "pregnant and nonpregnant animals both eat nearly 120 kcal/kg/day" (monkeys weights averaged 5.0 kg). While inter-animal variability may account for the finding of similar caloric intakes in these two diverse states, Portman (19) has estimated an increase of 25% in protein requirements in pregnancy based on extrapolations from human data.

Hamilton and his colleagues, (6,7) (#5) studying a group of monkeys over many years and using relatively imprecise methods of measuring caloric intake, documented the voluntary (*ad lib*) intake of monkeys. Monkey chow pellets were counted when placed in the cup, and leftovers were counted again the next morning. His estimates of daily caloric intake ranged from 54 kcal/kg for a 16 kg monkey to 136 kcal/kg for a 4.7 kg rhesus monkey, and are compared in Table II to findings of the present laboratory where a liquid diet was fed.

a. Liquid Diet: Liquid formulas (Table I) have been used occasionally because of the ease of caloric intake measurement and because they can be modified with ease. Jones *et al.* (10) and Lofland *et al.*, (17) found that a liquid diet containing 35.9% protein was able to maintain the body weight of adult squirrel monkeys. Rahlmann *et al.*, (20) reported the use of Enfamil with Iron[R] (#6) in two adult pigtailed macaques. Monkeys remained in positive nitrogen balance for up to four months while receiving no other source of food or water. All continued to gain weight and were described as in excellent health throughout. "Slightly diarrheic feces" were observed during the first week on the liquid diet, but for the remainder of the study stools were soft and formed. The National Research Council publication on nutritional requirements of nonhuman primates (16) comments that the major disadvantages of liquid diets are that "they tend to produce loose stools and are often poorly accepted by adult animals."

Since 1969 over 80 monkeys have been maintained entirely on liquid diets in the course of studies of feeding regulation by the present investigators. Prior to the commercial introduction of Ensure[R] (Ross Laboratories) (#7, Table I) either Enfamil with Iron[R] (Mead Johnson) or Similac[R] (Ross Laboratories) were used. Occasionally significant diarrhea occurred with the latter two diets, and one monkey totally refused the diet. It appears now that these problems were due to the severity of lactose intolerance in rhesus monkeys (29). Since 1973 all monkeys have been maintained on Ensure[R], a liquid diet formulated for use in tube-fed patients. Monkeys readily adapt to the diet and continue to grow normally.

TABLE I Composition of Diets Used for Feeding Pigtailed or Rhesus Macques

Diet ref.	Estimated % of calories Protein	carbohydrate	fat	Calorie Concentration (kcal)	Energy intake (kcal/kg/day)	Monkey Weights (kg)
22[a]	16.0	79.0	5.0	4.2/g	Restricted	4.2 - 9.6
9[b]	16.0	77.0	7.0	4.09/g	Limited to 70	7.4 - 8.4
21[c]	13.4	66.0	20.5	4.0/g	120	3.7 - 9.2
22[d],23[d]	16.4	79.0	4.5	3.47/g	Restricted	4.2 - 12.1
6[e],7[e]	16.9	70.1	13.0	4.18/g	112[h]	4.1 - 16.0
20[f]	8.9	41.3	49.8	1.0/ml	100	6.0 - 7.0
27[g]	14.0	54.5	31.5	1.06/ml	72[h],107[i]	4.0 - 8.0

[a] Caesin
[b] Diet B
[c] Control diet
[d] Okatie
[e] Purina[R] Chow 15
[f] Enfamil with Iron[R], Mead Johnson
[g] Ensure[R], Ross Laboratories
[h] 7.0 - 7.9 kg monkey
[i] 4.0 - 4.9 kg monkey

With the use of liquid diet and a suction-activated automated feeder, from which data can be directly fed into a data logger, precise daily food intake and feeding patterns have been obtained in this laboratory.

Several other powdered diets were previously tested for possible use in monkey feeding, but were found to be undesirable for *ad lib* feeding due to the occurrence of settling and clogging of the feeder pumps. Commercial homogenization is optimal. Souring of the Ensure[R] over the 8-hour feeding period is not a problem. However, when 24-hour feeding is needed, changing the diet twice daily or continuous refrigeration of the diet is necessary to prevent spoilage.

When data obtained from this feeding device are compared with those reported by Hamilton (6) (Table II), it is apparent that monkeys fed Ensure[R] consumed fewer calories per kg of body weight. The discrepancy may be attributed to the fact that dry pellets were used in Hamilton's experiment which make the estimate of food intake less accurate. On the other hand, the higher digestability, and hence higher feeding efficiency of the liquid diet used in this laboratory, may also be a factor in the reported differences in *ad lib* intake.

b. Other Considerations: In addition to the data reported above concerning energy requirements of monkeys, several others have estimated caloric requirements based upon expenditure. Portman (19) for example, estimated the basal metabolic rate of a 4 kg rhesus adult at 49 kcal/kg based on an assumed mean basal metabolic rate of 650 kcal/m^2 (2). Portman then estimated (generously) the energy requirements for the adult rhesus monkey to be 100 kcal/kg.

TABLE II Ad Lib Caloric Intakes of Rhesus Monkeys

Monkey body weight	*Purina Chow 15* (N=6) kcal/kg/day*	*Ensure[R] (N-16) kcal/kg/day*
4.0 - 4.9 kg	*136*	*107*
5.0 - 5.9 kg	*127*	*93*
6.0 - 6.9 kg	*115*	*84*
7.0 - 7.9 kg	*112*	*72*

**ref. 6*

Estimates of caloric requirements should specify the conditions under which the monkey is maintained together with the criteria used to determine the estimate. Variations in ratios of dietary nutrients, activity levels of the monkeys, environmental temperature, rate of growth, and other factors affecting basal metabolism or energy expenditure will all affect caloric need estimates. In addition there remains the question of what is an "optimal" level of energy intake, that is, one which provides for the most healthful state. This level may or may not be the intake which produces maximal rate of weight gain, as the nature of the added tissue must be considered. Some diets may lead to high rates of weight gain in adult animals which may be primarily adipose tissue or fluid volume and this may be less than optimal.

Because weight gain has been, and remains, almost the sole criterion of adequacy of a diet, further consideration of the "norms" for body weight change and its variability within and between monkeys is appropriate. Of importance is the consideration of intra-animal variability in the rate of weight gain. For example, in Figure 1 each square represents

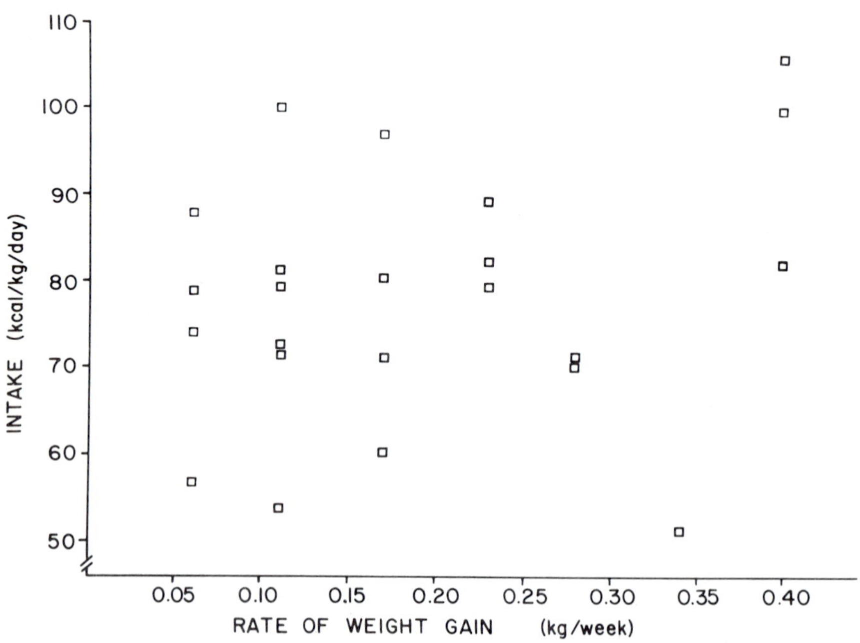

Fig. 1. Intra-monkey variation for rates of weight gain at a given caloric intake are demonstrated for a single monkey. Each square represents the average intake during a period of 1 week and the measured rate of weight gain.

a seven-day period of stable intake, and the respective weight change for that period in a single monkey. This monkey gained a highly variable amount of weight (from .06 kg/week to .40 kg/week) with virtually the same amount of food intake (about 80 kcal/kg/day). Conversely, a similar rate of weight gain was achieved by ingesting of a wide range of calories. For example, the monkey in Figure 1 had gained .11 kg/week while ingesting a constant number of calories ranging between 54 kcal/kg/day to 101 kcal/kg/day. This intra-animal variability between intake and rate of weight gain varies widely between monkeys.

c. Feeding Schedules and Weight Gain: The effect of different feeding regimens on daily food intake and rate of weight gain also have been investigated. Feeding schedules of 1 meal, 2 meals, 3 meals, 8 hours and 24 hours *ad lib*

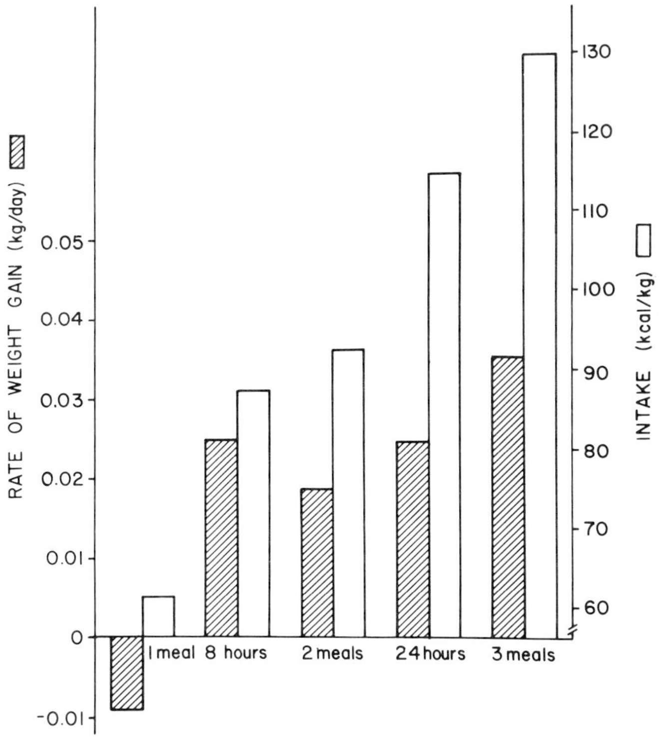

Fig. 2. The effect of food availability on daily food intake and rate of weight gain are demonstrated for a group of 12 monkeys over a 3-year period.

feeding per day were randomly assigned to 12 rhesus monkeys over a 3-year period. As illustrated in Figure 2, the longer food was available for feeding, the greater amount of food ingested. However, the rate of weight gain did not correlate well with food intake per kg body weight per day. With one 30 minute meal per day, monkeys were not able to ingest enough calories to maintain their body weights. With at least 2 meals or 8 hours feeding a day, steady weight gain was achieved. Although daily caloric intake per kg of body weight varied widely, there was no difference in the rate of weight gain between these feeding schedules. This suggests some adaptation mechanisms may have developed in response to different feeding schedules in order to maintain a constant rate of weight gain.

III. BODY WEIGHT CHANGE IN ADULT RHESUS

The Van Wagenen and Catchpole (25) average chronological weight gain for rhesus monkeys is shown in Figure 3 and was used as a reference to predict the age of our own monkeys. The monthly weight gain of monkeys maintained on Ensure[R] in this laboratory has also been plotted in Figure 3. There was no difference in rate of weight gain between these 2 groups of monkeys. The rates of weight gain also compare favorably with those of Hamilton's monkeys fed Purina chow (6). The food intake per kg of body weight was, however, lower for monkeys taking Ensure[R] (Table II). This suggests that monkeys ingesting this liquid diet may have higher feeding efficiencies in terms of body weight gain. This does not rule out other factors which may contribute to this difference, such as reduced energy expenditure for monkeys in primate restraint chairs and overestimates of intake for pelleted diets.

IV. EXPERIMENTAL CONSIDERATIONS

Since caloric requirements of adult rhesus monkeys are poorly documented, body weight gain has been the usual criterion used to judge the nutritional status of the monkeys. While the symptoms of undernutrition are easily identified, overnutrition has seldom been considered in estimating nutritional needs. King (14) reported that there is a tendency to feed foods which are too calorically concentrated or contain excessive vitamins. Percent of caloric intake as protein has also varied in different diets from 8.9% to 28.8% (Table I). The effects of overnutrition by excessive intake of calories, protein, vitamins or other nutrients require further examination.

In order to establish a "norm" for adult monkey growth or weight change, the accuracy of measuring this change becomes extremely important. The amount consumed by each monkey during each "meal" varies widely. It is therefore suggested that the early morning body weight before food is offered to the monkey be used as the most stable and representative body weight. Even if food is continuously available, the early morning weight still is less affected by gut content than any other time during the day due to the diurnal eating habit in primates.

Besides accurate body weight measurements, food intake also requires precise measurement. When dry chow pellets are offered to the monkeys, the difference between the amount given and amount left has been used as the food intake of the animals. The spillage and the pellets soaked in the urine pool generally make such measurements somewhat inaccurate. It is recommended that only the "dry" weight be used to estimate spillage. When a liquid diet is adopted, commercial homogenizing is advantageous to prevent layering and clogging and to permit accurate intake assessment. The feeding device should be stable enough to avoid spillage due to cage or restraint chair shaking by the monkeys. The feeding device also needs daily cleaning to prevent spoilage.

Both Hamilton's (6) and our data indicate that the caloric requirement for rhesus monkeys varies widely for individual monkeys over a long period of time. The inter-monkey variation is even greater than that within monkeys (6). It is therefore more meaningful to use each individual monkey as its own control when dealing with any experimental regimen related to feeding and growth. This finding also demonstrates that only over a relatively long period of time can any changes in feeding patterns due to experimental manipulation be established.

V. CONCLUSIONS

The data of Van Wagenen and Catchpole (25), Hamilton (6) and this laboratory (Figure 3) demonstrate that in rhesus monkeys "normal" growth continues well into adulthood, but at a slower rate compared to infant or adolescent monkeys. The body composition and the upper limits of this continuous growth have not yet been determined. There is some evidence that the weight gain in adulthood is mainly due to increased adiposity (8), but data from other species also indicate that this increased body weight may not be due to adiposity (26).

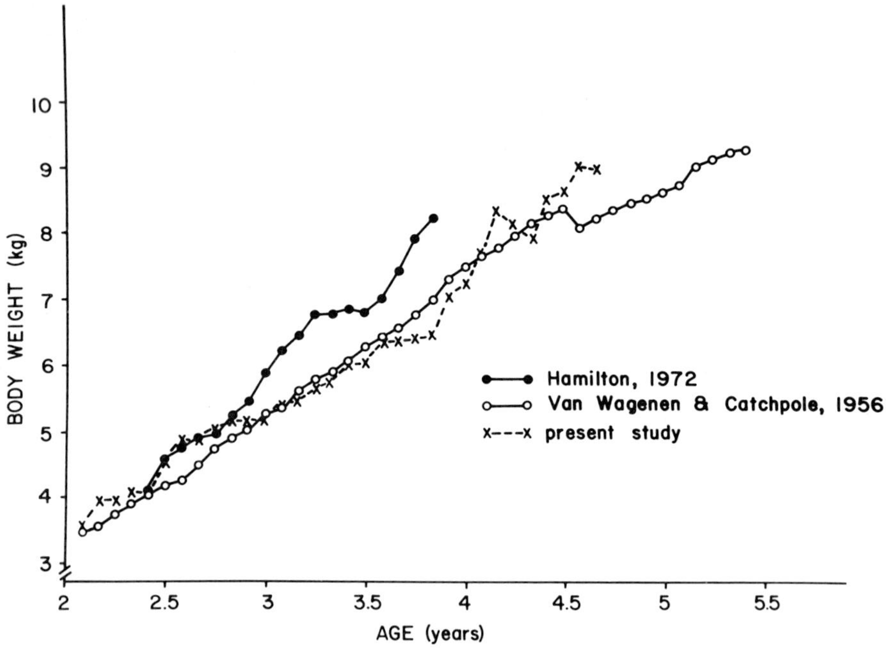

Fig. 3. Comparison of weight gain in 3 studies of rhesus monkeys. a) Van Wagenen and Catchpole, 1956, diet not described and animals caged. b) Hamilton, 1972, Purina Chow 15 ad lib, animal caged. c) Present study, Ensure[R] 8 hours/day, animals in chronic chair restraint.

Although caloric requirements at certain body weights can be generally calculated, the specific nutritional requirement will depend on the quality of protein in the diet, its digestibility, caloric density of the diet, age of the animal, environmental surroundings, health/physiological state of the animal, the feeding schedule, and the adaptive mechanisms developed by each individual animal. Therefore, any "normal" nutritional requirement reported should be considered in the context of the laboratory of origin and experimental conditions. Moreover, the individual differences of unknown nature in the efficiency of energy utilization and the variation of energy utilization within each individual monkey over time make the "optimal" nutritional requirement even more specific to each animal. Finally, the "optimal" nutritional requirement can only be established when food intake, energy expenditure and energy balance have been accurately estimated. This information will not be obtained until a calorimeter has been constructed and used to assess the short and long term changes in feeding efficiency and body weight in the nonhuman primate.

REFERENCES

1. Ausman, L.M., Hayes, K.C., Lage, A., and Hegsted, D.M. (1970) Nursery care and growth of Old and New World infant monkeys. *Lab. Anim. Care 20: 907-913.*
2. Bruhn, J.M. (1934) The respiratory metabolism of infrahuman primates. *Am. J. Physiol. 110: 477-484.*
3. Crampton, E.W., and Lloyd, L.E. (1959) Fundamentals of nutrition. San Francisco: W. H. Freeman and Co.
4. Deo, M.G., and Ramalingaswami, V. (1960) Production of periportal fatty infiltration of the liver in the rhesus monkey by a protein deficient diet. *Lab. Invest. 9: 319-329.*
5. Durnin, J.V.G.A., Edholm, O.G., Miller, D.S., and Waterlow, J.C. (1973) How much food does man require? *Nature 242: 418.*
6. Hamilton, C.L. (1972) Long term control of food intake in the monkey. *Physiol. and Behav. 9: 1-6.*
7. Hamilton, C.L., Ciaccia, P.J., and Lewis, D.A. (1976) Feeding behavior in monkeys with and without lesions of the hypothalamus. *Am. J. Physiol. 230: 818-830.*
8. Hamilton, C.L., Kuo, P.T., and Feng, L.Y. (1972) Experimental production of syndrome of obesity, hyperinsulinemia and hyperlipidemia in monkeys. *Proc. Soc. Exp. Biol. Med. 140: 1005-1008.*
9. Hoffman, R.A., Dozier, E.A., Mack, P.B., Hood, W.N., and Parrott, M.W. (1968) Physiologic and metabolic changes in *Macaca nemistrina* on two types of diets during restraint and non-restraint. *Aerospace Medicine. 39: 693-708.*
10. Jones, D.C., Lofland, H.B., Clarkson, T.B., and St. Clair, R.W. (1975) Plasma cholesterol concentrations in squirrel monkeys as influenced by diet and phenotype. *J. Food Sci. 40: 2-7.*
11. Kerr, G.R. (1972) Nutritional requirements of subhuman primates. *Physiol. Rev. 52: 415-467.*

12. Kerr, G.R., Allen, J.R., Scheffler, G., and Waisman, H.A. (1970) Malnutrition studies in the rhesus monkeys. I. Effect on physical growth. *Am. J. Clin. Nutr. 23: 739-748.*
13. Kerr, G.R., Scheffler, G., and Waisman, H.A. (1969) Growth and development of infant *Macaca mulatta* fed a standardized diet. *Growth 33: 185-199.*
14. King, G.J. (1978) Comparative feeding and nutrition in captive, non-human primates. *Br. J. Nutr. 40: 55-62.*
15. Lofland, H.B., Clarkson, T.B., St. Clair, R.W., and Lehner, N.D.M. (1972) Studies on the regulation of plasma cholesterol levels in squirrel monkeys of two genotypes. *J. Lipid Res. 13: 39-47.*
16. National Research Council (NRC) (1978) Nutrient requirements of non-human primates. National Academy of Sciences, Washington, D.C.
17. Pace, N., Hansen, J.T., Rahlmann, D.F., Barnstein, N.J., and Cannon, M.D. (1964) Preliminary observations of some physiological characteristics of the pigtailed monkey, *Macaca nemistrina*. *Aerosp. Med. 35: 118-121.*
18. Pickering, D.E., Scott, F.S., Vangenen, G., and Fisher, D.A. (1953) Growth and metabolism in normal and thyroid-ablated infant rhesus monkeys (*Macaca mulatta*). *Am. J. Dis. Child. 86: 1-10.*
19. Portman, O.W. (1970) Nutritional requirements of non-human primates. In R.S. Harris, ed. Feeding and Nutrition of non-human primates. New York: Academic Press.
20. Rahlmann, D.F., Pace, N., and Grunbaum, B.W. (1968) Adequacy of liquid diets for maintenance of pigtailed monkeys. *Lab. Anim. Care. 18: 631-636.*
21. Riopelle, A.J., Hill, C.W., Li, S.C., Wolfe, R.H., Siebold, H.R., and Smith, J.L. (1974) Protein deprivation in primates. *Am. J. Clin. Nutr. 27: 13-21.*
22. Robbins, R.C., and Gavan, J.A. (1965) Influence of diet on the qualitative amino acid patterns in the urine of rhesus monkeys *(Macaca mulatta)*. *Human Biol. 37: 137-147.*
23. Robbins, R.C., and Gavan, J.A. (1966) Utilization of energy and protein of a commercial diet by rhesus monkeys (*Macaca mulatta)*. *Lab. Anim. Care. 16: 286-291.*
24. Samonds, K.W., and Hegsted, D.M. (1973) Protein requirements of young cebus monkeys *(Cebus albifrons and apella)*. *Am. J. Clin. Nutr. 26: 30-40.*
25. Van Wagenen, G., and Catchpole, H.R. (1956) Physical growth of the rhesus monkey *(Macaca mulatta)*. *Am. J. of Phys. Anthrop. 14: 245-273.*

26. Walike, B.C., Goodner, C.J., Koerker, D.J., Chideckel, E.W., and Kalnasy, L.W. (1977) Assessment of obesity in pigtailed monkeys. *J. Med. Primatol. 6: 151-162.*
27. Walike, B.C., Kribbs, P.J., and Kalnasky, L.W. (1974) Feeding behavior in rhesus monkeys *(Macaca mulatta)*. Fifth International Conference on the Physiology of Food and Fluid Intake, Jerusalem.
28. Waisman, H.A., Rasmussen, A.F., Jr., Elvehjem, C.A., and Clark, P.F. (1943) Studies on the nutritional requirements of the rhesus monkey. *J. Nutr. 26: 205-218.*
29. Wen, C.P., Antonowicz, I., Tovar, E., McGandy, R.B., and Gershoff, S.N. (1973) Lactose feeding in lactose-intolerant monkeys. *Am. J. Clin. Nutr. 26: 1224-1228.*

FOLIC ACID NEEDS FOR BREEDING AND GROWTH OF NONHUMAN PRIMATES

Kathleen Maher Rasmussen

Division of Nutritional Sciences
Cornell University
Ithaca, New York

I. INTRODUCTION

Nonhuman primates were first used as animal models for the study of folate deficiency in the early 1930's (1), more than a decade before the isolation (2), and chemical characterization (3), of the vitamin. The similarity of the characteristics of folate deficiency in nonhuman primates to those in human beings argues for the continued use of these animals as models for studies of various aspects of folate metabolism. This review characterizes folate deficiency and estimates the folate requirement of nonhuman primates. The use of nonhuman primates as models for problems

ISBN 0-12-334250-3

in folate nutrition is discussed. The particular example of folate needs during pregnancy is considered along with the more general topic of our current knowledge of nutrient requirements for optimal reproduction in nonhuman primates.

II. CHARACTERISTICS OF FOLATE DEFICIENCY

Experimental folate deficiency has been studied in young, growing animals of many species including the chick (4), pig (5), cat (6), mouse (7-9), rat (10-12), and guinea pig (13-15) (Table I). When fed a folate deficient purified diet without the use of sulfonamide drugs (as intestinal antiseptics) or vitamin antagonists, only guinea pigs developed megaloblastic anemia (14). Slungaard and Higgins (14) observed anemia in only some of their animals with bone marrow megaloblastosis, but other workers have been unable to confirm the simultaneous appearance of both megaloblastosis and anemia in the guinea pig (13-15). Bone marrow megaloblastosis without anemia was observed in cats fed a folate-deficient diet (6). Investigators have produced megaloblastic anemia in the pig (5), mouse (8), and rat (10) by feeding a folate-deficient diet with intestinal antiseptics and/or folate antagonists.

Experimental folate deficiency has been studied without the use of sulfonamides or folate antagonists in a number of nonhuman primate species including the rhesus monkey (16), baboon (17), cebus monkey (18), squirrel monkey (19), and marmoset (20) (Table II). Weight loss and leukopenia were more prominent findings in these nonhuman primates than in man (21). Anemia resulting from folate deficiency has been reported in the marmoset and rhesus, cebus, and squirrel monkeys, while only the baboon and cebus and squirrel monkeys displayed bone marrow megaloblastosis. The cytologic characteristics of megaloblastic anemia in the cebus monkey (Figure 1) were very similar to those seen in man (18).

In their survey of Old World primates, Huser and Beard (22) concluded that the normal rhesus monkey resembled man most closely in serum and red blood cell (RBC) folate values. Unfortunately, neither they nor other investigators have reported values for these parameters in folate-deficient rhesus monkeys. A comparison of blood and liver folate values and urinary excretion of formiminoglutamic acid (FIGLU) (a biochemical index of folate status) in several nonhuman primate species is presented in Table III. In those species

TABLE I Comparison of the Characteristics of Experimental Folate Deficiency in Young Animals of Various Species

Characteristic	Chick[a]	Pig[*b]	Cat[c]	Mouse[def]	Rat[ghi]	Guinea Pig[jkl]
Weight loss or growth failure	+	+	−	+	+*	+
Diarrhea		+	−		+*	+
Gingivitis		−	−		+*	−
Anemia	+	+	−	+*	±*	+
Leukopenia	+	+	−	+*	+*	+
Thrombocytopenia	+	±				
Low serum folate			+		+	+
Low RBC folate			+		+*	+
Low liver folate			+	−	+	+
High urinary FIGLU			+			−*
Macroovalocytosis	+	+	−		−	−*
Megaloblastic bone marrow		+	+	+*	+*	+

*Sulfonamide drugs and/or folate antagonists were used.

[a] Campbell et al. (4)
[b] Cartwright et al. (5)
[c] Thenen and Rasmussen. (6)
[d] Nielson and Black. (7)
[e] Weir et al. (8)
[f] Shaw et al. (9)
[g] Franklin et al. (10)
[h] Chanarin et al. (11)
[i] Tigner and Roe. (12)
[j] Reid et al. (13)
[k] Slungaard and Higgins. (14)
[l] Thenen. (15)

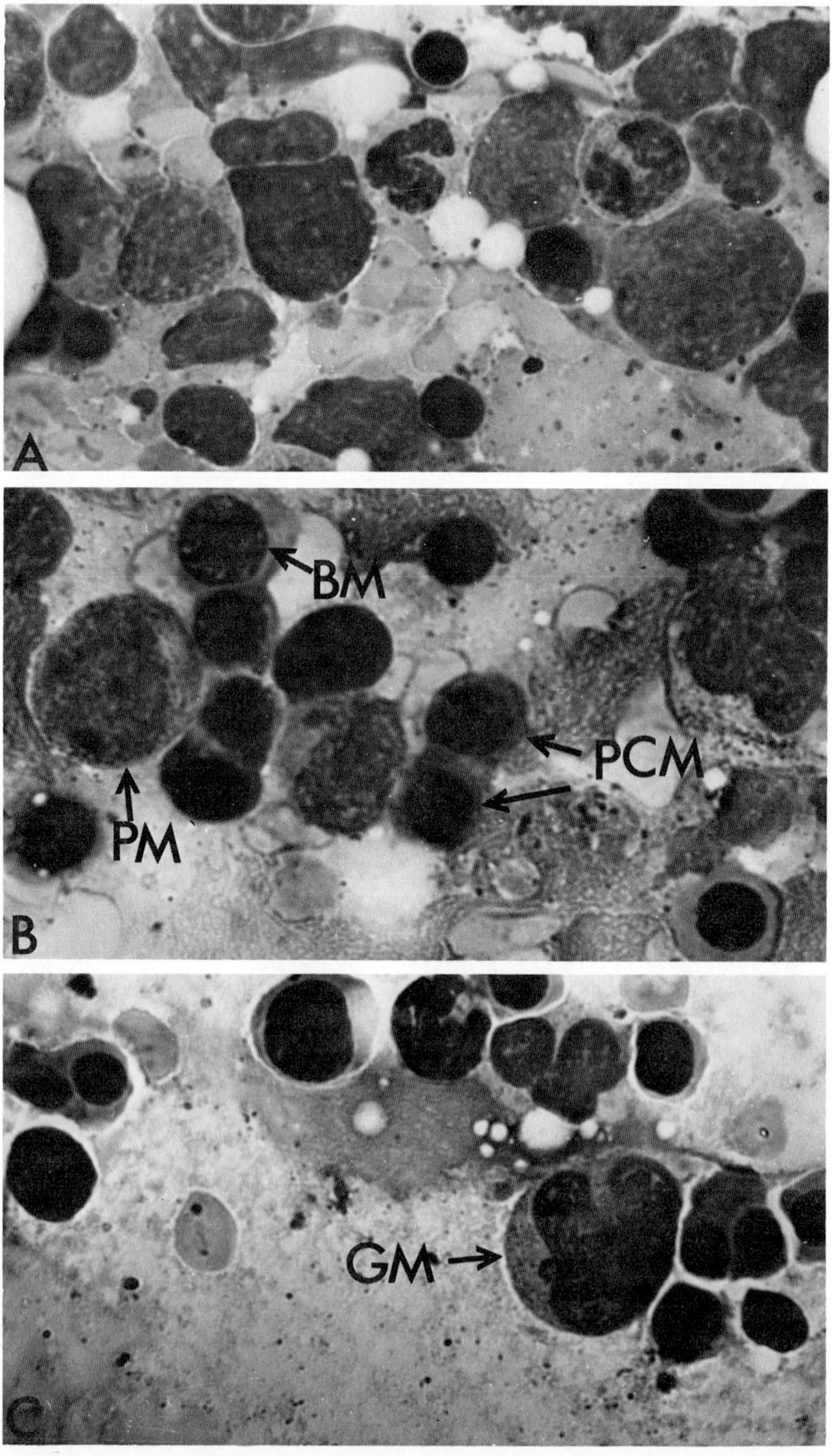

Fig. 1. Bone marrow smears obtained from cebus monkeys

for which the data exist, plasma and liver folate values were reduced and urinary excretion of FIGLU was increased during experimental folate deficiency. In the folate-deficient baboon, RBC folate values were reduced in comparison with animals fed the control diet (17). This reduction was not as clear in cebus (18) and squirrel (19) monkeys because the control diets used contained only marginally adequate amounts of folic acid.

III. ESTIMATES OF FOLATE REQUIREMENTS

Folate requirements have been estimated for three nonhuman primate species, namely the rhesus (23, 24), cebus (18), and squirrel (19) monkeys. The estimated folate requirement of the rhesus monkey of 40 μg/kg body weight/day was derived from studies of the ability of various purified fractions containing the vitamin to prevent the development of nutritional cytopenia. Recently a more mathematical approach for estimating the folate requirement of cebus and squirrel monkeys has been used (18,19). Figures 2 and 3 illustrate the application of this method to data from female squirrel monkeys. Using this method the authors found that 28 and 35 μg of total folate/kg body weight/day were required for weight maintenance in nearly adult squirrel and cebus monkeys, respectively. It is noteworthy that the folate requirement for weight maintenance varied considerably between individual animals and that in animals of this age, the requirement for weight maintenance was less than half the amount needed for maintenance of normal hematologic parameters and cytologic features in the bone marrow (>75 μg/kg body weight/day).

---

fed diets containing various levels of added folic acid. Wright's stain X 1,000. *A, bone marrow smear from monkey fed control diet throughout the experiment; normal cell size and cytologic features are evident.* *B, bone marrow smear from monkey fed the deficient diet; many cytologic abnormalities characteristic of a megaloblastic process are present including the open nuclear chromatin of the promegaloblast (PM) and the clumped nuclear chromatin of the basophilic* (BM) and polychromatophilic megaloblasts (PCM) producing a *"clockface" pattern.* *C, bone marrow smear from another monkey fed the deficient diet; a giant metamyelocyte (GM) and two orthochromic megaloblasts with double nuclei are visible. From Rasmussen (28).*

TABLE II Comparison of the Characteristics of Experimental Folate Deficiency in Man and Five Nonhuman Primate Species

Characteristic	Man[a]	Nonhuman Primates				
		Old World		New World		
		Rhesus[b]	Baboon[c]	Squirrel[d]	Cebus[e]	Marmoset[f]
Weight loss	±	+	+	+	+	+
Diarrhea	−	+	+	±	−	+
Gingivitis	−	+	+	−	−	+
Anemia	+	+	±	+	+	+
Leukopenia	±	+	+	±	+	+
Thrombocytopenia	±	±	+	±		
Low serum folate	+		+	+	+	
Low RBC folate	+		+	+	+	
High urinary FIGLU	+			+	+	
Neutrophil hypersegmentation	+		−			
Macroovalocytosis	+		+	+	+	
Megaloblastic bone marrow	+	±	+	+	+	

[a]Herbert. (21)
[b]Day. (24)
[c]Siddons. (17)
[d]Rasmussen et al. (19)
[e]Rasmussen et al. (18)
[f]Dreizen and Levy. (20)

TABLE III Values for Blood and Liver Folate and FIGLU Excretion in Four Nonhuman Primate Species Fed Folate-Deficient (DD), Control (CD) or Stock (SD) Diets

Species	Plasma Folate ng/ml			RBC folate ng/ml			Liver folate µg/g			FIGLU excretion µmoles/8 hr		
	DD	CD	SD	DD	CD	SD	DD	CD	SD	DD	CD	SD
Baboon	1.0[a]	7.8[a]	18[b]	25[a]	67[a]	42[b]	1.5[a]	6.8[a]	--	2000[a] (max)	285[a]	--
Rhesus	--	--	10[b]	--	--	200[b]	0.2[c*]	1.1[c]	--	--	--	--
Cebus	1.0[d]	1.1[e] 1.8 2.7	2.4[e]	24[d]	23[e] 27 23	18[e]	--	--	--	88[e]	4.4[e]	--
Squirrel	0.6[d]	1.6[f]	7.8[f]	49[d]	43[f]	103[f†]	--	0.2[f]	--	4[f§]	0.1[f§]	--

[a]Siddons. (17)
[b]Huser and Beard. (22)
[c]May et al. (1952) Blood, 7: 978
[d]Rasmussen, Thenen and Hayes (unpublished statistical analyses)
[e]Rasmussen et al. (18)
[f]Rasmussen et al. (19)
*Fed diet deficient in both folate and vitamin C.
†Fed stock diet supplemented with folic acid.
§24-hr. collection period used.

These estimates indicate that, for example, a 2-kg cebus monkey would need more than 150 µg of folic acid/day. This figure allows little margin for individual differences, yet it exceeds the recent NRC recommendation (25) by 50%. Since it is likely that folate requirements are even higher in animals younger than the nearly adult monkeys studied and since some allowance should be made for destruction of this vitamin during storage, purified diets for nonhuman primates should be formulated to provide at least 200 µg of folic acid/day.

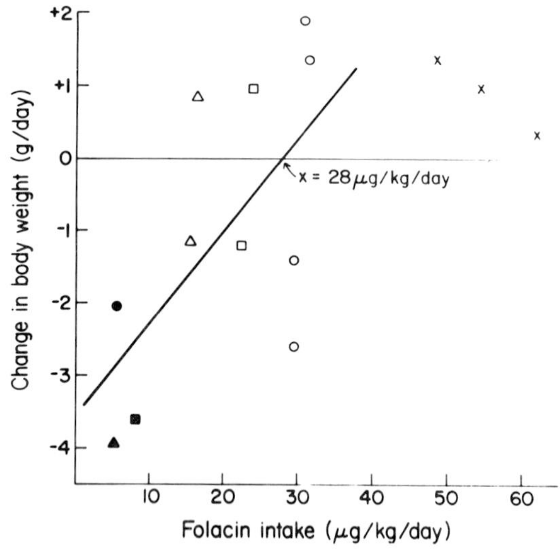

Fig. 2. Effect of folate depletion on body weight in squirrel monkeys. Regression lines are shown for each period of consistent weight change used in calculation of the folate requirement of these animals. The magnitude of the weight change (the slope of the regression line) is indicated with the mean food intake during the same period shown in parentheses. Each arrow denotes an injection of 500 µg of folic acid. From Rasmussen et al. (19).

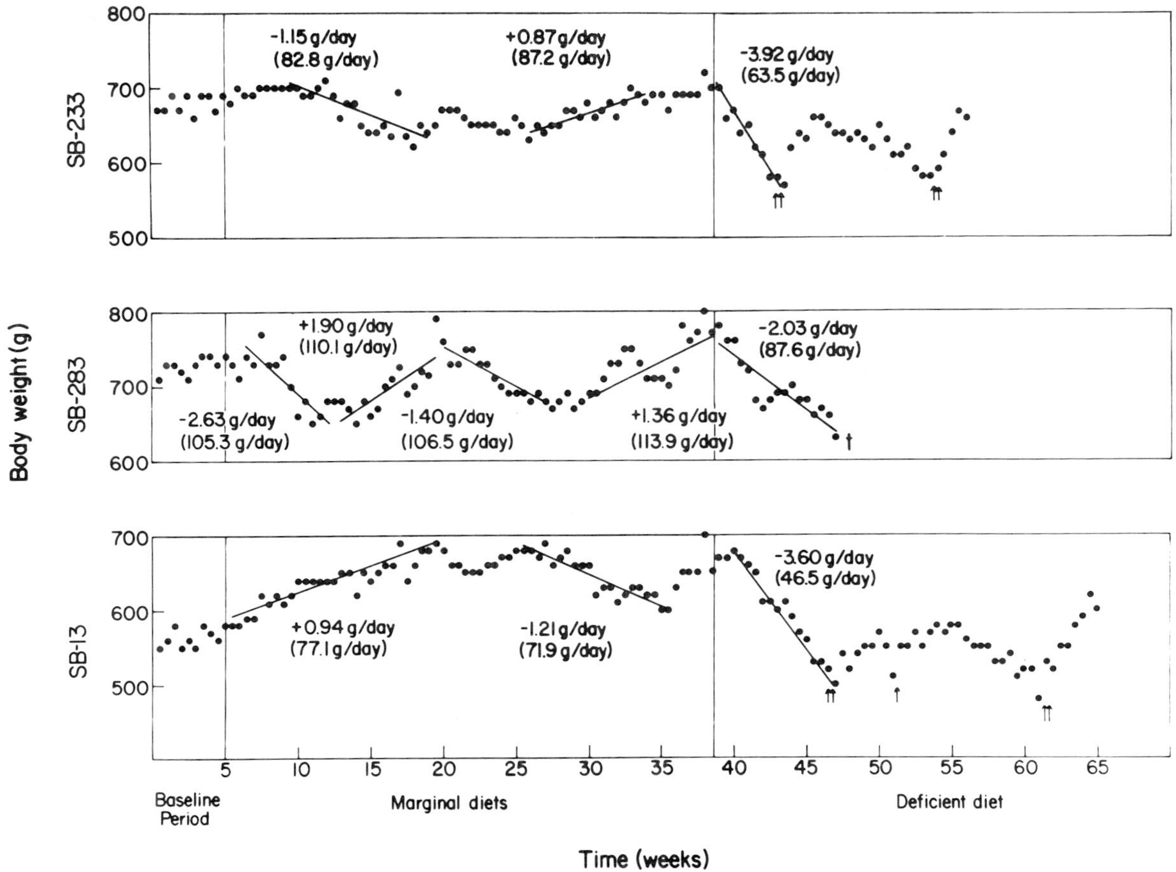

(See bottom of next page for legend).

IV. NONHUMAN PRIMATE MODELS FOR HUMAN FOLATE METABOLISM

Nonhuman primates represent the animal model of choice for studies of folate deficiency in which the development of megaloblastic anemia is an important criterion for assessing the degree of folate depletion. Such studies would include the effect of folate deficiency during pregnancy, the mechanism of the production of megaloblastic anemia, and the interaction of folic acid with vitamins B_{12} and C. Other areas of folate metabolism currently under investigation for which the use of nonhuman primates as animal models would be valuable include the effects of alcoholism on folate metabolism (particularly during pregnancy), the interaction of folate with iron and zinc, and the neurological consequences of folate deficiency.

The effect of folate deficiency during pregnancy is an example of current research in folate metabolism where studies with nonhuman primates have the potential to resolve the conflicts present in the literature. Folate deficiency has been associated with a variety of obstetric problems including megaloblastic anemia, abruptio placentae, placenta previs, "accidental" hemorrhages, toxemia, and impaired immunocompetence (reviewed in 26-28). In addition, folate deficiency during pregnancy has been associated with a variety of adverse fetal outcomes such as anoxia, abortion, stillbirth, low birth weight, prematurity, congenital anomalies, and impaired immunocompetence (reviewed in 26-28). Debate about the role of maternal folate deficiency in the etiology of most of these conditions exists in the literature.

Fig. 3. Folate requirement for maintenance of body weight in squirrel monkeys. Points represented by crosses were obtained from animals fed control diet, open symbols from animals fed diets marginally deficient in folate, and solid symbols from animals fed folate-deficient diet (see legend for Fig. 2). A regression line is shown for points obtained from monkeys fed the marginally deficient and deficient diets and indicates that 28 μg of total folate/kg body weight/day are required for maintenance of body weight in nearly adult animals. Considerably higher folate intakes are required for maintenance of normal hematologic parameters and cytologic features in the bone marrow. From Rasmussen et al. (19).

In view of the high prevalence of folate deficiency among pregnant women (especially those in developing contries) (29), it is important to clarify the effects of folate deficiency during pregnancy on both the mother and the developing fetus.

V. FOLATE NEEDS FOR SQUIRREL MONKEY REPRODUCTION

Both the usefulness of one nonhuman primate species as a model for studies of folate nutrition during pregnancy and their folate requirement during this critical period have been explored in a recent study (30). Letitian squirrel monkeys maintained in the breeding colony at the Harvard School of Public Health were divided into two groups during the first of that two-year study. One group received their usual stock diet (Purina Monkey Chow 5045, Ralston-Purina, St. Louis, Mo.) supplemented with folic acid (FAS) (added to the drinking water, providing a supplement of about 60 μg/day) while the other group received only the usual stock diet (NS). Observations were made on changes in maternal body weight during gestation, the outcome of pregnancy, infant size at birth and, in subgroups representing half of the FAS and NS groups, changes in various hematologic parameters before, during and after pregnancy.

Rasmussen and coworkers (30) demonstrated that supplemental folic acid produced significant changes in mean cell volume, mean cell hemoglobin, mean cell hemoglobin concentration, and RBC folate values in non-pregnant animals over a three-month period (Table IV). At the end of a year and following an intervening pregnancy for most of the animals, RBC folate values in the FAS group remained significantly higher than they were before folic acid supplementation was begun. It was concluded that the supplement improved the folate status of these monkeys.

Monkeys receiving supplemental folic acid in this study (30) began to gain weight about 10 days earlier in gestation and gained significantly more weight during pregnancy (Figure 4). As one would anticipate from this difference between the groups in maternal weight gain, infants born to FAS dams were significantly heavier than those born to NS dams (Table V).

The folic acid supplement also produced significant differences between the groups during pregnancy and the post partum period in mean cell volume (Figure 5) and RBC folate (Figure 6) values in the study of Rasmussen and associates

TABLE IV Short- and Long-term Effects of Folic Acid Supplementation On Hematologic Parameters and RBC Folate Values in Non-Pregnant Squirrel Monkeys. All Comparisons Presented Showed Significant ($P < 0.05$) Differences by Analysis of Variance.

	Duration of Supplementation			
	Before	5 Weeks	11 Weeks	13 Months
	(26)[1]	(26)	(21)	(22)
MCV, fl	62.7 ± 0.6[1]	63.3 ± 0.4	61.8 ± 0.5	62.8 ± 0.5
MCH, pg	20.5 ± 0.2	19.8 ± 0.2[a]	18.6 ± 0.2[a]	20.6 ± 0.3
MCHC, g/dl	32.7 ± 0.3	31.4 ± 0.2	30.2 ± 0.3[a]	32.7 ± 0.3
RBC folate, ng/ml	69.9 ± 4.8	55.1 ± 3.2[a]	91.0 ± 7.0[a]	96.1 ± 4.6[a]

[1]Mean ± SE; values in parentheses represent number of animals.
[a]Significantly different ($P < 0.05$) from values before supplementation.
From Rasmussen et al. (30)

TABLE V Effect of Folic Acid Supplementation on Birth Weight of Squirrel Monkeys During Year 1 of a Two-Year Study at the Harvard School of Public Health

	Birth Weight, g		
	Females	Males	All animals
Non-supplemented	96.1 ± 6.1[1] (7)	106.9 ± 4.5 (19)	104.1 ± 3.8 (26)
Folic acid-supplemented	105.2 ± 6.6 (6)	121.3 ± 4.0[a] (15)	116.7 ± 3.7[a] (21)

[1] Mean ± SE; values in parentheses represent number of animals.
[a] Significantly different ($P < 0.05$) from the non-supplemented group.
From Rasmussen et al. (30).

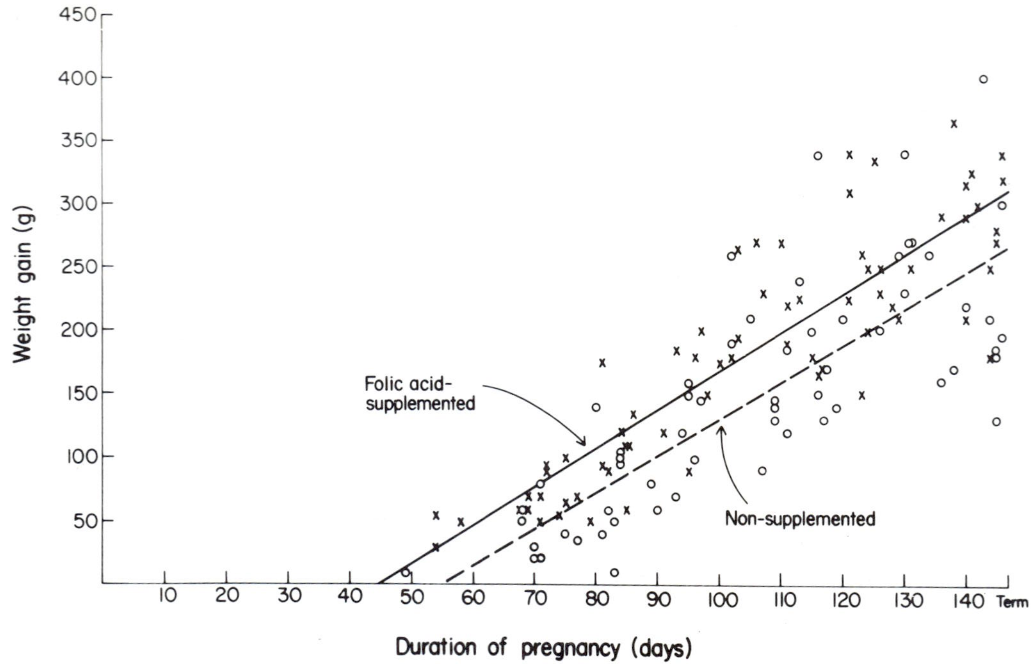

Fig. 4. Effect of folic acid supplementation on maternal weight gain during pregnancy in squirrel monkeys. Data points are restricted to healthy dams that delivered fullterm live-born infants; the length of gestation was assumed to be equal in this selected group of monkeys. The duration of gestation was computed by working backward from the date of parturition. Crossess represent monkeys receiving the vitamin supplement; open circles represent animals fed stock diet alone. Regression lines for both groups are shown and differ significantly by analysis of covariance. From the first year of a two-year study at the Harvard School of Public Health by Rasmussen et al.(30)

(30). These results showed that folic acid supplementation improved both hematologic and folate status during pregnancy and the post partum period in these squirrel monkeys. The authors concluded that the stock diet lacked sufficient folate for optimal reproduction in this species. Their analyses indicated that these breeding animals were ingesting approximately 110 μg of folate/day or about 125 μg of total folate/kg body weight/day. The drop in RBC folate values seen in the FAS monkeys during pregnancy suggested that the amount of folic acid in the supplement was still less than would be desirable. The authors recommended that at least 150 μg of total folate/kg body weight/day be provided for squirrel monkeys used for breeding. Based on the characteristics of folate deficiency in this species (19) and their response to the folic acid supplementation during gestation, it was concluded that squirrel monkeys represented a suitable animal model for future studies in dams and their infants of the effects of folate deficiency during pregnancy (30).

VI. NUTRITIONAL CONSIDERATIONS FOR NONHUMAN PRIMATE REPRODUCTION

With importation severely curtailed and demand increased for nonhuman primates as animal models for biomedical research, it is crucial that our domestic breeding programs be successful. Over long-term periods we must be able to produce healthy infants with minimal maternal and fetal morbidity and mortality. To ascertain if the diets fed to nonhuman primates maintained in breeding colonies are adequate for optimal reproduction over both short- and long-term periods, data concerning colony productivity, morbidity, mortality, and birth weight are required. Little information

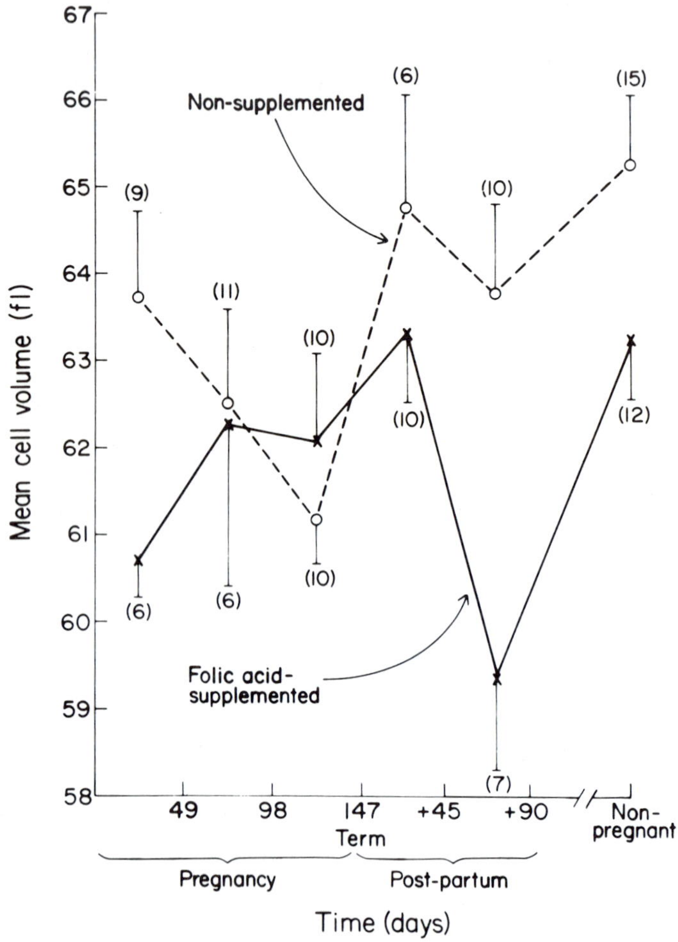

Fig. 5. Effect of folic acid supplementation on mean cell volume during pregnancy and the post partum period in squirrel monkeys. Data are from the first year of the study of Rasmussen et al. (30). Crosses represent the group receiving the vitamin supplement; open circles represent monkeys fed stock diet alone. Mean ± SE is illustrated by the points and the vertical bars with the number of animals shown in parentheses. Mean cell volume was significantly ($P < 0.05$) lower (by analysis of variance) in the supplemented monkeys.

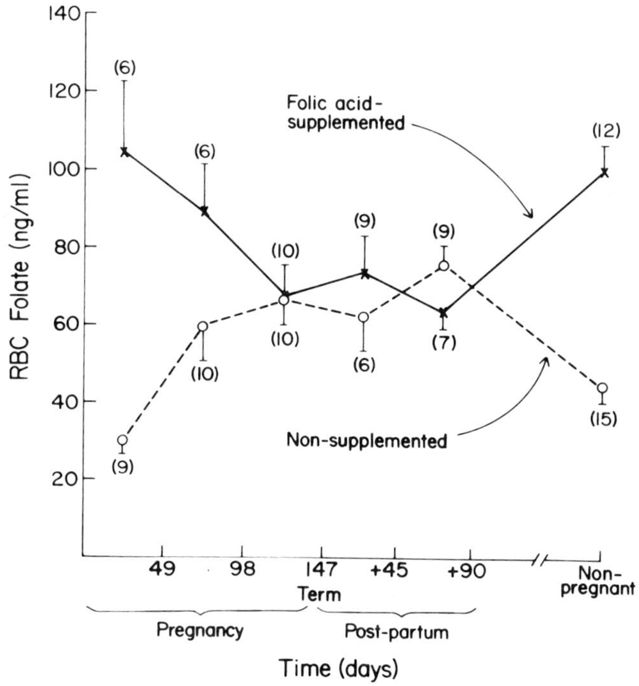

Fig. 6. Effect of folic acid supplementation on RBC folate during pregnancy and the post partum period in squirrel monkeys. Data are from the first year of the study of Rasmussen et al. (30). Crosses represent the group receiving the vitamin supplement; open circles represent monkeys fed stock diet alone. Mean ± SE is illustrated by the points and the vertical bars with the number of animals shown in parentheses. RBC folate values were significantly ($P < 0.05$) higher (by analysis of variance) in the supplemented animals.

is available in the literature for making this assessment. Reports summarizing longitudinal data from rhesus monkeys (31,32) dealt largely with colonies of free-ranging animals and this provided no data on birth weight or fetal wastage. Similarly, this type of information was not available from a recent report (33) on squirrel monkeys.

We have compiled data on 513 pregnancies over a nine-year period in the squirrel monkey breeding colony maintained at the Harvard School of Public Health (manuscript in preparation). Data for weight at birth among infants born to these animals are shown in Table VI. The 11-g difference in birth weight between fetuses aborted or stillborn and infants born alive that lived less than one week was statistically significant. The similar difference in birth weight between this latter group and infants born alive that lived more than one week also was significant, indicating that the larger infants had an improved chance of survival. A downward trend over time in birth weight was observed in this colony which could not be accounted for by aging of the monkeys in the colony (data not shown). When birth weight

TABLE VI. Analysis of Birth Weight by Pregnancy Outcome in Squirrel Monkeys, Harvard School of Public Health, 1969-1977. Values vary significantly (P < 0.00001) by Analysis of Variance

Outcome of pregnancy	*Number*	*Birth weight (g)*
Abortions or stillbirths	*43*	*89.1 ± 4.3*
Neonatal deaths	*41*	*100.7 ± 2.8**
Livebirths	*375*	*110.7 ± 0.8***

*Significantly different (P < 0.05) from value for abortions and stillbirths.
**Significantly different (P < 0.0001) from values for other two groups.
From Rasmussen, Ausman and Hayes (manuscript in preparation).

data were analyzed by the number of years that the dam had been in the colony, a pronounced downward trend was observed until the institution of folic acid supplementation (Figure 7). It is clear from the study of Rasmussen and coworkers (30) that the folate status of these monkeys was not adequate for optimal reproduction at the time supplementation with the vitamin was begun. One must ask, however, what other nutrients also might be limiting in diets fed to nonhuman primates used for breeding purposes?

Requirements for few nutrients other than protein and energy have been studied during gestation in nonhuman primates. The study of Rasmussen et al. (30) included analyses of plasma vitamin B_{12} concentrations in pregnant squirrel monkeys. Values generally were higher in the FAS than in the NS group (Figure 8). More important were the low values observed in these animals (relative to monkeys fed a purified diet providing a similar amount of vitamin B_{12} daily) (19) and the downward trend of the plasma vitamin B_{12} values toward the end of pregnancy and in the post partum period. These data suggest that the amount of vitamin B_{12} provided by the stock diet may not have been adequate for maintenance of optimal vitamin B_{12} nutriture during reproduction in squirrel monkeys.

Vitamin A and zinc have been shown to be essential for optimal reproduction in nonhuman primates. The effects of

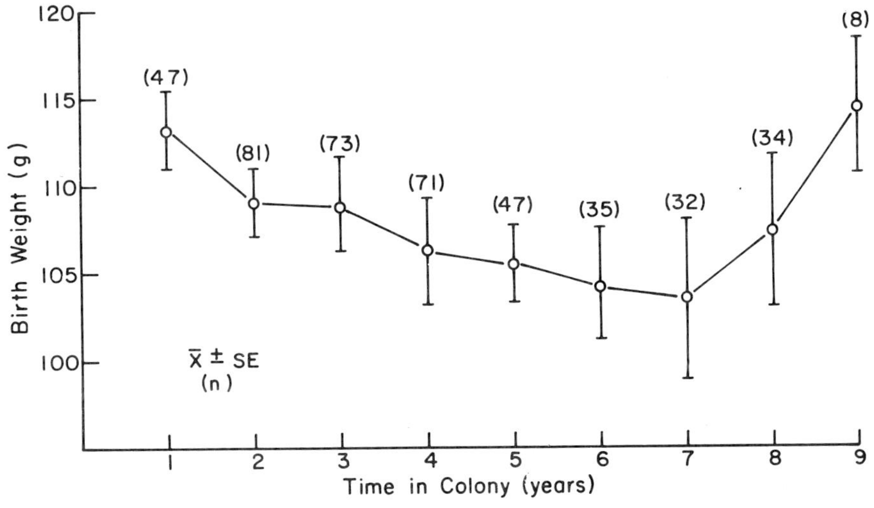

Fig. 7. Effect of time in colony on infant birth weight among squirrel monkeys, Harvard School of Public Health, 1969-1977. Rasmussen, Ausman and Hayes, manuscript in preparation.

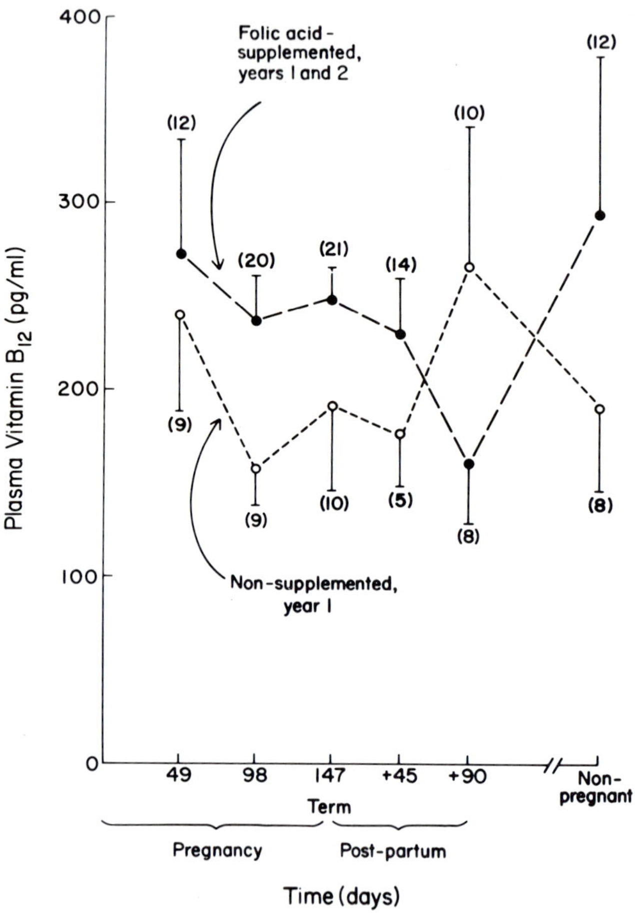

Fig. 8. Effect of folic acid supplementation on plasma vitamin B_{12} values during pregnancy and the post partum period in squirrel monkeys. Data from monkeys fed stock diet (open circles) are compared with values from animals fed stock diet supplemented with folic acid (closed circles) during both years of the two-year study of Rasmussen et al. (30). Mean ± SE is illustrated by the points and the vertical bars with the number of animals shown in parentheses. Although plasma vitamin B_{12} levels generally were higher in monkeys receiving the folic acid supplement, the differences between the groups was significant only among non-pregnant animals.

deficiencies of vitamin A (34) and zinc (35,36) have been studied and are listed in Table VII. Preliminary values for plasma zinc during pregnancy and the post partum period in squirrel monkeys compared with values for monkeys fed control and zinc-deficient diets (37) suggests that this nutrient also may be limiting in the stock diets used for these monkeys (30).

TABLE VII Summary of Literature Reports of Vitamins and Minerals Required for Optimal Reproduction in Nonhuman Primates

Nutrient	*Species*	*Probable Deficiency Signs*
Vitamin A	*M. mulatta*[a]	↓ *plasma retinol* *congenital xeropthalmia*
Vitamin B_{12}	*M. mulatta*[b]	↓ *plasma vitamin* B_{12} *in dams and infants*
Folic Acid	*S. sciureus*[c]	↓ *maternal plasma and RBC folate* ↓ *birth weight*
Zinc	*M. radiata*[de]	↓ *plasma zinc* *dermal lesions, alopecia* *small ovaries* ↓ *number of pregnancies* *impaired behavioral development of infants*

[a] *O'Toole et al. (34)*
[b] *Oxnard (1964) Nature, 201: 188*
[c] *Rasmussen et al. (30)*
[d] *Swenerton and Hurley. (35)*
[e] *Sandstead et al. (36).*

VII. SUMMARY AND CONCLUSIONS

Nonhuman primates represent excellent animal models for studies of various aspects of folate metabolism. The characteristics of folate deficiency have been studied in several species. Recent estimates of the folate requirement of two New World species suggest that current NRC recommendations should be increased to about 200 μg/day.

Investigation of the effects on reproduction in squirrel monkeys of folic acid supplementation has shown that a stock diet commonly used for breeding lacks adequate folate and also is probably too low in vitamin B_{12} and zinc for optimal reproductive performance. Little information is available on needs for other nutrients during gestation except for protein, energy, and vitamin A. If the objective of production of large numbers of healthy nonhuman primate infants in captivity over long periods of time is to be met, nutritional requirements for optimal reproductive performance must be assessed for a number of these species.

ACKNOWLEDGMENTS

Supported by USPHS NIH Grants HD-08353 and HL-10098.

REFERENCES

1. Wills, L., and Bilimoria, H.S. (1932) Studies in pernicious anaemia of pregnancy. V. Production of a macrocytic anaemia in monkeys by deficient feeding. *Ind. J. Med. Res. 20: 391-402.*
2. Pfiffner, J.J., Brinkley, S.B., Bloom, E.S., et al. (1943) Isolation of the antianemia factor (vitamin B_c) in crystalline form from liver. *Science. 97: 404-405.*
3. Angier, R.B., Boothe, J.H., Hutchings, B.L., et al. (1946) The structure and synthesis of the liver L. casei factor. *Science. 103: 667-669.*
4. Campbell, C.J., Brown, R.A., and Emmett, A.D. (1944) Influence of crystalline vitamin B_c on hematopoiesis in the chick. *J. Biol. Chem. 152: 483-484.*
5. Cartwright, G.E., Tatting, B., Ashenbrucker, H., and Winthrob, M.M. (1949) Experimental production of a nutritional macrocytic anemia in swine. *Blood. 4: 301-323.*
6. Thenen, S.W., and Rasmussen, K.M. (1978) Megloblastic erythropoiesis and tissue depletion of folic acid in the cat. *Amer. J. Vet. Res. 39: 1205-1207.*
7. Nielson, E., and Black, A. (1944) Biotin and folic acid deficiencies in the mouse. *J. Nutr. 28: 203-207.*

8. Weir, D.R., Heinle, R.W., and Welch, A.D. (1948) Pteroylglutamic acid deficiency in mice: Hematologic and histologic findings. *Proc. Soc. Expt. Biol. Med. 69: 211-215.*
9. Shaw, W., Schreiber, R.A., and Zemp, J.W. (1973) Perinatal folate deficiency: effects on developing brain in 67BL/6J mice. *Nutr. Rept. Intern. 8: 219-228.*
10. Franklin, A.L., Stokstad, E.L.R., Belt, M., and Jukes, T. H. (1947) Biochemical experiments with a synthetic preparation having an action antagonistic to that of pteroylglutamic acid. *J. Biol. Chem. 169: 427-435.*
11. Chanarin, I., Smith, G.N., and Wincour, V. (1969) Development of folate deficiency in the rat. *Br. J. Hawmatol. 16: 193-195.*
12. Tigner, J., and Roe, D.A. (1977) Folacin deficiency in rat tissues measured by competitive protein binding assay. *Fed. Proc. 36: 1120 (Abstr.).*
13. Reid, M.E., Martin, M.G., and Briggs, G.M. (1956) Nutritional studies with the guinea pig. IV. Folic acid. *J. Nutr. 59: 103-119.*
14. Slungaard, R.K., and Higgins, G.M. (1956) Experimental megaloblastic anemia in young guinea pigs. *Blood. 11: 123-142.*
15. Thenen, S.W. (1978) Blood and liver folacin activity, formiminoglutamic acid excretion, growth and hematology in guinea pigs fed a folacin deficient diet with and without sulfonamides. *J. Nutr. 108: 836-842.*
16. Day, P. L. (1944) The nutritional requirements of primates other than man. *Vit. Horm. 2: 71-105.*
17. Siddons, R.C. (1974) Experimental nutritional folate deficiency in the baboon (*Papio cynocephalus*). *Br. J. Nutr. 32: 579-587.*
18. Rasmussen, K.M., Thenen, S.W., and Hayes, K.C. Megaloblastic anemia and the requirement for folacin in the Cebus monkey. Manuscript submitted for publication.
19. Rasmussen, K.M., Thenen, S.W., and Hayes, K.C. Folacin deficiency and requirement in the squirrel monkey *(Saimiri sciureus). Am. J. Clin, Nutr. (in press).*
20. Dreizen, S., and Levy, B.M. (1969) Histopathology of experimentally induced nutritional cheilosis in the marmoset (*Callithrix jacchus*). *Arch. Oral. Biol. 14: 577-582.*
21. Herbert, V. (1962) Experimental nutritional folate deficiency in man. *Trans. Assoc. Amer. Phys. 75: 307-320.*

22. Huser, H.J., and Beard, M.E.J. (1969) Studies on folate and vitamin B_{12} metabolism in primates. I. Blood and bone marrow morphology, folate and vitamin B_{12} levels. *Folia Primatol. 10: 172-180.*
23. Cooperman, J.M., Elvehjem, C.A., McCall, K.B., and Ruegamer, W.R. (1946) "Folic acid" active compounds in the nutrition of the monkey. *Proc. Soc. Expt. Biol. Med. 61: 92-97.*
24. Day, P.L., and Totter, J.R. (1947) The bioassay of the vitamin M group. *Biol. Symp. 12: 313-338.*
25. NRC, Panel on Nonhuman Primate Nutrition (1978) Nutrient requirements of nonhuman primates. National Academy of Sciences, Washington, D.C.
26. Kitay, D.Z. (1969) Folic acid deficiency in pregnancy. *Amer. J. Obstet. Gynecol. 104: 1067-1107.*
27. Rothman, D. (1970) Folic acid in pregnancy. *Amer. J. Obstet. Gynecol. 108: 149-175.*
28. Rasmussen, K.M. (1978) Studies of folacin nutrition in primates. Sc.D. Thesis, Harvard University.
29. Sauberlich, H.E. (1977) Detection of folic acid deficiency in populations. *Folic Acid: Biochemistry and physiology in relation to the human nutritional requirement. pp. 213-223, Natl. Aca. Sci., Washington, D.C.*
30. Rasmussen, K.M., Thenen, S.W., and Hayes, K.C. Effect of folic acid supplementation on pregnancy and hematologic and other nutritional parameters in the squirrel monkey *(Saimiri sciureus). J. Med. Primatol. (in press).*
31. Drickhamer, L.C. (1974) A ten year summary of reproductive data for free-ranging *Macaca mulatta. Folia Primatol. 21: 61-80.*
32. Hird, D.W., Hendrickson, R.V., and Hendricks, A.G. (1975) Infant mortality in *Macaca mulatta:* neonatal and post-neonatal mortality at the California Primate Research Center, 1968-1972. *J. Med. Primatol. 4: 8-22.*
33. Taub, D.M., Adams, M.R., and Auerbach, K.C. (1978) Reproductive performance in a breeding colony of Brazilian squirrel monkeys *(Saimiri sciureus). Lab. Anim. Sci. 28: 562-566.*
34. O'Toole, B.A., Fradkin, R., Warkany, J. et al. (1974) Vitamin A deficiency and reproduction in rhesus monkeys. *J. Nutr. 104: 1513-1524.*
35. Swenerton, H., and Hurley, L.S. (1972) Zinc deficiency in the bonnet monkey *Macaca radiata). Fed. Proc. 31: 667 (Abstr.).*

36. Sandstead, H.H., Strobel, D.A., Logan, Jr., G.M., et al. (1978) Zinc deficiency in pregnant rhesus monkeys: effects on behavior of infants. *Amer. J. Clin. Nutr. 31: 844-849.*

37. Macapinlac, M.P., Barney, G.H., Pearson, W.N., and Darby, W.J. (1967) Production of zinc deficiency in the squirrel monkey *(Saimiri sciureus)*. *J. Nutr. 91: 499-510.*

SEMIPURIFIED DIETS FOR NONHUMAN PRIMATES

Paul M. Newberne
Department of Nutrition and Food Science
Massachusetts Institute of Technology
Cambridge, Massachusetts

K. C. Hayes
Department of Nutrition
Harvard School of Public Health
Boston, Massachusetts

I. INTRODUCTION

Nonhuman primates have gained in popularity and use during the past twenty five years with earlier uses restricted mainly to selected types of research, particularly the field of cardiovascular disease. The establishment of the Primate

ISBN 0-12-334250-3

Research Centers by the Division of Research Resources, National Institute of Health has facilitated research with nonhuman primates and expanded their use to a variety of investigations cutting across such diverse areas of biomedical research as basic nutritional biochemistry, food and drug safety and efficacy evaluations, carcinogenesis, heart disease and others. Interest in and use of nonhuman primates has, as a seeming paradox, increased during a period when supplies of animals have plateaued or, in the case of some species, sharply decreased. The relative scarcity of nonhuman primate species and the need to more effectively and accurately compare data among and between laboratories forces us, as responsible investigators, to ensure adequate dietary quality control.

Experimentation with nutritionally deficient animals due to improperly formulated or imbalanced diets leads to misinterpretation of data and erroneous conclusions from which serious consequences can arise if attempts are made to extrapolate such conclusions to humans. The careful selection and formulation of diets with known quality and quantity of ingredients will preclude many of the potential problems and reduce the number of discrepancies in data from different laboratories. The use of semipurified diets in a considerable segment of nonhuman primate research is therefore justified. This review will briefly define the three major types of diets and focus on the development and use of semipurified diets in nonhuman primate research. Particular attention will be accorded the variability in source of protein, fat and carbohydrate, the wide discrepancies in mineral and vitamin content of diets reported in the literature, the lack of dietary control, and failure to provide sufficient information about the diet in research publications for the study to be duplicated. This review is not intended to be exhaustive, but to provide key references and examples of the development and use of semipurified diets in nonhuman primates.

II. MAJOR TYPES OF DIETS

The three major types of diets for nonhuman primates are classified in general according to the degree of refinement of the various ingredients used in the formulation. The classifications are: a) natural ingredient, b) semipurified, and c) chemically defined. A brief description of each follows:

a) Natural Ingredients: Natural ingredient diets are economical and widely used in nonhuman primate research. They are formulated from ingredients with a limited degree of refinement. For example, protein may be derived from one or more sources such as meat meal, soybean meal, peanut meal, fish meal, or dried milk solids. A limited amount of protein and fat, but significant proportions of carbohydrate, can be derived from corn, oats, wheat, barley, or rice and on a more refined basis, from flour derived from these ingredients. The natural fats in the various ingredients are supplemented with crude or refined soybean oil, cottonseed oil, animal fat, or in rare cases, more refined oils such as corn oil prepared for human consumption. Minerals and vitamins may be added to supplement and balance those occurring naturally in the various ingredients. While these natural ingredients are convenient, and generally considered "safer bets" for proper growth and reproduction due to their diverse source of ingredients and their economy, there are a number of disadvantages, particularly with the closed-formula type (see page 121 for discussion of open- vs. closed-formula diets): There is a lack of control for nutrient concentration and quality from one lot to another (Table I); contamination with pesticides, heavy metals, mycotoxins, nitrates and nitrosamines is common (1); it is impractical to study effects of a specific nutrient since the capacity for altering a single component while maintaining the others is not possible, and natural products contain enzyme inducers and inhibitors (2) which can modify the response to the entity under study. Particular attention must be given to those diets which contain fish products because of nitrosamines contamination in many lots (3).

b) Semipurified Diets: Semipurified diets are more widely used now than a decade ago, but still lack acceptance on a broad scale because of expense, equipment requirements, lack of investigator familiarity with formulations, and reduced acceptability by some species. However, the advantages of semipurified diets formulated from refined ingredients outweigh disadvantages. Protein sources may be casein, lactalbumin or soy protein isolate; carbohydrate may be derived from sucrose, dextrose, dextrin or starch; fat may be derived from lard, tallow, corn oil or cottonseed oil; vitamins and minerals are provided by chemically pure materials; and fiber can be supplied by cellulose. A high degree of control can thus be exerted over any single ingredient or additive to permit an evaluation of a specific nutrient effect. Contamination of these diets is minimal.

TABLE I Variation in Selected Components of Randomly Sampled Closed-formula Natural Product Diets for Rats[a]

Component		Concentration by lot			Rat requirement[b]
Calcium, %	(source A)	0.5	0.27	0.89	0.56
	(source B)	1.67	0.82	2.10	
Phosphorus, %	(source A)	0.20	0.79	0.45	0.44
	(source B)	1.90	0.68	0.13	
Selenium, mg/kg	(source A)	0.002	0.007	0.17	0.04
	(source B)	0.05	0.21	0.001	
Iodine, mg/kg	(source A)	0.10	0.73	1.67	0.17
	(source B)	0.08	1.35	2.00	
Methionine, %	(source A)	0.13	1.21	0.22	0.67
	(source B)	0.75	0.18	0.40	
Vitamin A, mg/kg	(source A)	0.28	0.90	0.50	0.67
	(source B)	2.10	0.31	3.75	
DDT, ppm	(source A)	0.17	5.0	-	
	(source B)	2.10	0.0	0.0	
Nitrates, ppm	(source A)	0	23	3	
	(source B)	90	5	0	

[a]Adapted from Newberne: Fed. Proc. 34: 209, 1975.

[b]Based on NAS recommended levels for rats (Board of Agriculture and Renewable Resources, 1972).

c) Chemically Defined Diets: Chemically defined diets are comprised of chemically pure ingredients; such formulations may include amino acids, sugars, essential fatty acids, triglycerides, vitamins, and reagent grade inorganic salts as mineral sources. While too expensive and troublesome from a technical point of view for general use, such diets are useful in studies where strict control of concentration of specific nutrients is required. In addition to expense, an additional potential problem with chemically defined diets is modified availability of nutrients as a result of oxidation and interactions among the various dietary ingredients.

Solid vs. Liquid Diets: The form in which the three types of diets is presented to the animal will depend to some extent on the requirements of the study and the type of diet. Natural ingredient type diets are usually offered in a pelleted form, but a ground mash type is useful where additions are to be made and where the heat of pelleting may be harmful. Semipurified diets are used as the dry powdered form or they may be pelleted. They can also be used as an agar cake, gelatin gel, or as a liquid. The choice of physical form will depend on the nature of the study, but generally the pelleted or gel types are readily accepted by the nonhuman primate. Liquid formulations are useful, for example where additions of another liquid such as alcohol is used in the study and where quantification of caloric sources is a factor in the experimental design.

Chemically defined diets are usually pelleted, but may be given in a mash, gel, or liquid form. The less physical stress this type of formulation incurs the better the result because of the danger of interactions, destruction, or reduced availability of the various ingredients.

III. NATURAL INGREDIENT DIETS

Closed - Formulas: Unfortunately, the most widely used diets for many nutritional experiments in monkeys have been, and more disturbingly continue to be, slightly modified commercial formulas usually fed as biscuits, the exact formulas of which are not available. These diets are generally supplied with a proximate analysis; they may be of excellent quality and result in good health of the monkey. As suggested earlier, however, such diets are not appropriate for controlling specific ingredients and thus cannot be used to factor out specific dietary variables bearing on metabolism. Closed-formula type diets may also vary in their caloric density and protein or mineral concentration, depending on their ingredients and the method of formulation. Nonetheless, they are fed as a short-cut for study of dietary supplements such as extra fat and cholesterol.

One of the first such experimental monkey diets was published by Cox et al. (4) using rhesus monkeys. The diet represented a modified closed-formula type referred to as a ground monkey-food "mash" to which water was added to make a thick, uniform "porridge" that was accepted, but not relished, by the monkeys. It was a low-fat, low-cholesterol preparation made by a commercial feed mill with ingredients listed as wheat flour, wheat germ, soybean oil meal, dired skim milk, alfalfa-leaf meal, sucrose, bone meal, and vitamin mix. The specific concentration of each was not provided so it could never be reproduced experimentally. The authors noted that extra supplements of vitamin K, ascorbic acid, folic acid and pyridoxine were given, but it was not clear as to the quantities. Further, two or three prunes were given daily as "treats" for good behavior and for supplemental iron. The discrepancies and difficulty in interpretation of results are apparent if one is attempting to define precisely effects of specific nutrients.

Malinow, Maruffo and Perley described studies with squirrel monkeys wherein a number of dietary permutations were used to examine effects of fat and cholesterol on atherosclerosis (5). Although sources of calories were very carefully calculated and attempts made to relate supplemental margarine, butter and cholesterol to atherogenesis, the diet was based on a commercial natural product diet, the composition of which was not known. In addition, fat supplements diluted the vitamin and mineral concentration provided by the basal ration, making these microconstituents dangerously low or deficient. The authors further indicate that "animals were fed twice a week with a few leaves of lettuce, grapes, or small pieces of apples." Bananas were also added to concentrations of 8.5-9.0% of the diet, further complicating the interpretation of results. A similar diet was recently fed to a number of primate species while varying the cholesterol content (6). Interpretation of the data is complicated by the fact that it is impossible to assess species differences in a possible cholesterol interaction with unknown dietary components, such as specific fats or proteins, in the basal ration.

The same basic approach utilizing a closed-formula commercial ration with supplements of casein, salts (Hegsted IV), vitamins (unlisted), butter, tallow, and cholesterol was used to compare the cholesterol metabolism of three primate species (7). Many of the same criticisms apply, except the dilution of vitamins and minerals was compensated by supplementation of these nutrients. However, the continued use of Hegsted IV salts is inappropriate since it is apt to be low or deficient in magnesium, zinc, selenium, chromium, and fluoride and high in iron and iodine as originally formulated.

Wissler and Vesselinovitch (8) described an interesting approach to the study of atherosclerosis by feeding rhesus monkeys table-prepared human diets. The diets were made up of food ingredients (25 for the "average" American diet, 17 for the prudent ration) selected on the basis of food consumption tables. The proximate analysis and consumption rate of these diets is shown in Table II. While this approach has considerable practical implications and has yielded useful information, attempts to reproduce the exact formula would be severely handicapped; and control of various nutrient interactions can never be factored out with such an approach. In addition, differences in vitamin and mineral concentration, caloric density, texture, and palatability between the two diets led to appreciable discrepancies in their consumption, which considerably complicates interpretation of the results.

A unique category of essentially closed-formula diets are the commercially available human infant liquid formulas fed to infant nonhuman primates(9). Whereas these diets are technically superb and highly standardized, experiments based on them are flawed by the fact that the exact formulation is known only to the producer, and the formula may be low in, or lack, certain nutrients generally included in monkey diets such as choline, iron or selenium. Nor do they contain dietary fiber, which should be a consideration for long term consumption. Furthermore, an inflexible bias is introduced by the type of fat or amount and kind of protein included which limit experimental manipulations considerably. For example, the protein content (approximately 10% calories) can be inadequate for infant New World species, and other unexplained peculiarities may develop after prolonged consumption (Ausman and Hayes, unpublished data).

Open-Formulas: Gresham et al. reported on arterial lesions in baboons associated with an open-formula rat cake (ingredients and amounts were listed) and enriched with butter and egg yolk or butter and cholesterol. These diets were fed for 18 months along with daily supplements of greens or carrots. A number of dietary indiscretions can be pointed out in addition to the complex nature of the "rat cake":
1) The bran, oats, wheat, barley, as well as maize, meat and

TABLE II Composition and Consumption Data for Natural Ingredient Diets Fed to Rhesus Monkeys During Studies of Atherogenesis by Wissler and Vesselinovitch (8)

Ingredient	Average American diet % dry wt	Average American diet g/day consumed	Prudent American diet % dry wt	Prudent American diet g/day consumed
Fat	22.6	26.9	18.0	14.0
Protein	25.0	29.8	31.5	24.4
Carbohydrate	45.6	54.3	44.5	34.5
Ash (salts)	2.2	2.6	2.5	1.9
Fiber	3.2	3.8	3.3	2.6
Cholesterol	136 mg	175 mg	90 mg	80 mg
Total food consumption/day-wet wt		261		224
Daily calories/monkey		560		360

bone meal all contained some fat which apparently was not accounted for in dietary calculations of fat content. 2) The salts mix (minerals) was contributed by the rat cake and was not described. Again, supplementation with fat diluted the total mineral and vitamin content. 3) The total vitamin content was unknown except for vitamin C and was crudely supplied by dried yeast and cod liver oil. The vitamin A and D content of cod liver oil varies considerably from one lot to another and the fatty acids in the oil are subject to oxidation and development of rancid diets.

An example of a true open-formula diet was fed by Ackman and Loew (11) to cynomolgus monkeys to study the effect of rapeseed oil and hydrogenated fish oils on skeletal and cardiac muscle. All ingredients were listed with percent contribution from rolled oats, ground wheat, wheat bran, soybean meal, dried skim milk, alfalfa meal, brewers' dried yeast, experimental oil, mineral and vitamin premixes (completely listed). In addition, the complete fatty acid profile of each ingredient was provided since this was a major variable in the study. Aside from the general criticism that most ingredients, such as brewers' yeast or alfalfa meal, have numerous unknown components that complicate interpretation, this represents adequate documentation of an open-formula diet.

Semi-natural, Open - Formulas: Of decidedly better classification among open-formulas are semi-natural diets that comprise a mixture of purified and crude ingredients. The Bowman-Gray group has successfully utilized such diets in squirrel monkeys, African green monkeys, and stump-tailed macques, both in solid and liquid form (12). A perusal of ingredients in Table 3 indicates a potential for high calcium concentration from dried milk solids that was balanced by the salts mix (USP XIV), which has a low calcium and copper content and no zinc, chromium, or selenium. Dry milk solids (presumably skim milk) is a variable source of mixed ingredients and is low in iron and copper, but iron would be added via the salts mix. As in studies by many investigators, the cholesterol was 2-4 times higher than we (humans) consume and in terms of nutritional modeling, could be lowered to a more realistic level. Applesauce, added for sweetness and possibly fiber, is a complex mixture and not apt to be reproducible from lot to lot. The source of the vitamin and mineral pre-mixes is indicated, but a listing of their ingredients would be preferential. Wheat flour binds the ingredients together and allows the diet to be baked as a cake; it also represents a crude mixture of starch and protein. The gelatin in the liquid diet was presumably to keep the ingredients in suspension, but it represents an additional

TABLE III Composition of Semi-natural, Open-formula Solid and Liquid Diets Successfully Fed to Old and New World Primates (12)

Ingredient	*Solid diet (g/kg)*	*Liquid diet (g/kg dry mix)*
Dry milk solids	*300.0*	*425.6*
Casein	*130.0*	-
Gelatin	-	*20.3*
Lard	*250.0*	-
Corn oil	-	*189.2*
Wheat flour	*200.0*	-
Sucrose	-	*304.1*
Applesauce[a]	*78.0*	-
Vitamin mix (with D_3)	*22.0*	*30.4*
Salts mix (USP XIV)	*20.0*	*30.4*
Total	*1000.0*	*1000.0*
Calories	*5120*	*4700*
Water	-	*3000*

[a]*0.5g cholesterol replaced 0.5g applesauce in atherogenic diet.*

source of poor quality protein that could be replaced by an appropriate suspending agent (see Table I , page 42). Also, the vitamin and mineral concentrations per calorie are approximately 50% higher in the liquid diet. The latter diet might also tend to produce loose stools since no fiber is included.

IV. SEMI-PURIFIED DIETS

It has been pointed out that development of semi-purified diets prepared for nonhuman primates was prompted for the most part by studies related to cardiovascular disease. Ironically the first major nutritional study in monkeys (13) actually utilized semi-purified diets to investigate the effects of diet on experimental atherosclerosis. Table IV lists a basal diet fed for study of experimental atherosclerosis in cebus monkeys. This series of diets was satisfactory for the design of the study and moved the semipurified diet concept ahead

TABLE IV Examples of Original Semi-purified Diets Fed to Cebus Monkeys by Mann et al. (13)

	Diet number	
Ingredients	*840*	*820*
Casein (vitamin free)	*25.0*	-
Alpha-protein	-	*20.0*
Sucrose	*60.3*	*60.3*
Corn oil	*8.0*	*8.0*
Salt mix (Hegsted IV)	*4.0*	*4.0*
Cod liver oil, USP	*2.0*	*2.0*
Choline chloride	*0.5*	*0.5*
Inositol	*0.1*	*0.1*
P-aminobenzoic acid	*0.1*	*0.1*
Cholesterol	-	*5.0*

Each day 5 ml. of a 20 percent ethanol solution containing the vitamins was added to the food portion: Thiamine HCl 1 mg., riboflavin 1 mg., pyridoxine HCI 1 mg., niacin 4.9., calcium pantothenate 3 mg., folacin 0.1 mg., biotin 0.02 mg., and ascorbic acid 25 mg.

considerably. A few aspects of the diets may be pointed out to indicate the progress that has been made during the 25 years since this work was published. Two types and levels of protein were used: Casein was fed at either 10 or 25% of the diet and soybean protein (alpha-protein) was included at either 10 or 20%. Corn oil was the source of fat, included at either 8 or 15% with an additional increment of 2% cod liver oil as the source of vitamins A and D; no vitamin E was added eventhough corn oil was heated to facilitate incorporation of cholesterol. The mineral mix (Hegsted IV) was the original formula developed and has already been critiqued. Other comments are as follows: 1) The alpha-protein prepared at this time was not intended as a source of protein for animals, but rather was used as an industrial mucilage. It did however, provide the low concentration of sulfur-containing amino acids desired by the investigators. It had not been purified and, unless water washed several times and reground by the investigator to remove the sulfite residue, thiamine and possibly other vitamins were destroyed to the point of near or real deficiency. 2) Cod liver oil as a supply of vitamins A and D has already been criticized. The highly poly-unsaturated nature of this fish oil would also increase the requirement for vitamin E, which was not even included in the vitamin mix. 3) Corn oil at 15% of the diet should have supplemental vitamin E, especially since it was heated to facilitate solubilization of cholesterol. 4) The addition of water-soluble vitamins to the daily food portion was added effort without justification since the vitamins should be uniformly mixed into an entire diet as it is prepared. The stability of the 25 mg ascorbic acid was not stated. Para-aminobenzoic acid was probably not necessary since folic acid was added as the pure compound in sufficient amounts for adequate folate nutrition. 5) The addition of 5% cholesterol was excessive and actually interferred with the interpretation of the data. Despite these problems, the study, and the semipurified diet was considerably ahead of most contemporaries.

Strong et al. (14) fed semi-purified diets to baboons and controlled dietary variables considerably better than the open-formula diet of Gresham et al. (9) referred to previously. The semi-purified diets were casein-sucrose based and vitamins and minerals were added in known quantities along with fats, the quantify and quality of which was determined. However, the salt mix was Hegsted IV and all the carbohydrate was fed as sucrose which is less than desirable in metabolic studies involving lipid metabolism and atherogenesis.

Scott et al. (15) used a semi-purified diet with a number of modifications in studies designed to examine different types and amounts of fat on atherogenesis in rhesus monkeys. (Table V lists the composition of the diet indicating the extraordinarily high fat content (30% by weight), cholesterol (5%), and salt, fed with the unphysiological insults of propylthiouracil and sodium cholate. The casein-sucrose basal diet that was described can not be accurately duplicated because the composition of neither the salt mix nor vitamin mix was given. Furthermore, the depressed growth rate in all experimental groups underscores the inadequacies of the diet.

TABLE V Various High-fat Diets Fed to Rhesus Monkeys During Studies on Atherogenesis by Scott et al. (15)

Ingredients	*Butter diet*	*Peanut oil diet*	*Corn oil diet*
Casein	*20.0*	*20.0*	*20.0*
Sucrose	*21.4*	*21.4*	*18.4*
Butter	*30.0*	-	*5.0*
Peanut oil	-	*30.3*	*5.0*
Corn oil	-	-	*20.0*
Vitamin mix	*6.0*	*6.0*	*6.0*
Salt mix	*6.0*	*6.0*	*6.0*
Choline chloride	*1.0*	*1.0*	*1.0*
Cellulose	*15.0*	*15.0*	*18.6*
Propylthiouracil	*0.2*	*0.2*	-
Sodium cholate	*0.4*	*0.4*	-

Continuing studies in the Department of Nutrition, Harvard University, have been published during the past several years (16-18) which define limitations and advantages of selected semi-purified diets in nonhuman primates. Robinson et al. (16) utilized a semi-purified diet based on casein, dextrin, and vegetable oils to investigate the effect of dietary fat on blood lipids and aortic ultrastructure in squirrel monkeys. The diet was consumed in quantities that permitted weight gains from 60-90 grams over a six-month period. This diet has since been modified (Table VIII). The original diet can be faulted for use of Hegsted IV salts, having 25% calories as protein whereas 12-16% is sufficient, feeding all carbohydrates as dextrin which reduces palatability, including an unusually high fiber content (24%), and supplementing folic acid at a level below that now known to be required; the latter criticism is true for most all vitamin mixes used to date for New World monkeys. The diet was fed as an agar cake to facilitate weighing and determination of food intake. With modifications referred to later this has proved to be a convenient diet that can be utilized or altered slightly to assure reasonable nourishment for most nonhuman primate species.

Special Diets: Development of specialized semi-purified diets deficient or limiting in specific dietary ingredients can be a challenge, such as those required for study of trace element or vitamin deficiencies and protein-calorie malnutrition. An example of the former is the diet prepared by Sandstead et al. (19) for study of zinc deficiency in pregnant rhesus (Table VI), where a low-zinc protein of high quality such as egg albumin was utilized. However, the fiber content of the diet (solka floc) was relatively low.

Enwonwu et al. (20) studied protein-calorie deprivation in macaques feeding different levels of protein (Table VII). Unfortunately, the reduced protein content was replaced by fiber such that the caloric density and, therefore, the vitamin and mineral concentration of the 2 diets differed appreciably due to the difference in fiber content. Thus, the 2% protein group had to consume 23% more diet than those receiving the 20% protein diet to acquire the same number of calories. In doing so, the vitamin and mineral intake was increased by the same proportion. These differences in fiber, vitamins and minerals necessarily bias any interpretation. The Hawk-Oser salt mix is also lacking copper, selenium, zinc, and chromium and is low in manganese.

Diet Formulation: From this rather selective, but varied background on diets it is possible to formulate a semi-purified

TABLE VI Zinc-deficient Semi-purified Diet Fed to Pregnant Rhesus Monkeys (19)

Ingredient	g/100 g
Sprayed egg white	20
Dextrose	64
Solka floc (SW 140)	2
Corn oil	8
Salt mix*	4
Vitamin mix	2

*Prepared zinc free, but neither chromium nor fluoride were included.

TABLE VII Control and Protein Deficient Diets Fed to Young Pigtail Macaques by Enwonwu et al. (20)

Ingredients	20% protein diet	2% protein diet
Casein	20.0	2.0
Sucrose	61.0	61.0
Cottonseed oil	8.0	8.0
Salt mix (Hawk-Oser)	4.0	4.0
Vitamin mix	1.0	1.0
Calcium stearate	1.0	1.0
Cellulose	5.0	23.0
Approx/kcal/1000g	3950	3200

diet for nonhuman primates that should prove reliable for most investigative nutritional purposes. Consideration has been given to protein quality and known nutrient requirements including vitamins and minerals. Table 8 lists ingredients and proportions which are thought to be satisfactory for monkeys.

This formula is suggested for general use in juvenile or adult monkeys based on previous experience with several primate species. Good quality proteins should be fed to provide approximately 12 to 16% of the calories depending on species size and rate of growth. Smaller rapidly growing monkeys with high metabolic rates require a relatively high caloric density (higher fat content) as well as a higher concentration of good quality protein, most readily obtained from animal sources. Equivalent sources of vegetable protein are not well characterized in terms of their utilization by primates. The carbohydrate source generally includes some sucrose to increase palatability since a diet with only dextrin is poorly consumed (see Figure 1, page 190). The fat source can vary widely from vegetable oils to animal fats, and a range from 20 to 40% of the calories has been fed successfully with this formula. Adjustments in fat calories or fiber content should also be accompanied by changes in vitamins and minerals to assure that they are supplied as a constant percent of calories, not simply weight percentages. The vitamin mix and mineral mix are as complete and up-to-date as current knowledge allows, and both can be purchased as pre-mixes as indicated in Table 8. Choline should be mixed into the diet directly since it tends to oxidize B-vitamins if stored in the vitamin pre-mix. Cellulose is generally added at 5 to 10% to provide necessary bulk, but may be increased substantially (up to 20%) to reduce diarrhea or constipation, if necessary. The agar gel serves to form cakes which can be refrigerated until fed. Alternatively, the entire diet can be purchased in pelleted form as indicated, providing the fat content does not exceed 10 to 12% by weight.

Liquid formula semi-purified diets currently in use are described elsewhere (page 39). Preparation of the protein source and solubility of the mineral mix are important considerations that need re-emphasis. A currently utilized water-soluble mineral pre-mix is included in Table 9. This pre-mix tends to prevent settling out of minerals which plagues many liquid diets.

TABLE VIII. Semipurified Primate Diet[a]

Ingredients	g/100 g	% kcal
Casein (or lactalbumin)	20.00	20.5
Dextrin	23.00	21.8
Sucrose	26.45	25.2
Corn oil (or lard)	15.00	32.5
Vitamin mix[b]	0.50	
Mineral mix (Ausman-Hayes)[c]	4.75	
Choline chloride	0.30	
Cellulose	10.00	
Agar (1.25% solution)	80.00 ml	

[a]Available in pelleted form from BioServ, Inc., Frenchtown, N.J. (Bio-Mix #1351).

[b]The vitamin mix contained (in g/kg): thiamine HCl, 0.80; riboflavin, 1.60; niacinamide, 8.0; folic acid, 0.8; biotin, 0.04; cyanocobalamin, 0.03; menadione, 0.1; D,L-α-tocopheryl acetate, 20.0 (500 IU/g); vitamin A acetate, 5.0 (500,000 IU/g); vitamin D_3, 1.25; (200,000 IU/g); ascorbic acid beadlets, 120; inositol, 100; taurine, 50; made to 1kg with 687 g dextrin.

[c]The mineral mix contained (in g/kg): calcium carbonate, 290.5; calcium phosphate (dibasic), 72.6; potassium phosphate (dibasic), 327.9; sodium chloride, 162.4; magnesium sulfate, 98.7; magnesium oxide, 32.0; ferric citrate, 13.3; manganese sulfate, 1.2; zinc chloride, 0.915; cupric sulfate, 0.290; potassium iodide, 0.077; chromium acetate, 0.044; sodium fluoride, 0.023; sodium selenite, 0.004. Available as premix (Bio-Mix #1321), BioServ, Inc., Frenchtown, N.J. and Teklad Mills, Madison, Wis. (mineral mix #TD79033).

TABLE IX. Ausman Water-Soluble Salt Mix I for Liquid Diets

Salt	g/kg
Calcium phosphate tribasic·$Ca_3(PO_4)_2$	356.51
Potassium carbonate·K_2CO_3	260.19
Sodium chloride·NaCl	162.37
Magnesium sulfate·$MgSO_4$ (anhydrous)	143.89
Potassium phosphate·K_2HPO_4	67.71
Ferrous sulfate·$FeSO_4$	6.74
Manganese sulfate·$MnSO_4 \cdot H_2O$	1.22
Zinc chloride·$ZnCl_2$	0.91
Copper sulfate·$CuSO_4 \cdot 5H_2O$	0.29
Potassium iodide·KI	0.077
Chromium acetate·$Cr(C_2H_3O_2)_3 \cdot H_2O$	0.048
Sodium fluoride·NaF	0.023
Sodium selenite·Na_2SeO_3	0.004

V. DISCUSSION AND CONCLUSIONS

The development and use of semi-purified diets for investigations with nonhuman primates has been painfully slow. Their acceptance by investigators is still a problem because relatively little is known about the needs of the various species, using such diets is time consuming, expensive, and may result in unanticipated deficiency disease, funding agencies are reluctant to fund long-term investigations for their research and development, and acceptance of semi-purified diets by the animals themselves varies considerably. Nevertheless, considerable progress has been made during the past decade and results are now emerging from some of the more significant efforts. In reviewing the literature over the past 30 years a number of points are evident: 1) Natural product diets have been, in general, preferred by both investigators and animals, but adequate semi-purified diets for most nonhuman primates now exist. 2) Investigators have not voluntarily described, nor have journal editors required, dietary regimens in sufficient detail to permit accurate reproducibility. 3) A large proportion of results in the literature using semi-purified diets have been derived from studies related to cardiovascular disease. 4) Minerals, particularly trace elements, appear to be an area of limited knowledge concerning needs for nonhuman primates.

A standardized approach for utilization of semi-purified diets would greatly enhance our understanding of nutrient requirements in the various nonhuman primate species and facilitate interpretation of experimental results between investigators.

REFERENCES

1. Newberne, P.M., and Fox, J.G. (1978) Chemicals and toxins in the animal facility. Laboratory Animal Housing. NAS/NRC, Washington, D.C.
2. Wattenberg, L., Page, M., and Leong, J. (1968) Induction of increased benzopyrene hydroxylase activity by flavones and related compounds. *Cancer Res. 28: 934.*
3. Edwards, G., and Fox, J.G. (1979) Significance of nitrosamines in animal diets. *Science 203: 6.*
4. Cox, G.E., Taylor, C.B., Cox, L.G., and Counts, M.A. (1958) Atherosclerosis in rhesus monkeys. I. Hypercholesteremia induced by dietary fat and cholesterol. *Arch. Path. 66: 32.*

5. Malinow, M.R., Maruffo, C.A., and Perley, A.M. (1966) Experimental atherosclerosis in squirrel monkeys *(Saimiri sciurea). Path. Bact. 92: 491.*
6. Srinivasan, S.R., Radhakrishnamurthy, B., Smith, C.C., Wolf, R.H., and Berenson, G.S. (1976) Serum lipid and lipoprotein responses of six nonhuman primate species to dietary changes in cholesterol levels. *J. Nutr. 106: 1757.*
7. Eggen, D.A. (1974) Cholesterol metabolism in rhesus monkey, squirrel monkey, and baboon. *J. Lipid Res. 15: 139.*
8. Wissler, R.W., and Vesselinovitch, D. (1973) The effects of feeding various dietary fats on the development and regression of hypercholesterolemia and atherosclerosis. *Adv. Exp. Med. Biol. 60: 65.*
9. Kerr, G.R., Scheffler, G., and Waisman, H.A. (1969) Growth and development of infant *Macaca mulatta* fed a standardized diet. *Growth 33: 185.*
10. Gresham, G.A., Howard, A.N., McQueen, J., and Bowyer, D. E. (1965) Atherosclerosis in primates. *Brit. J. Exp. Path. 46: 94.*
11. Ackman, R.G., and Loew, F.M. (1977) The effects of high levels of fats rich in erucic acid (from rapeseed oil) or cetoleic and cetolaidic acids (from partially hydrogenated fish oil) in a short-term study in a nonhuman primate species. *Sette-Seifen-Anstrichmittel 79: 1.*
12. Bullock, B.C., Lehner, N., Clarkson, T.B., Feldner, M., Wagner, W., and Lofland, H. (1975) Comparative primate atherosclerosis. *Exp. Mol. Path. 22: 151.*
13. Mann, G.V., Andrus, S.B., McNally, A., and Stare, F.J. (1953) Experimental atherosclerosis in cebus monkeys. *J. Exp. Med. 98: 195.*
14. Strong, J.P., and McGill, H.C., Jr. (1967) Diet and experimental atherosclerosis in baboons. *Am. J. Path. 50: 669.*
15. Scott, R.F., Morrison, E.S., Jarmolych, J., Nam, S.C., Kroms, M., and Coulston, F. (1967) Experimental atherosclerosis in rhesus monkeys. *Exp. Mol. Path. 7: 11.*
16. Robison, R.L., Hayes, K.C., McCombs, H., and Faherty, T. (1971) Effect of dietary fat and cholesterol on circulating lipids and aortic ultrastructure of squirrel monkeys. *Exp. Mol. Path. 15: 281.*
17. Ausman, L.M., Hayes, K.C., and Hegsted, D.M. (1972) Protein deficiency and carbohydrate tolerance of the infant squirrel monkey *(Saimiri sciureus). J. Nutr. 102: 1519.*

18. Corey, J.E., Hayes, K.C., Dorr, B., and Hegsted, D.M. (1974) Comparative lipid response of four primate species to dietary changes in fat and carbohydrate. *Atherosclerosis 19: 119.*
19. Sandstead, H.H., Strobel, D.A., Logan, G.M., Jr., Marks, E.O., and Jacob, R.A. (1978) Zinc deficiency in pregnant rhesus monkeys: effects on behavior of infants. *Am. J. Clin. Nutr. 31: 844.*
20. Enwonwu, C.O., Stambaugh, R.V., and Jacobson, K.L. (1973) Protein - energy deficiency in nonhuman primates: biochemical and morphological alterations. *Am. J. Clin. Nutr. 26: 1287.*

OPEN FORMULA NATURAL INGREDIENT DIETS FOR NONHUMAN PRIMATES

Joseph J. Knapka
Martin L. Morin

U.S. Department of Health, Education and Welfare
Public Health Service
National Institutes of Health
Division of Research Services, Veterinary Resources
Bethesda, Maryland

I. INTRODUCTION

Natural feed ingredients are used in a large percentage of the nonhuman primate colonies maintained throughout the biomedical research community. Natural ingredients are appropriately processed whole grains such as wheat, barley or corn and commodities that have been subjected to limited amounts of refinement such as fish meal, soybean meal or wheat bran (1). The most readily available natural ingredient diets for nonhuman primates are products manufactured by commercial feed manufacturing companies. These are referred to as "closed formula" diets because the amounts of ingredients used in the diet formulation is privileged company

ISBN 0-12-334250-3

information. The manufacturers of these diets supply the list of ingredients used in the formulation and the guaranteed analysis. Upon request, the mean concentrations of all nutrients are generally provided. In general, the performance of various nonhuman primate colonies fed commercial closed formula diets has been acceptable, particularly for Old World species. The production of closed formula diets is geared to the manufacture of large quantities of a diet for most efficient production. It is impractical for most commercial feed manufacturing companies to provide relatively small amounts of closed formula diets with nutrient concentrations to accommodate specific requirements.

Open formula diets can supply the nutrient requirements of specific nonhuman primate species or provide graded concentrations of nutrients for nutrition experiments. The quantitative ingredient composition of these diets is readily available, allowing for adjustments in formulations to alter nutrient concentrations. In the past, natural ingredient open formula diets have been used in nonhuman primate colonies on a relatively limited basis (2,3,4). However, with the increased interest in nonhuman primate nutrition throughout the research community, open formula diets are a useful research tool in identifying the nutrient requirements of nonhuman primates or for developing animal models of diseases caused by either deficient or excess concentrations of nutrients. The objective of this presentation is to provide information regarding the ingredients, formulation, manufacture and application of open formula diets for nonhuman primates.

II. NATURAL INGREDIENTS FOR NONHUMAN PRIMATE DIETS

Diets made with natural ingredients are fed because they are readily available, economical to procure, and accepted by most of the species maintained in laboratories. The disadvantages in using them for experimental animal colonies is the variation in nutrient concentrations among production batches and the potential for contamination with pesticide residues, heavy metals, or other agents that might alter the response of animals to experimental treatments.

The data presented in Table I indicate variation in nutrient concentration observed among production batches of a commercial nonhuman primate diet during a two year period. Representative samples of diet manufactured for the National Institutes of Health (NIH) were collected by the manufacturer and sent to a private laboratory for analyses of proximate

TABLE I Variation in Percent Nutrient Concentration Among 30 Samples of a Commercial Monkey Diet During a Two Year Period

Item		Crude Protein	Crude Fiber	Ash Max	NFE Min	Calcium	Phosphorus
Stated Concentration		15.00	3.00	5.00	60.00	0.84	0.45
Mean		17.80	2.61	5.27	60.77	0.95	0.53
Standard Deviation		1.47	0.56	1.16	2.90	0.13	0.12
Range	high	20.30	3.66	9.20	65.54	1.24	1.00
	low	14.60	1.00	3.76	54.53	0.74	0.36

nutrients, plus calcium and phosphorus. It should be recognized that the variation observed in analytical results from a series of feed samples includes inherent errors associated with product sampling, sample preparation, sub-sampling, and the analytical procedures utilized. Results presented in Table 1 indicate that nutrient concentrations in a large percentage of the samples analyzed were in excess of the levels indicated by the manufacturer, and that the observed degree of nutrient variation would have little effect on animals under most practical conditions. The ranges in nutrient concentrations indicate that occasional batches of feed either contain an excess or deficiency of nutrients which may justify concern where animals are subjected to stress or in experiments where the dietary concentration of a specific nutrient is critical.

The primary source of nutrient variation among production batches of natural ingredient diets is the inherent variation in the nutrient concentrations in feed ingredients (5). Through careful management of the ingredients and diet manufacture, it is possible to control the degree of nutrient variation to the extent that natural ingredient diets have been used to determine the nutrient requirements of various farm animal species. Procedures that tend to decrease the variation include diet formulation utilizing a large number of ingredients, the selection of ingredients based on the variability in nutrient concentrations, procurement of ingredients with specified nutrient concentrations, and the frequency of product manufacture. For instance, the variation in dietary nutrients for a particular experiment can be practically eliminated by freezer storage of a sufficient quantity of feed from a single production batch to last the entire experiment.

Chemical contamination of natural ingredient diets is a factor which should not be overlooked (6,7). There are only limited published data (8) to document the concentrations of various contaminants in nonhuman primate diets. An analytical program designated to detect chemical contaminants in natural ingredient diets is essential if trace concentrations of a particular contaminant could compromise experimental results. The concentrations of chemical contaminants in diets can be minimized by using dietary ingredients that have a low probability of being contaminated. The data in Table II indicated that the source of a large percentage of the lead found in a rodent diet was in the mineral premix. Subsequent lead analysis of the components of this premix indicated that practically all the lead was in the limestone used as a source of calcium. A considerable decrease in the dietary lead concentration was noted when limestone was obtained from a different source.

TABLE II Lead Concentrations of Ingredients Used in an Open Formula Rat and Mouse Diet

Ingredients	*PPM Lead*
Fish meal	*1.6*
Soybean meal	*<0.2*
Alfalfa meal	*1.2*
Corn gluten meal	*0.6*
Whole wheat	*0.2*
Shelled corn	*<0.2*
Whole oats	*0.6*
Wheat middlings	*0.3*
Brewer's years	*<0.2*
Salt	*<1.0*
Vitamin premix	*0.2*
Mineral premix	*16.1*
Total diet	*0.4*

III. FORMULATION OF NATURAL INGREDIENT DIETS

The objective of diet formulation is to determine the ratios of various ingredients and levels of vitamin and mineral supplementation which will produce a diet containing the required concentrations of essential nutrients. The process of formulating natural ingredient diets is rather complex because each ingredient may contribute some percentage of all the approximately 50 required nutrients. It is necessary to account for the essential nutrients in each ingredient to determine total dietary nutrient concentrations. Consequently, a systematic procedure for diet formulation must be established to insure the required concentrations and ratios of nutrients are provided.

Nutrient Concentrations: The initial step in the process of diet formulation is to establish nutrient concentrations for the planned diet or diets. Guidelines for the dietary requirements of nonhuman primates (1) have been published and are included elsewhere in this monograph. Planned dietary nutrient concentrations can be based on experimental objectives and, in some cases, observations regarding the dietary requirements of specific nonhuman primate species. Dietary nutrient concentrations must be adjusted for losses occurring during the manufacturing process, storage, and for factors that affect feed consumption. For example, animals consume less feed and total nutrients when dietary energy is increased or when compounds that affect palatability are added to the diet. Actual dietary nutrient concentrations are not included in this discussion of diet formulation because these procedures apply regardless of the planned nutrient concentrations.

Ingredients: The second step in diet formulation is to select the major ingredients to be used in the formula. The primary basis for selection is the ingredient nutrient composition, but factors such as the ability to readily obtain the ingredient, the palatability for nonhuman primates, and the contamination potential are also important considerations. Information regarding the nutrient composition of the most readily available feed ingredients throughout the world has been published in tabular form (9,10,11,12). Only readily available ingredients should be selected to preclude the need to alter formulations during periods when certain ingredients may be difficult to obtain. In general, diet quality increases as the number of ingredients used in the formulation increases. This tends to minimize the effect that variation in nutrient composition of any one ingredient has on the total dietary nutrient concentrations, and it increases the probability that all trace nutrients are provided in adequate quantities.

A typical diet for nonhuman primates will include one or two ingredients as the primary source of each nutrient class as well as supplemental vitamins and minerals (Table III). Crude protein is supplied by a combination of ingredients of animal and plant origin, such as fish meal, milk products, meat meal, corn gluten meal, or soybean meal. A source of crude fiber may be alfalfa meal, oat hulls, or dried beet pulp. The primary source of carbohydrates is a combination of feed grains such as corn, wheat, barley, oat groats. The kinds of feed grains selected for a particular diet formulation is influenced by the procedures to be used in the manufacture of the diets. Heat treated or gelatinized starches are more efficiently utilized by nonhuman primates than raw starches. Therefore, when diet manufacturing procedures do not include a process involving heat, such as extruding or baking, heat-processed feed grains are included in diet formulations. When manufacturing processes such as baking are involved, it is necessary to include wheat flour whose gluten will serve as a binder to improve biscuit quality. An unsaturated fat such as soybean oil or corn oil is used as a

TABLE III Examples of Nutrient Sources from Major Ingredients in Nonhuman Primate Diets

Nutrient	
Crude Protein	
Animal Source	*Fish meal, meat meal, dried milk products*
Vegetable Source	*Soybean meal, corn gluten meal*
Crude Fiber	*Alfalfa meal, beet pulp, oat hulls, wheat bran*
Crude Fat	*Soybean oil, corn oil, animal fat*
Carbohydrates	*Whole ground wheat, ground corn, ground barley, wheat middlings*

source of essential fatty acids and to increase the dietary energy concentration. Brewer's dried yeast is a major source of water-soluble vitamins while the addition of sugar increases palatability and acceptance. Ingredients such as limestone, dicalcium phosphate, calcium carbonate, and sodium chloride are used as mineral sources. Vitamin and mineral premixes are formulated to provide the difference in concentrations between the total amount of nutrients supplied by the ingredients and the levels required in the diet. Premixes are stored separately until added to minimize the destruction of vitamins through mineral-catalyzed reactions.

Calculating Ingredient Concentrations: The final step in the process of diet formulation is to determine the amount of each ingredient required to produce a diet with the desired nutrient concentrations. This involves a large number of calculations using nutrient data, from ingredient tables, to determine the concentration of each nutrient in a specific combination of ingredients. The amount of each ingredient is expressed as a percentage of weight and the formula must total 100%.

An initial "working" formula is made to permit adjustments in the amounts of various ingredients until a final formula containing the planned nutrient concentrations is established. Since the crude protein and crude fiber concentrations cannot be readily increased by supplementation in natural ingredient diets, the initial formula is based on the planned concentrations of these nutrients. In the initial formula 20 to 25% of the planned dietary crude protein should be of animal origin, 25 to 30% from a combination of ingredients rich in plant protein, and the remainder from other ingredients. Including crude protein from these sources in the indicated ratios usually results in a good quality protein for mammals. The ingredient included as a primary source of crude fiber should be added at a level that will supply approximately 80% of the planned crude fiber concentration. The amounts of feed grains included in the working formula are based on the planned concentrations of dietary crude protein and crude fiber. This is accomplished by adjusting the amounts of the feed grains with the highest and lowest concentrations of crude protein. Minimal adjustments in the amounts of ingredients included in the formula as the primary sources of crude protein and crude fiber may be necessary. The total amount of these ingredients in the initial formulations should not total 100%. This allows for the addition of ingredients such as sugar, brewer's yeast, vitamin and mineral premixes, supplemental fat, and sources of major minerals. These ingredients may account for up to 8 to 10% of the final formulation. The essential amino acid concentrations in the working formula are calculated to verify adequate or planned concentrations (Table IV). Based on these calculations, it may be necessary to readjust ratios of the ingredients included in the formula as primary sources of crude protein. However, when sulfur containing amino acid concentrations are deficient, appropriate amounts of methionine hydroxy-analogue can be included in the vitamin premix.

The total amount of crude fat, sodium, calcium and phosphorus in the ingredients are calculated. Based on these data, the final formula (with the exception of the premixes) is established by adding sufficient quantities of the appropriate ingredients to ensure that the planned concentrations of these particular nutrients are provided. Slight adjustments in the amount of feed grain lowest in crude protein may be necessary so the amount of ingredients total 100%. The composition of the mineral premix is established by calculating the total concentrations of cobalt, copper, iron, magnesium, manganese, potassium, zinc, and iodine in the formula. The appropriate mineral salts are then added at concentrations equal to the difference between these values and the planned dietary concentrations. In most cases, trace minerals are not a concern in natural ingredient diets because the ingredients ordinarily contain adequate amounts.

TABLE IV. Approximate Estimates for Minimal Allowances of Essential Amino Acids When Formulating Open-Formula Diets*

	% of diet		*% of diet*
Arginine	*0.2*	*Phenylalanine*	*0.9*
Histidine	*0.3*	*Threonine*	*0.5*
Isoleucine	*0.6*	*Tryptophan*	*0.15*
Leucine	*0.8*	*Valine*	*0.6*
Lysine	*0.8*	*Cystine*	*0.1*
Methionine	*0.5*		

*Essentiality of the various amino acids has not been established for nonhuman primates, but inferences have been made from the human experience. Cystine is not essential when methionine is adequate, except possibly in infants.

Similar calculations are made for vitamins A, B_{12}, E, D, K, C, biotin, choline, folic acid, niacin pantothenic acid, pyridoxine, riboflavin, and thiamin to establish their concentration required in the vitamin premix. The mineral and vitamin premixes are prepared with a major ingredient as a carrier to insure that the required concentration of each nutrient is provided. The carrier is used in order to obtain a maximum distribution of the vitamins and minerals throughout the diet.

Diets with graded concentrations of nutrients can be formulated using natural ingredients. To formulate diets with graded concentrations of crude protein and crude fiber, the quantities of major ingredients used in the formula are adjusted. Examples of nonhuman primate diets having graded concentrations of crude fiber are those of Morin et al. (13) shown in Table V. To achieve the diets with increased concentrations of crude fiber, the amount of oat hulls and alfalfa meal used in each formula was increased and the amount of wheat flour and middlings decreased. Adjustments in the amount of fish meal was necessary to maintain protein quality because the protein supplied by oat hulls contains

TABLE V. Formulation of Experimental Open Formula Baked Diets for Nonhuman Primates

Ingredients	Percent dietary crude fiber		
	2.4	7.0	9.8
Wheat flour	50.00	39.00	28.00
Wheat middlings	6.30	4.75	3.50
Wheat bran	4.20	3.50	3.00
Alfalfa meal (17% protein)	2.25	6.25	10.00
Oat hulls	1.00	10.00	19.00
Soybean meal (48% protein)	7.50	7.50	7.50
Skim milk (dried)	2.75	2.50	2.50
Fish meal (60% protein)	4.00	4.50	5.00
Brewer's dried yeast	2.50	2.50	2.50
Sugar	3.00	3.00	3.00
Cornstarch	8.40	8.60	8.40
Soybean oil	3.70	3.60	3.50
Calcium carbonate	1.70	1.50	1.30
Monosodium phosphate	0.40	0.50	0.50
Salt	0.30	0.30	0.30
Mineral premix	1.00	1.00	1.00
Vitamin premix	1.00	1.00	1.00
	100.00	100.00	100.00

Ingredients	Percent dietary crude fiber 2.4	7.0	9.8
Mineral premix provided (mg/kg diet)			
Cobalt	0.11	0.22	0.33
Copper	2.86	3.30	3.30
Iron	59.47	44.05	35.24
Magnesium	99.12	99.12	99.12
Manganese	8.81	13.22	17.62
Potassium	1751.10	870.04	-
Zinc	1.65	3.85	3.74
Iodine	1.43	1.43	1.43
Vitamin premix provided (per kg diet)			
Vitamin A (IU)	24,200	13,200	2,200
Vitamin D_3 (IU)	6,400	6,400	6,400
Vitamin E (mg)	44.00	39.60	35.20
Vitamin K (mg)	8.80	8.80	8.80
Vitamin C (g)	2.95	2.95	2.95
Biotin (mg)	0.05	0.05	0.05
Choline (mg)	220.00	312.43	407.04
Vitamin B-12 (ug)	3.52	2.20	1.10
Folic acid (mg)	2.64	2.64	2.20
Niacin (mg)	44.00	50.60	55.00
Pantothenic acid (mg)	57.20	57.60	55.00
Pyridoxine (mg)	6.60	6.60	7.70
Riboflavin (mg)	5.50	4.95	3.96
Thiamin (mg)	17.60	18.70	18.70
Methionine hydroxyanalogue (g)	0.50	0.50	0.50

Adapted from Morin et al. (13).

an adequate amount of essential amino acids. Diets containing graded concentrations of crude protein are formulated by making similar adjustments in the quantities of all major ingredients so that the quantity of protein is altered, but not the quality. Changes in the quantity of major ingredients used in a formula required adjustments in the quantities of vitamins and minerals included in the respective premixes for each diet. The limitation to formulating diets with graded concentrations of nutrients other than crude protein and crude fiber, is the minimum quantity of a specific nutrient supplied by the major ingredients. Diets differing only in crude fat concentrations are attained by incorporating an ingredient, such as corn starch in the initial formula then substituting the source of crude fat for corn starch at increments that will result in the planned crude fat concentrations in the final formula. Altering the fat content alters the caloric density of the diet, and the mineral and vitamin premixes should be adjusted to maintain these micronutrients as a constant percent of calories. Diets with graded concentrations of vitamins or minerals (above the concentrations in the major ingredients) are formulated by making the appropriate increase in the respective premix. It is not practically possible to formulate natural ingredient diets with graded concentrations of amino acids.

IV. DIET MANUFACTURE

Almost all commercial natural ingredient diets for nonhuman primates are extruded biscuits manufactured by subjecting the diet in meal form to high temperature in a pressurized chamber and forcing it through a die. This process results in a low density product that is readily accepted and highly digested by most species of nonhuman primates. Ascorbic acid, which is unstable in the extruding process, is applied as the final coating on the biscuits. However, extruders are designed for the efficient production of a single diet in large quantities, not for experimental diets that are usually required in small quantities.

In general, high density pelleted products are not readily consumed by most nonhuman primate species. Diets fed in meal form or as a liquid have limited application for nonhuman primates because it is impractical to maintain large numbers of animals with these kinds of diets.

Baking is an efficient method for manufacture of experimental diets for nonhuman primates because relatively small amounts of experimental diets can be produced. This process results in a product which most species readily accept and appear to utilize quite well. The digestion coefficients presented in Table VI are from a total collection digestibility study at our facility (14). A series of metabolism trials were conducted using 12 rhesus *Macaca mulatta* monkeys to compare the digestibility of a commercial extruded diet and an experimental baked diet. The diets were comparable in nutrient composition and the study was designed so the digestibility of both diets was measured by analysis of feces in the same 12 animals. These data indicate the utilization of baked diets was comparable to extruded diet in these non-human primates.

TABLE VI. Digestibility of a Commercial Extruded Diet and an Experimental Baked Diet in Rhesus Monkeys

Item	*Extruded diet*	*Baked diet*
No. Animals	*12*	*12*
X feed consumption (g)[1]	*735*	*822*
Dry matter		
% of diet	*89.5*	*93.5*
% digested	*89.0*	*88.6*
Crude protein		
% of diet	*18.0*	*18.0*
% digested	*90.0*	*88.3*
Ether extract		
% of diet	*5.0*	*4.5*
% digested	*85.7*	*91.5*
Crude fiber		
% of diet	*2.2*	*3.4*
% digested	*47.0*	*38.8*
Nitrogen-free extract		
% of diet	*61.9*	*61.6*
% digested	*91.3*	*94.5*

[1]*Total consumption during the 5 day collection period.*

V. APPLICATION OF OPEN FORMULA DIETS

As indicated previously, the use of open formula diets for nonhuman primates has not been as widespread as in other species. The American Institute of Nutrition has recommended open formula diets as standard reference diets for rat and mouse experiments (15). Open formula diets have proven to be valuable research tools for thses species. Recently published results from studies using open formula diets with mice have shown apparent differences in crude protein and crude fat requirements among inbred strains (16).

A similar approach for nonhuman primate diets should be considered for research with these species. Open formula diets, with graded nutrient concentrations, could be used to generate the data essential to ascertain the quantitative nutrient requirement of nonhuman primate species or for establishing animal models of diseases caused by either deficient or excess dietary nutrients. An example of this is the previously mentioned study (13) where the effect of dietary crude fiber concentrations on intestinal disorders in rhesus monkeys during quarantine was examined. Diets containing three crude fiber levels were used (Table V), 2.4% - a relatively low fiber level similar to the level of some commercial diets; 7.0% - an intermediate crude fiber level; and 9.8% - the maximum crude fiber level convenient to manufacture. A commercial diet was also fed with a 2.2% crude fiber level. These diets were fed a newly arriving rhesus monkeys for the first 60 days of a quarantine period when the stress on the monkeys was the greatest. The expected incidence of intestinal disorders was about 20% for monkeys fed a standard, commercial complete nonhuman primate diet during this period. The diets were evaluated on the basis of incidence of intestinal disorders and the number of days the monkeys required treatment for the disorder. The data presented in Table VII show the dramatic decrease in the incidence of intestinal disorders occurring in monkeys fed the diet containing 7% crude fiber.

Based on these data it appears that optimum fiber concentration for the rhesus monkeys during guarantine is approximately 7% as compared to the 2 to 2.5% that has been supplied in some commercial nonhuman primate diets. When this higher level of crude fiber is fed, the animals fecal colume increases approximately two-fold, based on visual observation, as compared to the volume for monkeys fed the 2% fiber level standard commercial diet. It may be that the low fiber level in commercial diets was established for the convenience of caring for animal colonies and not for the monkeys' requirements. These results indicate a need for additional data on these specied dietary requirements so their management can be improved.

TABLE VII. Application of Open Formula Diets With Varying Levels of Dietary Crude Fiber on Intestinal Disorders of Quarantined Rhesus Monkeys

Criterion	Dietary crude fiber level 2.4%*	7.0%*	9.8%*	2.2%†
Number of monkeys	72	72	72	72
Morbidity rate	11.1%	1.4%[a]	12.5%	20.8%
Total number of treatment days	66	1	39	104
Mean treatment days per monkey	0.9	0.014[a]	0.5[b]	1.4

*Diets listed in Table 5.
†Commercial Diet.
[a]Significantly different from all other groups.
[b]Significantly different from 2.2% Dietary crude fiber level group.

Another recent application of open formula diets occurred in Iquitos, Peru, in a newly established breeding colony of squirrel monkeys and marmosets. These colonies were fed a gruel consisting of fresh fruits, whole eggs, immature corn, soybean meal, and small amounts of various protein and vitamin supplements. A large number of animals developed alopecia and the new hair of juvenile and mature animals was grey. An evaluation of the mineral content of the ingredients in the gruel indicated a relatively low level of minerals was being supplied. Most of the dietary ingredients were harvested in the Iquitos region of heavy rainfall where minerals are readily leached from the soil. When a mineral premix (Table VIII) was added to the gruel there was a dramatic decrease in the number of abnormal monkeys. This mineral mix was manufactured in the United States and shipped to Peru. A small amount of methionine was included in this premix because there was some concern regarding the amount of this essential amino acid in the diet being used.

Subsequently, a diet was formulated with ingredients available in the Iquitos area for manufacture by baking in one of the local establishments (Table VIII). Raw whole eggs are included as a source of protein because they are readily

TABLE VIII Formulation of Baked Primate Diet for Iquitos, Peru

Ingredient	*Baked Diet g/kg*
Soybean meal (44% protein)	*380*
Raw eggs (dry weight, assumed 70% water)	*70*
Alfalfa meal (15% protein)	*90*
Dried skim milk	*100*
Wheat flour	*180*
Wheat bran	*70*
Ground whole brown rice	*70*
Sucrose	*30*
Vegetable fat	*40*
Mineral mix 1-77[1]	*20*
Vitamin mix 1-77[2]	*20*
	1070

[1]*The mineral mix contained (in g/kg): calcium phosphate, 413: magnesium sulfate•$7H_2O$, 121.5: sodium chloride, 104.5; iron gluconate, 40; zinc sulfate, 2.25; manganese sulfate•H_2O, 2.55; copper sulfate•$5H_2O$, 1.0: cobalt sulfate•$7H_2O$, 0.025; potassium iodide, 0.13; sodium selenite, 0.035; methionine, 50, added to 265 g cornstarch.*

[2]*The vitamin mix contained (in g/kg): vitamin A (650,000 IU/g), 1.846; vitamin A and D_3 (325,000 IU each/g), 0.922; vitamin E 50% (500 IU/g), 6.042; Hetrogen K 63% MSB, 0.456; vitamin B_{12} (2.206 mg/g), 1.137; thiamine mononitrate USP, 1.0; riboflavin 50%, 2.0; niacin 98%, 5.262; calcium pantothenate USP, 7.318; pyridoxine HCl USP, 1.0; folic acid USP, 0.75; biotin 1%, 2.5; ascorbic acid USP coated, 100.0; choline chloride* 50%, 50.23; added to 820 g dextrin.

available. However, the diet is being formulated on the basis of 1070 grams rather than a kilogram because the moisture in the eggs and some of the other ingredients evaporates during the baking process. A vitamin premix is also being manufactured for this diet (Table VIII). The diet is being supplemented with fresh citrus fruit to insure an adequate amount of vitamin C.

These are but two examples of how open formula nonhuman primate diets can be used to solve specific problems involving the management of nonhuman primate colonies. Even though sufficient data are not yet available concerning the long term performance of nonhuman primates fed open formula nonhuman primate diets, these diets are essential to better identify the nutrient requirements of the various species in order to improve the quality of all research involving nonhuman primates.

REFERENCES

1. National Research Council (NRC) (1978) Nutrient Requirements of Nonhuman Primates. Committee on Animal Nutrition, National Academy of Sciences, Washington, D.C.
2. Hodson, H.H., Jr., Mesa, V.L., and Van Riper, D.C. (1967) Protein requirements of the young, growing chimpanzee. *Lab. An. Care 17: 551-562.*
3. Hummer, R.L. (1970) Observations of the Feeding of Baboons. In: Feeding and Nutrition of Nonhuman Primates, ed. by R.S. Harris, New York: Academic Press, pp. 183-203.
4. Schmidt, L.H., Cramer, D.V., Rossan, R.N., and Harrison, J. (1977) The characteristics of *Plasmodium cynomolgi* infections in various Old World Primates. *Am. J. Trop. Med. Hyg. 26: 356-372.*
5. National Research Council (NRC) (1978) Control of Diets in Laboratory Animal Experimentation. Committee on Laboratory Animal Diets, Assembly of Life Sciences, National Academy of Sciences, Washington, D.C.
6. Fox, J.G., Aldrech, F.D., Boglen, G.W., Jr. (1976) Lead in animal feeds. *J. Toxicology and Environmental Health 1: 461.*
7. Newberne, P.M. (1975) Influence on pharmocological experiments of chemicals and other factors in diets of laboratory animals. *Fed. Proc. 34: 209.*

8. Yang, R.S.H., Mueller, W.F., Grace, H.K., Goldberg, L., and Coulston, F. (1976) Hexachlorobenzene contamination in laboratory monkey chow. *J. Agric. Food Chem. 24: 563.*
9. National Research Council (NRC) (1971) Atlas of Nutritional Data on United States and Canadian Feeds. Committee on Animal Nutrition, Agricultural Board, National Academy of Sciences, Washington, D.C.
10 Nutrient Composition of Some Philippine Feedstuffs (1976) Department of Animal Science, College of Agriculture, College Laguna. Leopoldo S. Castillo, Amelia L. Gerpacio.
11. FAO Agricultural Studies. (1975) *Tropical Feeds*: Feeds Information Summaries and Nutritive Values. Bo Gohl, FAO Feeds Information Centre, Animal Production and Health Division, Food and Agriculture Organization of the United Nations, Rome.
12. Latin American Tables of Feed Composition. (1974) University of Florida, Institute of Food and Agricultural Sciences, Center for Tropical Agriculture, Department of Animal Science, Gainesville, Florida.
13. Morin, M.L., Renquist, D.M., Knapka, J., and Judge, F.J. (1978) The effect of Dietary Crude Fiber Levels on Rhesus Monkeys During Quarantine. *Lab. An. Sci. 28: 405.*
14. Barnard, D., Renquist, D.M., Morin, M.L., and Knapka, J. J. (1979) The Effect of Dietary Crude Fiber on the Digestibility of Crude Protein, Crude Fat, and Carbohydrates in *Macaca mulatta*. In preparation, *Lab. An. Sci.*
15. American Institute of Nutrition. (1977) Report of Ad Hoc Committee on Standards for Nutritional Studies. *J. Nutr. 107: 1340.*
16. Knapka, J.J., Smith, K.P., and Judge, F.J. (1977) The Effect of Crude Fat and Crude Protein on Reproduction and Weanling Growth in Four Strains of Inbred Mice. *J. Nutr. 107: 61.*

ROLE OF DIET IN NORMAL BILIARY PHYSIOLOGY AND GALLSTONE FORMATION

Oscar W. Portman

Department of Nutrition and Metabolic Disease
Oregon Regional Primate Research Center
Beaverton, Oregon

Department of Biochemistry
University of Oregon Health Sciences Center
Portland, Oregon

Manfred Alexander

Department of Nutrition and Metabolic Diseases
Oregon Regional Primate Research Center
Beaverton, Oregon

Naomi Tanaka
Toshiaki Osuga

Department of Medicine
University of Tsukuba School of Medicine
Sakamura, Niihari-gun, Ibaraki-ken, Japan

First Department of Internal Medicine
University of Tokyo
Tokyo 111, Japan

ISBN 0-12-334250-3

I. INTRODUCTION

Bile has been one of the most extensively studied of the body fluids of man and many animal species. Abnormalities in bile physiology have been linked to a number of human diseases. For example, the association of altered cholesterol metabolism with atherosclerosis has focused attention on the bile, since cholesterol as well as bile salts, which are the main catabolic products of cholesterol, are secreted into the bile (1). Another medical reason for emphasizing cholesterol in bile is the realization that it is the principal constituent of the most common type of gallstone in the United States and other Western countries (2). We will devote much of this paper to a consideration of the factors that determine cholesterol gallstone formation and the suitability of different animals for such studies.

There is also some evidence that nutrition is an important variable that influences the metabolism of the principal constituents of bile as well as the development of gallstones (3,4). Of course, the bile salts also have a more general interest for the nutritionist since they are necessary for the digestion of fats and the absorption of lipids and fat-soluble vitamins.

II. EPIDEMIOLOGY OF CHOLESTEROL GALLSTONES

The epidemiology of human cholesterol gallstones and of atherosclerosis are similar in many ways. Both diseases are common in the industrial countries and rare in developing countries. For example, cholesterol gallstones have been rare in several areas of Africa (5). The Japanese provide a particularly dramatic example of an increase in cholesterol gallstones and atherosclerotic heart disease with a change in environmental factors. People of Japanese origin in the United States had a greater incidence of atherosclerotic heart disease and probably of cholesterol gallstones than Japanese in Japan. The changes in lifestyle that occurred in Japan after World War II were also associated with an increase in cholesterol gallstones and decrease in pigment stones (6,7). The Japanese way of life resembled that in the United States and Western Europe in many respects such as industrialization and urbanization even prior to the war. However, one dramatic recent change has been the Westernization of the diet, i.e., an increase in the level and saturation of fats, and this may explain the increase in cholesterol gallstones and the incidence of atherosclerotic heart disease.

Although the correlation between the rates of incidence of cholesterol gallstones and atherosclerotic heart disease is high in many populations, there are some exceptions. One of the most remarkable is the two-fold greater incidence of gallstones in women than in men whereas the incidence of ischemic heart disease is higher in men than in women of the same age group (8). Pregnancy, and the use of estrogens and progestins in birth control (9,10) and treatment of menopausal and post-menopausal signs and symptoms (11), also influence the incidence of cholesterol gallstones. Therefore, we are interested both in nutritional factors and in the mechanism of this apparent endocrine basis for susceptibility to gallstones and the possibility of producing animal models of the phenomena.

III. MICELLES IN BILE CHOLESTEROL STABILIZATION

The secretion of cholesterol by the liver and its transport into the bile may seem strange in view of its insolubility in water. Unlike the plasma, bile does not contain lipoproteins as transport vehicles for cholesterol. It had been realized for some time that the high concentrations of bile salts and lecithin must play some role in the stabilization of cholesterol in bile, but it is only in the last 10 years that substantial progress has been made in understanding and quantifying that process. Hoffman's (12) studies on the mixed micelle in digestion were important in the evolution of the present theory of a similar moiety in bile. The concept of single component micelles had been developed long ago from the behavior of soaps. Each micelle of this type is an aggregate of a large number of identical molecules that form a spherical or other ordered configuration such as the liposome or lipid vesicle. To form a micelle a compound must have both hydrophobic and hydrophilic properties; the former are responsible for the aggregation of molecules that occur with an oil in water, the latter for formation of an interface with the water phase. Lecithin, a major constituent of biological membranes and of bile, is an ideal amphipath or micelle-former. The bile salts are a very heterogeneous group of compounds, but all of them have the capacity to form micelles and true solutions, the extent for each form depending on their concentrations, the pH of the bile, and the pK of the molecule in question. Cholesterol has one hydroxyl group and one double bond, but it is predominantly hydrophobic and has little capacity to form micelles in water. The phenomenon of emulsification of oils in surface active materials like lecithin, as well as the general theories of membrane structure which visualize cholesterol solubilized in the hydrophobic portion of phospholipid bilayers, suggest the possibility of a mixed micelle that includes cholesterol. The mixed micelle of cholesterol, phospholipid, and the bile salts is sufficiently small to give a clear solution in water and to remain dispersed during low speed centrifugation. The quantity of cholesterol that can be stabilized depends to a large extent on the relative proportions of lecithin and total bile salts present, although absolute concentrations of the three constituents have some influence. The kinds of bile salts and, to a much lesser extent, the fatty acid constituents of the lecithin have some effect.

IV. LIMITS OF CHOLESTEROL STABILIZATION IN BILE

Small and his associates (13) made a major contribution to an understanding of the limits of cholesterol solubility in bile with the use of a three dimensional phase diagram like that indicated in Figure 1. The lower part of the triangle below the curved line defines concentrations of the three main constituents of bile in which only mixed micelles are present. Above that line one can expect cholesterol crystals to form and, if the composition is maintained, for cholesterol gallstones to form. Subsequent attempts to use this diagram of Admirand and Small (14) have failed to perfectly separate gallbladder biles of patients with

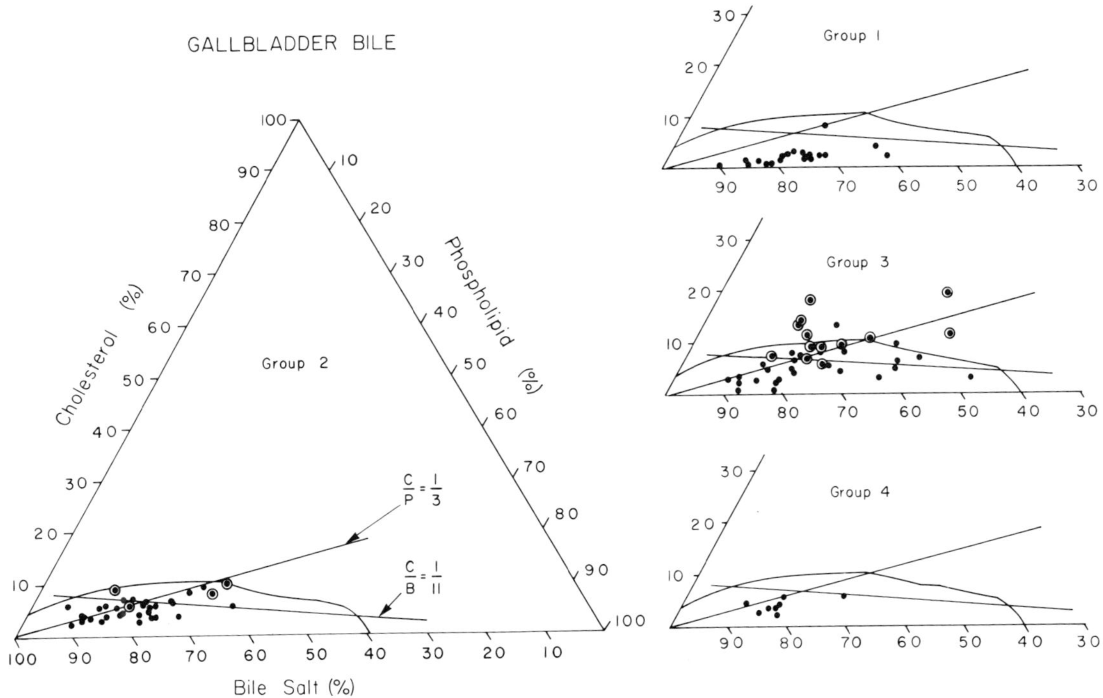

Fig. 1. Plots of the relative molar concentrations of cholesterol, bile salts, and phospholipids (the sum for three constituents - 100) in gallbladder biles of squirrel monkeys on a 3 dimensional graph to predict the presence of cholesterol gallstones. According to Admirand and Small (14) the area below the wavy line in the lower left hand corner of the triangle represents compositions in which all of the cholesterol is stabilized as a mixed micelle. The encircled points are for biles of monkeys with cholesterol gallstones. Group 1 is monkeys fed a commercial stock diet. Groups 2 and 4 are monkeys fed a semipurified diet with 8% (w/w) corn oil and no cholesterol. Group 3 was fed 20% butter and 0.9 mg cholesterol/kcal. From Osuga and Portman (56).

gallstones from those free of cholesterol gallstones. In particular, many people without gallstones have saturated bile. There are a number of reasons for this failure. After a overnight fast, the conditions under which bile is invariably obtained at surgery, both types of bile and particularly hepatic bile are frequently saturated with cholesterol whereas during the alimentary period they usually are not. A random sample of bile, particularly one taken after an overnight fast, will not be representative. Perhaps the proportion of the day that the bile is saturated is crucial. Another possible explanation for the failure of the 3 dimensional plot to segregate perfectly normals from people with gallstones is that most people in Western countries have an abnormally high cholesterol concentration in bile, but for some reason they have not or have not yet formed and retained cholesterol gallstones. Finally, there is some evidence that the original values of Admirand and Small (14) which defined the levels of cholesterol that are needed to saturate the bile were too high. This, of course, means that an even larger proportion of people without gallstones have fasting gallbladder bile that is saturated with cholesterol than was originally believed. Hegardt and Dam (15), Holzbach (16), and Carey and Small (17) have proposed new limits of cholesterol stability in the mixed micelle. Others have proposed simplifications of the methods of calculating and expressing the degree of saturation of cholesterol in bile. The procedure of Thomas and Hofmann (18) which is illustrated in Figure 2 is particularly useful. Here the maximum mole % cholesterol which is soluble in bile according to several experiments and conditions is plotted against the ratio of the molar concentration of phospholipid to that of phospholipid + bile acids. The percent saturation of bile with cholesterol was termed the lithogenic index of Metzger et al. (19) and is the ratio of the % cholesterol in bile to the maximum concentration that can be solubilized at the given concentrations fo bile salts and phospholipids X 100.

V. METABOLIC PATHWAYS FOR BILE LIPIDS

The saturation or supersaturation of bile with cholesterol can occur if the secretion of cholesterol is too high or the secretion of bile salts is low. The concentration of lecithin is nearly always proportional to that of bile salts in a given individual, and it has received little emphasis by investigators of the mechanisms of abnormal bile production.

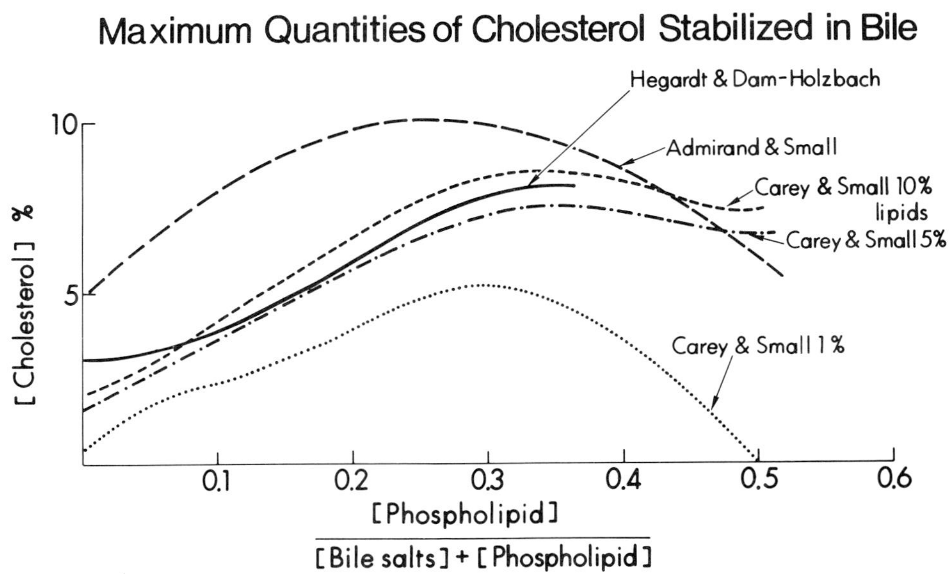

Fig. 2. Plots of the maximum mole % concentration of cholesterol which is stabilized in bile according to several investigators (14,17). The method of presentation is that of Thomas and Hofmann (18). The values 10%, 5%, and 1% after Carey and Small indicate the concentration (w/w) of cholesterol + phospholipid + bile salts in bile. 5-10% are typical values for gallbladder bile. The absolute concentration of the three constituents in hepatic bile is much less (about 1%).

Before we consider the factors responsible for abnormal bile production we should briefly summarize the metabolic pathways of the three principal lipid constituents of bile.

Biliary cholesterol is a mixture of nascent cholesterol formed in the liver and free cholesterol of lipoproteins. The latter in turn can be derived from the diet or be synthesized in the body. Recently Schwartz and coworkers (20) showed that only 20% of human biliary cholesterol is derived directly from newly synthesized cholesterol, the remaining 80% being obtained from lipoproteins. The same group (21) also showed that high density lipoprotein (HDL) cholesterol was secreted in preference to low density lipoprotein (LDL) cholesterol. This observation is consistent with the concept that HDL is involved in the net transport of cholesterol from peripheral tissues to the liver. Nearly one-half of biliary cholesterol is reabsorbed by the squirrel monkey (22), but it is transported via the lymphatics into the systemic circulation and is then distributed into a relatively large pool. Thus cholesterol secretion is not nearly as dependent as bile salt secretion on an intact enterohepatic circulation.

The bile salts, cholate and chenodeoxycholate, are formed in the liver from cholesterol and conjugated with taurine or glycine before their secretion into the bile (23). Schwartz et al. (21) concluded that 69% of bile salts secreted by humans was derived from lipoprotein cholesterol, presumably mainly HDL cholesterol. Part of the secreted bile, including the bile salts, passes down the common duct and directly through the sphincter of Oddi and into the duodenum, and part enters the gallbladder via the cystic duct. In the fasted baboon, more than half the hepatic bile enters the duodenum directly (24). The remainder enters the gallbladder where about 90% of the H_2O and electrolytes are removed. The gallbladder contracts in response to the hormone cholecystokinin which is released from the gut wall after a meal. Bile salts are very efficiently absorbed from the small intestine and colon. The more hydrophobic (e.g., dihydroxy bile salts conjugated with glycine) compounds tend to be reabsorbed throughout the gut whereas the more hydrophilic bile salts tend to be absorbed by active transport in the ileum (25). In the colon, the bile salts are deconjugated and the hydroxyl groups on the ring structure are modified by intestinal bacteria. Cholic is converted to deoxycholic acid and chenodeoxycholic is converted to lithocholic. The deoxycholic acid is well absorbed from the colon, the lithocholic acid less well. Since most of the bile salts that are secreted into the gut are reabsorbed, the rate of new bile salt synthesis is less than 5% of the rate of biliary secretion. Surgical or physiological interruption of the enterohepatic curculation of bile salts leads to a several-fold increase in new bile salt synthesis. There is a delay before this compensatory process comes into play, however, and new bile salt synthesis never totally replaces the normal rate of secretion (26).

Intact phospholipids undergo relatively little enterohepatic circulation; however, Scow et al. (27) have shown that some lysolecithin from administered lecithin is absorbed and acylated. Although most biliary lecithin is believed to come from the CDP choline:diglyceride pathway, Illingworth and Portman (28) showed that in the squirrel monkey, plasma lysolecithin was converted to biliary lecithin and that the quantity of plasma lysolecithin available (in part from plasma LCAT) and the rates of hepatic extraction of lysolecithin from blood and formation and secretion of lecithin could account for a large part of the biliary lecithin.

VI. EFFECTS OF INTERRUPTING THE ENTEROHEPATIC CIRCULATION

One of the most useful procedures to study the controls of bile secretion is the partial or complete interruption of the enterohepatic circulation with a common duct fistula. Figure 3 shows the cumulative excretion of bile acids from the common duct fistula of rats. There was a progressive drop in the rate of secretion, although the total volume of bile was rather well maintained. After 12 to 24 hours of continuous interruption of the bile duct, the bile salt secretion began to increase again. The bile salts secreted during the first period were largely preformed, whereas the secondary surge of bile salt secretion was due entirely to newly synthesized compounds. Dowling et al. (29) devised an ingenious method for sampling hepatic secretions of bile of the rhesus monkey and returning most of the bile to the duodenum. These were chronic preparations on animals in restraining chairs. Figure 4 shows the acute effect of complete interruption of the enterohepatic circulation in one monkey that had undergone surgery several weeks previously (30). There was a sharp drop in the secretion of bile salts and phospholipids and a less pronounced drop in cholesterol secretion. Thus bile became more saturated with cholesterol. After several hours when the feedback inhibition of bile salts that returned to the liver was at a minimum, new bile salt secretion increased, although the bile remained relatively saturated with cholesterol (Figure 5).

Although complete interruption of the enterohepatic circulation of bile salts rarely occurs spontaneously, partial interruption occurs during every nocturnal fast when some of the secreted bile is sequestered in an uncontracted gallbladder. The rate of secretion of cholesterol drops very little so that newly secreted hepatic bile contains relatively more cholesterol than does gallbladder bile and the gallbladder bile itself becomes relatively richer in cholesterol as the night progresses.

Iliectomy leads to increased loss of bile salts and to bile saturated with cholesterol (Figure 6). Cholestyramine is used clinically to bind the anionic bile salts and prevent their reabsorption from the gut and secondarily to lower plasma cholesterol. This agent also increases the saturation of bile with cholesterol. There is likewise some evidence that certain materials in food, particularly fiber, can bind bile salts and increase their loss from the enterohepatic circulation (32).

In order to determine how much bile salt can be lost from each enterohepatic circulation before the power of the liver

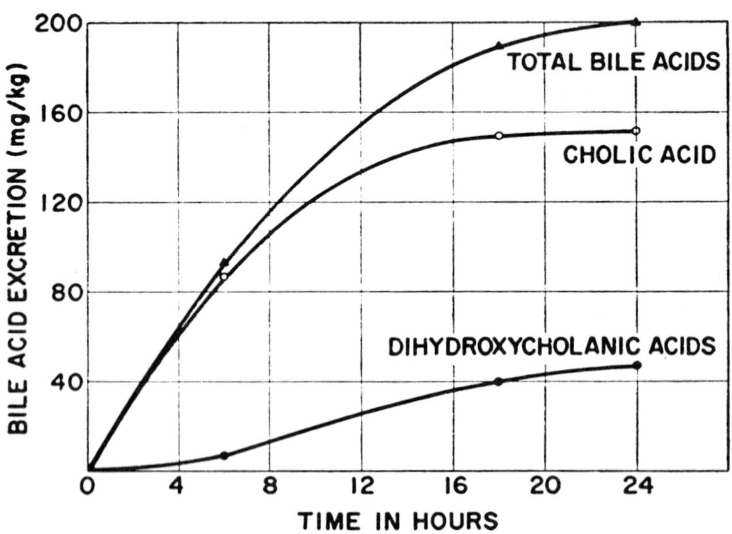

Fig. 3. The cumulative excretion of cholic, dihydroxycholanic and total bile acids from a rat with a bile duct fistula. (From Portman and Mann (75), 1955, J. of Biol. 213: 733.)

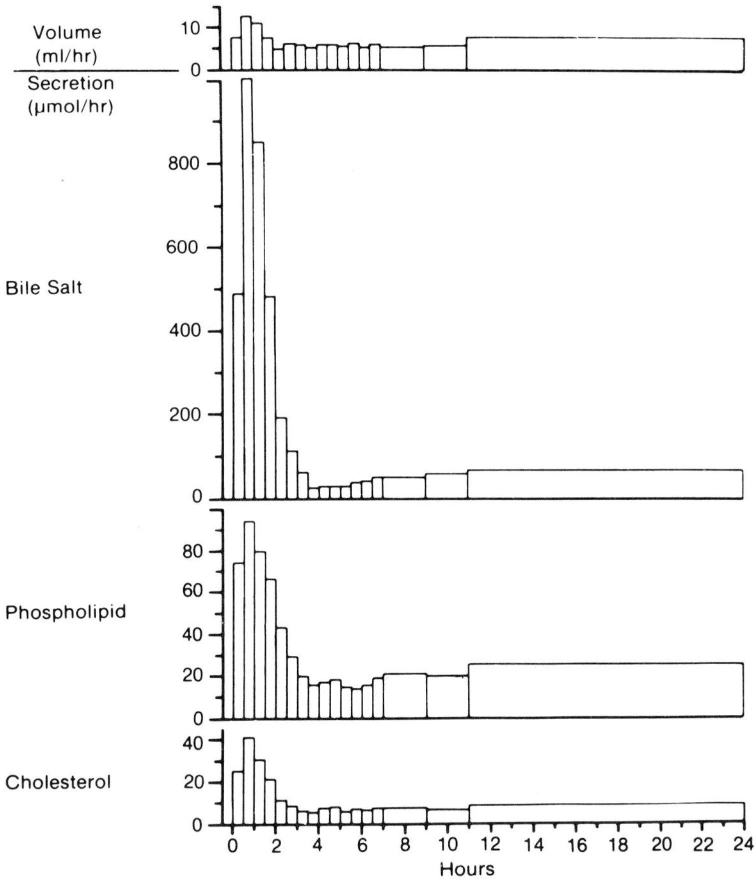

Fig. 4. The effect of acute interruption of the enterohepatic circulation on bile flow and secretion of bile salts, phospholipid and cholesterol by rhesus monkey. (From Dowling, Mack and Small (30), 1971, J. Clin. Invest. 50: 1917.)

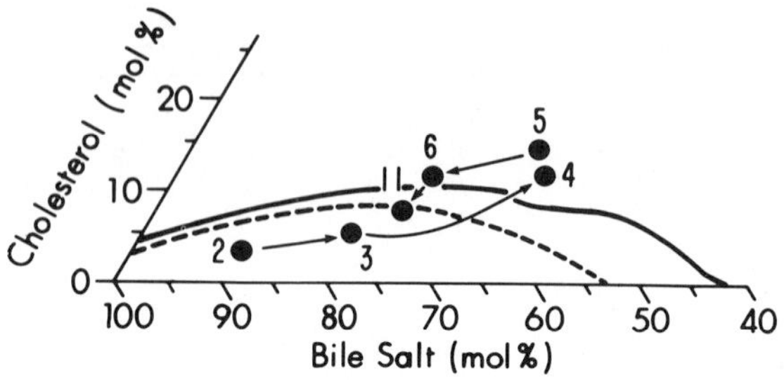

Fig. 5. The relative composition in a rhesus monkey at 2, 3, 4, 5, 6, and 11 hours after acute interruption of the enterohepatic circulation. (From Dowling, Mack, and Small (30). 1971 J. Clin. Invest. 50: 1917.)

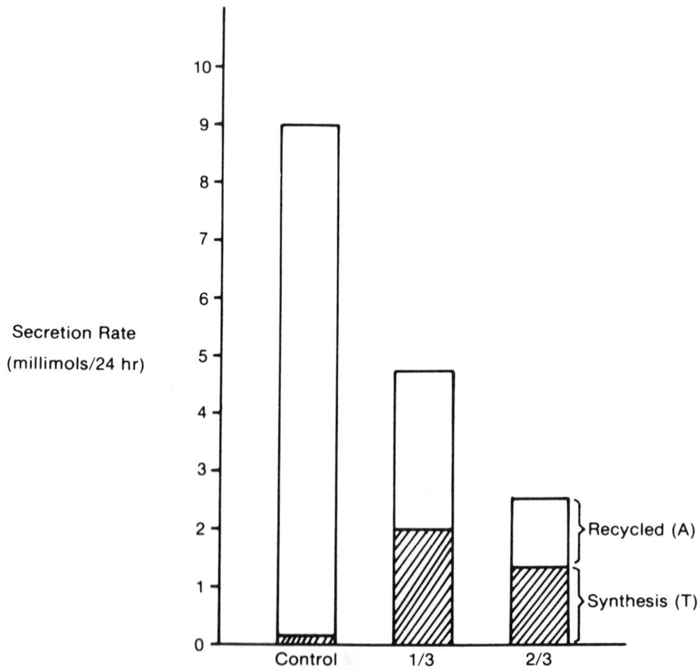

Fig. 6. Bile salt secretion and synthetic rates in rhesus monkeys with 1/3 or 2/3 of the distal small bowel resected. (From Small, Dowling, and Redinger (26). 1972 Arch. Int. Med. 130: 552.)

to replace the loss with new synthesis is surpassed, Dowling et al. (31) tested the chronic effects of varying the proportion of bile diverted from the rhesus monkey in their chair preparations. Figure 7 shows that the loss of more than 10% of the bile results in an absolute decrease in bile salt secretion and the loss of 20% resulted in maximum new bile salt synthesis. Thus an increase in the loss of bile salts from about 1 to 20% during each enterohepatic circulation results in a maximum increase in the catabolism of cholesterol. Greater losses such as may occur with ingestion of large doses of cholestyramine or after iliectomy only increase the hazard of saturation of bile with cholesterol and of its precipitation as crystals.

VII. FASTING AND CHOLECYSTECTOMY ON BILE

If the gallbladder is responsible for a physiological interruption of the enterohepatic circulation of bile salts and increase in cholesterol saturation of bile during the night, its removal should result in a more constant secretion of bile salts and composition of bile. Most studies (33) have shown that cholecystectomy causes hepatic bile to become unsaturated with respect to cholesterol. Shaffer and Small (34) found that the size of the bile salt pool tends to be somewhat smaller after cholecystectomy, although the secretion rate of this much more rapidly circulating pool was increased. The secretion of cholesterol was unaffected; therefore bile was less saturated with cholesterol than before cholecystectomy. Because of the more frequent circulation of bile salts the number of secondary or bacterially altered bile salts increases. The cholecystectomized patient does respond slightly to meals with increased bile salt secretion, but there is no response to cholecystokinin.

VIII. METHODS OF STUDYING BILE KINETICS

The ideal model for studies of biliary physiology should allow the investigator to obtain a continuous analysis of gallbladder and hepatic bile without significant interruption of the enterohepatic circulation in a subject with a functioning gallbladder and sphincter of Oddi. Some of this has been achieved in humans with multiple lumen tubes, one of which delivers formula meals and another samples duodenal contents.

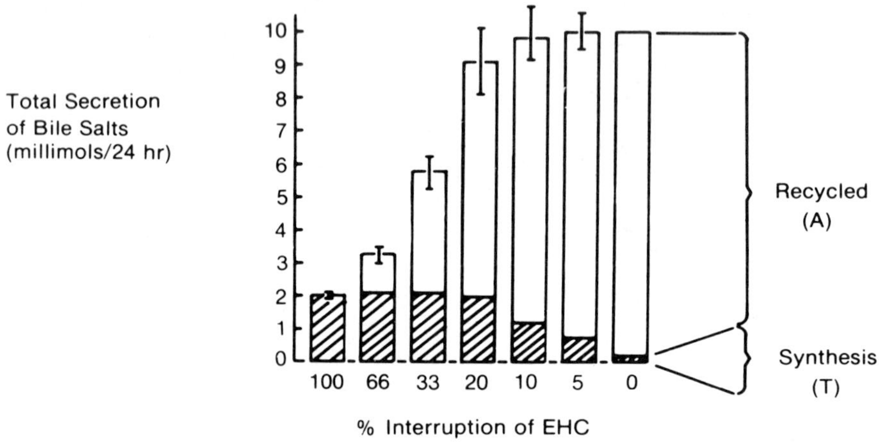

Fig. 7. Bile salt secretion and synthetic rates with chronic partial interruption of the enterohepatic circulation. The numbers under each bar represent the percentage of bile diverted during a test period. (From Small, Dowling, and Redinger (26), 1972 Arch. Int. Med. 130: 552.)

(35-36). The use of solid meals is difficult in intubated patients. Some investigators have used baboons with a cholecystopexy procedure (fundus of the gallbladder attached to the abdominal wall for percutaneous sampling of bile) and T-tubes in the common duct. An additional tool for understanding the metabolism of bile acids has been the use of labelled compounds which allow one to measure the pool of bile acids and their rate of turnover (38). Redinger and associates (39) have recently demonstrated that the continuous monitoring of expired air from the rhesus monkey or baboon which has been injected with [26-^{14}C]-cholesterol provides a very sensitive indicator of acute alterations in bile acid synthesis.

IX. CHOLESTEROL GALLSTONES IN NONPRIMATES

Few good animal models of human cholesterol gallstone disease have been described (40). Among nonprimates, only the mouse fed diets with cholesterol and cholic acid, (41) the prairie dog fed diets with cholesterol, and the hamster fed a diet low in polyunsaturated fatty acid and with sucrose or glucose as the source of carbohydrate (43) have received significant attention. The latter preparation has been thoroughly and continuously studied by Dam and his coworkers since 1952 (43). As shown in Table I, hamsters fed the basic lithogenic diet which was fat-free and contained glucose or sucrose had a very high incidence of cholesterol gallstones. The stones could be prevented by feeding rice or potato starch instead of glucose. Lactose was nearly as effective as starch. Small supplements of polyunsaturated fat with or without EFA activity to the glucose-containing diet were also preventative. Paradoxically, the addition of cholesterol or its precursor squalene to the lithogenic diet also prevented the formation of gallstones. Hyodeoxycholic acid prevented gallstones in hamsters on a lithogenic diet whereas chenodeoxycholic acid, which has been used to dissolve cholesterol gallstones in man, actually accentuated stone formation with lithogenic diets and induced gallstone formation in animals fed diets that were not otherwise lithogenic. There is little indication that the nutritional factors involved in the formation of gallstones in hamsters are involved in human cholelithiasis. On the other hand, we know little about nutritional factors in the human disease. The small size of hamsters and their relative cheapness are factors favoring this species for bile and gallstone research.

X. NUTRITIONAL EFFECTS IN RHESUS MONKEY BILE

The rhesus monkey maintained for long periods in a chair and prepared according to Dowling and coworkers, (29) with a mechanism for varying the degree of interruption of the enterohepatic circulation has provided us with a great deal of information about the control of bile formation and how nutritional factors could act. At least four laboratories have used this preparation, and in two, (45-46) the effects on biliary secretion of varying the diet have been studied. With the exception of a report concerning a single monkey, (47) we know of no reports of cholesterol gallstones in

TABLE I Diet and Cholesterol Gallstones in Hamsters (Adopted from Reports of Dam[4] and Associates).

	Principal Features of the Diet	% Incidence of Cholesterol Gallstones
1.	Rice starch; Lard 10%	0
2.	Rice starch; Fat Free	1.4
3.	Glucose; Fat Free	87
4.	Lactose; Fat Free	12.5
5.	Glucose-62%; Starch 12; Fat Free	84
6.	Glucose 62; Starch 12; Lard 2	60
7.	Glucose 62; Starch 12; Soybean oil 2	5
8.	Glucose 62; Starch 12; Cod liver oil 2	0
9.	Glucose 72; Lard 2	80
10.	Glucose 36; Starch 36; Lard 2	10
11.	Glucose; Butter fat 10	60
12.	Glucose; PFA rich margarine 10	5
13.	Glucose; Palm oil 7	39
14.	Glucose; Palm oil 7; Cholesterol 1	0
15.	Glucose; Palm oil 7; Squalene 1	0
16.	Glucose; Fat Free; Chenodeoxycholic 0.1	100
17.	Glucose; Fat Free; Hyodeoxycholic 0.1	4.2
18.	Starch; Fat Free; Chenodeoxycholic 0.1	91

rhesus monkeys. In addition to artificial interruption of the bile circulation, estrogen supplementation has been shown to produce bile that is saturated with cholesterol. Therefore, although the rhesus monkey is resistant to gallstone formation, it may be possible to devise the proper condition for induction of cholesterol gallstones. Campbell et al. (45) showed that supplements of polyunsaturated fat, but not of olive oil, medium chain triglycerides, or dextrose, increased biliary secretion of bile salts, cholesterol, and phospholipids. The bile acid increase was due to increased synthesis. Redinger and associates (46) agreed that a polyunsaturated fat increased bile secretion without altering its relative composition. Caloric restriction decreased bile secretion and increased the relative proportion of cholesterol.

XI. CHOLESTEROL GALLSTONES IN BABOONS

McSherry and his associates (48) have expanded on the original Russian observations (49) of the susceptibility of baboons to cholesterol gallstones. Baboons develop gallstones on natural diets, but pregnancy and estrogen treatment do not seem to accentuate the tendency (50). Cholesterol and bile salt secretion both drop but the bile becomes less saturated with cholesterol. This is in contrast to women treated with estrogens. The plasma cholesterol drops in baboons during pregnancy, whereas in women both pregnancy and estrogen treatment cause an increase in cholesterol, particularly in the HDL fraction. The large size of these animals makes them very suitable for preparations in which the gallbladder remains functional, and in which hepatic and gallbladder bile can be sampled simultaneously. So far they have not been used much for nutritional studies although they were very important in the development of the concept of intermittent lithogenic bile during the overnight fast. Although the development of cholesterol gallstones by baboons fed a stock diet has been used to support the proposition that baboons are ideal human models, one could argue that baboons are more susceptible to cholesterol gallstones than humans, since they form stones when they are fed diets that are based on a mixture of grains that contain low levels of fat and cholesterol and high levels of fiber. Certain groups of Africans that eat diets of similar composition rarely have gallstones (3).

XII. CHOLESTEROL GALLSTONES IN NEW WORLD MONKEYS

Among New World monkeys the owl monkey *(Aotus)* (51) is susceptible to spontaneous cholesterol gallstone formation, and Mann (52) has observed cholesterol gallstones in 2 cebus monkeys *(Cebus albifrons)* which were fed diets containing cholesterol. The Brazilian squirrel monkey *(Saimiri sciureus)* which we (53) as well as groups at Bowman-Gray (54) and the Harvard School of Public Health (55) have used for many years for studies of blood lipids, lipoproteins and atherosclerosis, is also susceptible to gallstones. (39,56 57,58). During that period, largely because of our interest in bile acids as catabolic products of cholesterol, we have analyzed the bile of more than 250 monkeys which were terminated as part of biliary physiology and atherosclerosis experiments and of over 300 gallbladder and hepatic biles obtained at non-terminal cholecystotomies and choledochostomies. Gallstones were never observed in squirrel monkeys fed our stock diet which is a vitamin-supplemented Purina Monkey Chow-25 unless cholesterol and fat were added. Even on one of the most lithogenic semipurified diets, which contained high levels of butter and 0.9 mg of cholesterol per kcal macroscopic stones did not appear in less than 3 months. Figures 8 and 9 illustrate the quantity and size of stones

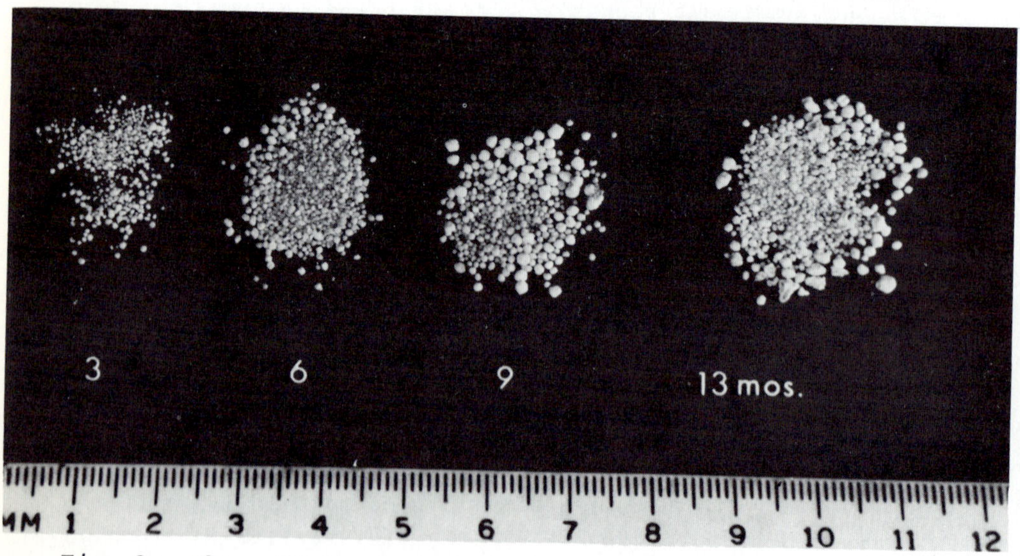

Fig. 8. The gallstones collected from individual squirrel monkeys fed a diet containing butter plus cholesterol for 3, 6, 9, or 13 months. (From Osuga and Portman (55).

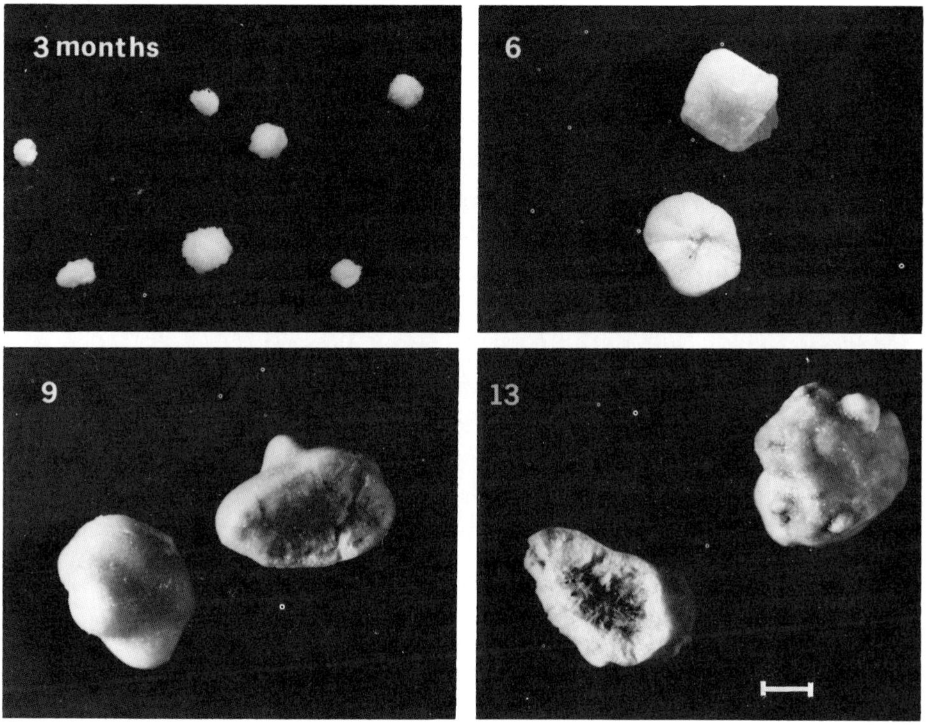

Fig. 9. Enlargement of stones seen at 3 (sandlike), 6 (simple radial), 9 (multilobular) and 13 months (mulberry) in squirrel monkeys fed the lithogenic diet. The stones at 6, 9, and 13 months have been fractured, and the fracture and external surfaces are shown. Line = 1 mm. From Osuga et al. (58).

that can be anticipated after different periods with this diet. The number and total mass of stones increase with time. The size and complexity of the largest stones also increase with time. Figure 10 shows the kind of crystals that appear very early in the hepatic bile of people and susceptible squirrel monkeys after initiation of a lithogenic diet. A prominent feature in the early growth of gallstones is the formation of platy units which result from layer on layer growth or the assembly of many crystals. Figure 11 shows a group of crystalline plates and also shows evidence of crystal dissolution. A continuous process of crystal growth and dissolution is probably necessary to change the essentially parallelogram-shaped crystals into structures that are arranged radially in a round gallstone. Figure 12 illustrates the type of rough surface structure that is

Fig. 10. Rhombic crystals which are frequently seen in the gallbladder and common duct bile of patients and squirrel monkeys with gallstones. Line = 5 μM. From Osuga et al. (59).

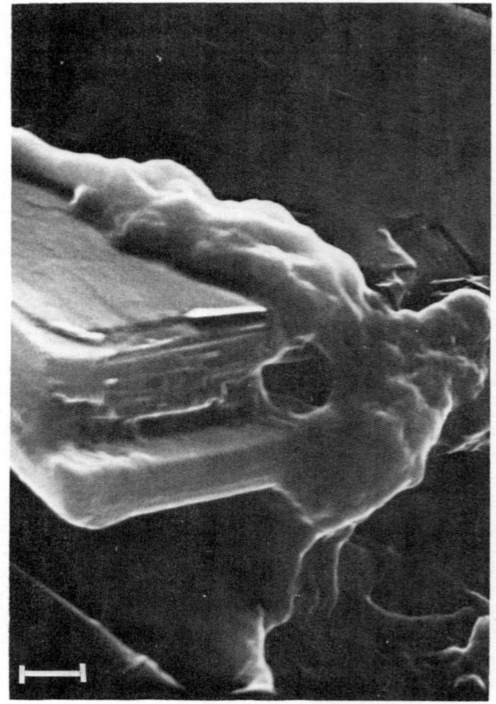

Fig. 11. A plate composed of laminated crystals which shows evidence of dissolution. Plate is from the center of a human gallstone. Line = 1 μM. From Osuga et al. (59).

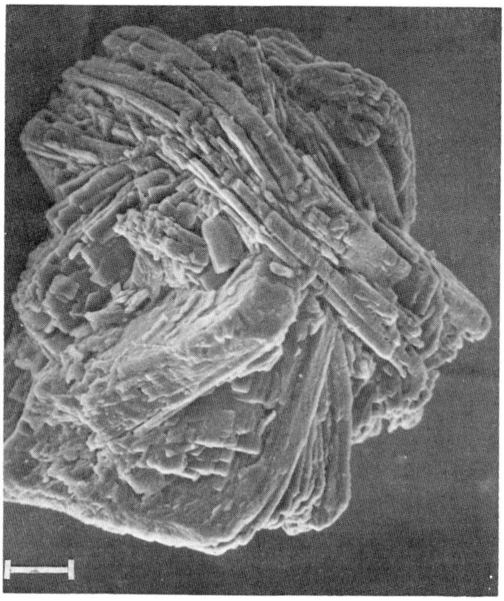

Fig. 12. The irregular external surface of a sandlike stone formed after 3 months on the lithogenic diet. There is random (subparallel) aggregation of smaller units. Line = 60 µM. From Osuga et al. (58).

apparent in the smallest macroscopic stones at 3 months. Those stones typically grow larger and smoother with time. A natural presumption is that cholesterol gallstones grow by extension from the center, and this has focused attention on the composition of the center as the original nidus for stone growth. Figure 13 is a high magnification of the center of a simple gallstone. The platy units at the center are somewhat irregular but if there was material other than these crystals, it has been lost. Comparative scanning electron microscopic examinations of crystals and intermediate microscopic forms in bile and of the fracture surface of gallstones indicated that simple radial stones may form by similar mechanisms in humans and squirrel monkeys (59,60). Larger gallstones in humans tend to have concentric rings which are really discontinuous grooves that separate aggregates of crystals with slightly different orientations. Irregularities in the shape of human gallstones seem to develop from new foci of crystal growth or accumulation on the surface of preexisting stones. Larger irregular stones in the squirrel monkey form invariably by the assembly of groups of smaller stones.

Fig. 13. The central cavity of a medium sized radial stone. There are no special features in the core. Line = 20 µM. From Osuga et al. (58).

XIII. STONE DEVELOPMENT AND INDIVIDUAL SUSCEPTIBILITY

Table II shows the incidence of gallstones and the mean weight of the gallstones in those animals involved as a function of time that the butter + cholesterol was fed and according to the sex of the monkey. Within 6 months the incidence of gallstones approaches 50% which is the approximate incidence during prolonged feeding of the lithogenic diet. There is no apparent sex difference. Three months is also the approximate time required for dissolution or expulsion of gallstones from the gallbladder of squirrel monkeys after they are changed to a nonlithogenic diet (Figure 14). We estimated this value by inserting approximately 90 mg of gallstones of mixed sizes into gallbladders of each of three monkeys which were fed modified Purina Monkey Chow-25. At one and 2 months after implantation, the total mass of stones recovered at cholecystostomy was reduced whereas after 4 months no stones remained.

TABLE II Rate of Development of Cholesterol Gallstones in Squirrel Monkeys Fed a Diet with Butter (20%) and Cholesterol (0.9 mg/kcal).

Time	Male	Female
0-3 months	0/9*	0/2
3-6 months	4/7 (14.5 mg)**	4/8 (24.6)
6-9 months	2/4 (51.0)	9/21 (89.9)
9-12 months	2/4 (62.5)	4/6 (36.9)
> 12 months	12/19 (108.0)	17/31 (53.4)

*Animals with stones/Total in group.
**( ) = mean weight of stone/afflicted animal.

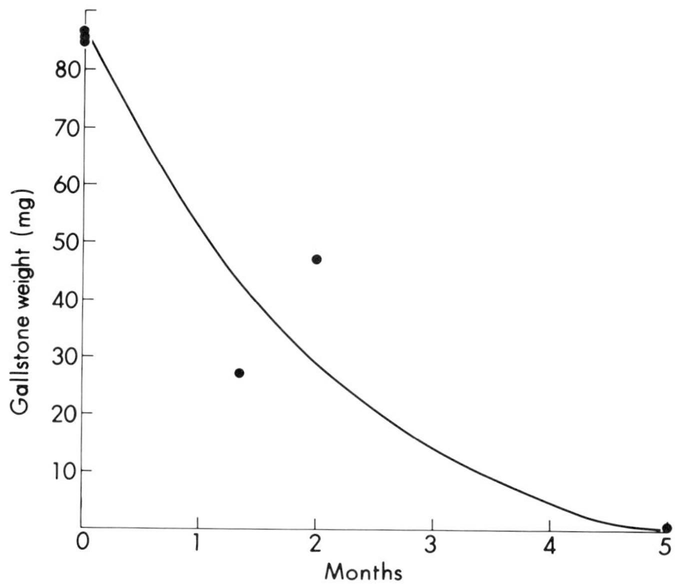

Fig. 14. The dissolution or expulsion from the gallbladder of gallstones as a function of time after their surgical implantation into squirrel monkeys which were fed a modified commercial monkey chow and had a low relative cholesterol concentration in bile. Approximately 90 mg of gallstones which had been formed in other squirrel monkeys were inserted into each recipient. The size of individual stones was reduced at 1 and 2 months.

We are particularly interested in the possibility of individual susceptibility or resistance to gallstone formation. We have, therefore, investigated the incidence of gallstones in squirrel monkeys which were subjected to two or more cholecystostomies while they were fed the same diet or when they were changed to a different diet after the first surgery. Table III shows that of 27 monkeys maintained on the same diet which was associated with a low incidence of stones during both test periods, 1 monkey which had stones during the first period also had stones during the second, whereas 26 remained free of stones for both periods. Of 40 monkeys fed a lithogenic diet during both periods 29 developed gallstones during the first period and 25 of these also had gallstones during the second period. Of 11 which did not form gallstones during the first period, 8 also were free of gallstones during the second period. Four monkeys were changed from a high to a low lithogenic diet. Three of 4 had stones during the first period and 2 of those 3 were free of stones during the second. The monkey without stones during the first period remained free during the second. In a fourth group, in which monkeys were changed from a low to a high lithogenic diet, 2 of 3 monkeys which were free of stones at the first cholecystostomy developed gallstones by the second. These results support the hypothesis that some monkeys are inherently susceptible and some resistant to cholesterol gallstone formation. They also indicate that repeated cholecystostomies and multiple experiments on the same monkey are possible. This is very important in light of the scarcity and cost of these animals.

One of our most often used semi-purified diets contained 8% w/w or 15% of calories as corn oil and no cholesterol. Cholesterol gallstones occurred in a much smaller percentage of this group than of the group fed the diet containing butter + cholesterol (Table IV). Also the mean weight of gallstones per afflicted animal was much less.

XIV. NUTRITIONAL FACTORS AND GALLSTONES IN SQUIRREL MONKEYS

Table V shows the fraction of monkeys fed the various diets for 12 months or more which had gallstones (57,61,63). Diets that contained cholesterol at 0.9 mg/kcal and high levels (20% w/w) of fat had a high incidence of gallstones regardless of whether the fat was butter or safflower oil or whether the butter and cholesterol were added to a semipurified diet based on casein and sucrose or to a natural

TABLE III Effect of Repeated Cholecystostomies on Presence of Gallstones. Values are numbers of Animals Conforming to the Diet Sequence in the Left Hand Column and the Gallstone Sequence Shown at the Top of the Table.

Period 1 Diet	Period 2 Diet	n	Stone → Stone*	Stone → Nonstone	Nonstone → Stone	Nonstone → Nonstone
Low	Low **	27	1	0	0	26
High	High	40	25	4	3	8
High	Low	4	1	2	0	1
Low	High	3	0	0	2	1

*Stone → Stone = Gallstones at first and second cholecystostomies.

**Low Low = Monkeys were on a diet with low incidence of stones for 9 months or more before the first cholecystostomy and on a similar diet for 9 months or more before the second cholecystostomy.

TABLE IV Rate of Development of Cholesterol Gallstones in Squirrel Monkeys with 8% Corn Oil and No Cholesterol.

Time	Male	Female
0-3 months	0/8*	0/2
3-6 months	0/2	2/9 (5.5mg)
6-9 months	0/0	1/6 (5.0)
9-12 months	0/0	0/4
> 12 months	4/13 (11.7)**	5/20 (20.5)

*Animals with stones/Total in group.
**( ) = mean weight of stones per afflicted animal.

TABLE V Diet and Cholesterol Gallstones in Squirrel Monkeys.

Principal Features of Diet	Incidence and Quantity of Stones		
	Males	Females	Male Castrates
Purina Monkey Chow-25-Modified	0/26*	0/25	
Chow + 20% + cholesterol	4/5 (35.9 mg)**		
20% Butter + 0.9 mg cholesterol/kcal	12/19 (108)	17/31 (53.4)	6/6 (96.8)
20% Butter	0.8	0/9	
20% Safflower oil	0.7	0/9	
8% Corn oil	4/13 (11.7)	5/20 (20.5)	
20% Safflower oil + 0.9 mg cholesterol	3/5	6/10 (76.2)	2/3 (63.6)
2% Safflower oil + 0.9 mg cholesterol	0/4		0/3
20% Butter + 0.3 mg cholesterol	1/4 (93.4)		
20% Butter + 0.9 mg cholesterol + cheno	4/4 (49.6)	1/5 (22.0)	
20% Safflower oil + cholesterol + cheno	6/6 (91.9)	5/5 (42.8)	

Except for the first 2, all diets were semipurified and contained casein at 21.2%. Animals were fed the diets indicated for 12 months or more.

Part of this data has been presented in another form. (56,60,61)

**Animals with stones/Total in group.*
***( ) = mean weight of stones per afflicted animal.*

commercial diet. When cholesterol was omitted from these high fat diets the tendency to form stones was lost, although one animal in 4 had gallstones when 0.3 mg cholesterol/kcal was used with butter. Reduction of the safflower oil content of cholesterol-containing diets to 2% resulted in a complete loss of lithogenicity. Although diets containing 20% w/w of fat and no cholesterol were also non-lithogenic, the diet with 8% corn oil was associated with some gallstones. Chenodeoxycholic acid which has been used to dissolve gallstones in humans had no effect in the prevention of gallstones in squirrel monkeys when it was added to lithogenic diets. Although there was no sex difference in gallstone incidence, there was a suggestion that male castrates might be particularly susceptible.

XV. MEASUREMENTS TO IDENTIFY SUSCEPTIBLE MONKEYS

One of our main interests has been to determine which chemical or biochemical measurements are most useful in separating the animals on a lithogenic diet from those on a non-lithogenic diet, as well as in separating monkeys with gallstones from those without on the same lithogenic diet.

The recognition of differences between lithogenic and non-lithogenic diet groups is relatively easy (Table VI). For example, the % saturation of bile with cholesterol which is determined from the relative concentrations of cholesterol, bile salts, and lecithin is greater and the pool of bile salts is less in monkeys fed lithogenic diets. The fraction of the total pool of bile salts outside the gallbladder during fasting is also less in the groups with gallstones.

Within the single group of monkeys on the high butter plus cholesterol diet it was also possible to detect differences between monkeys that had gallstones and those that did not (Table VII). The afflicted animals had higher mean plasma cholesterol in bile, but lower bile salt pools than animals free of stones. There was always some overlap, however, so that it was impossible to set limits for either of the three criteria that would invariably separate animals with from those without gallstones.

The consequences of a small pool of bile salts outside the gallbladder is illustrated in Figure 15 in which changes in the mole % composition of cholesterol, phospholipid, and total bile salts in hepatic bile are plotted against time after interruption of the enterohepatic circulation with a common duct fistula. In a monkey with gallstones and a small proportion of his bile salt pool outside the gallbladder the

TABLE VI Analyses of Plasma and Bile of Squirrel Monkeys from Different Diet Groups

Diet	Chow	8% Corn Oil	20% Butter	20% Butter + cholesterol
Plasma cholesterol, mg/dl	135 ± 7	158 ± 11	237 ± 13	373 ± 22
Biliary cholesterol, mole %	2.3 ± 0.3	4.5 ± 1.0	4.2 ± 0.9	7.5 ± 0.7
Lithogenic Index % (15)	37.9 ± 4.3	71.3 ± 13.5	57.3 ± 11.8	105.8 ± 8.7
Cholate Pool, mg/kg	82.2 ± 9.7	25.7 ± 6.0	28.4 ± 1.6	33.1 ± 4.0
Chenodeoxycholate Pool, mg/kg	28.1 ± 4.0	15.8 ± 1.2	16.3 ± 1.6	19.9 ± 2.5
Cholate Half Life, days	4.8 ± 0.8	2.2 ± 0.6	3.1 ± 0.3	1.6 ± 0.2
Chenodeoxycholate Half Life, days	2.7 ± 0.2	2.7 ± 0.5	4.9 ± 1.2	2.2 ± 0.3

Values are means ± standard errors of the mean.
Part of this data has appeared in another form. (61,62)

TABLE VII A Comparison of Plasma and Bile Analyses on Squirrel Monkeys with and without Gallstones. All Monkeys were Fed the Same Lithogenic Diet that Contained Butter and Cholesterol.

	Lithogenic Diet With Stones	Lithogenic Diet No Stones
Plasma cholesterol (mg/100 ml)	418 ± 27	322 ± 21
Lithogenic Index	133.3 ± 16.7	71.0 ± 7.4
Total Pool Bile Salts (mg/kg)	38.1 ± 3.2	73.0 ± 10.3
% Total Pool of Bile Salts outside gallbladder during fasting	5.9 ± 1.0	30.5 ± 6.0

Part of this data has appeared in another form. (61,62).

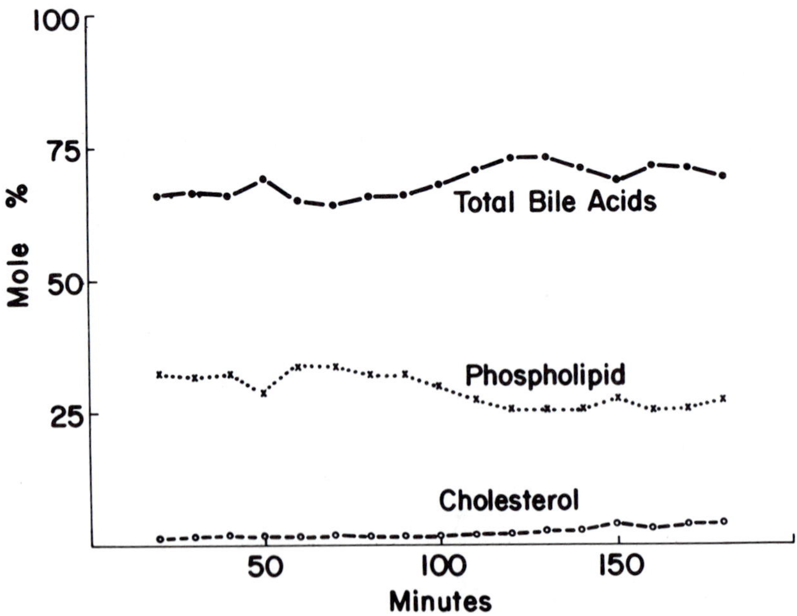

Fig. 15a.

relative saturation of bile with cholesterol increased rapidly, whereas the bile of a monkey without gallstones and comparatively large amounts of bile salts outside the gallbladder changed relatively little during 3 hours after common duct cannulation. Figure 16 shows the relative increase in mole % cholesterol of hepatic bile in several monkeys with gallstones and in a representative monkey fed the same lithogenic diet but without gallstones and in one fed a nonlithogenic diet.

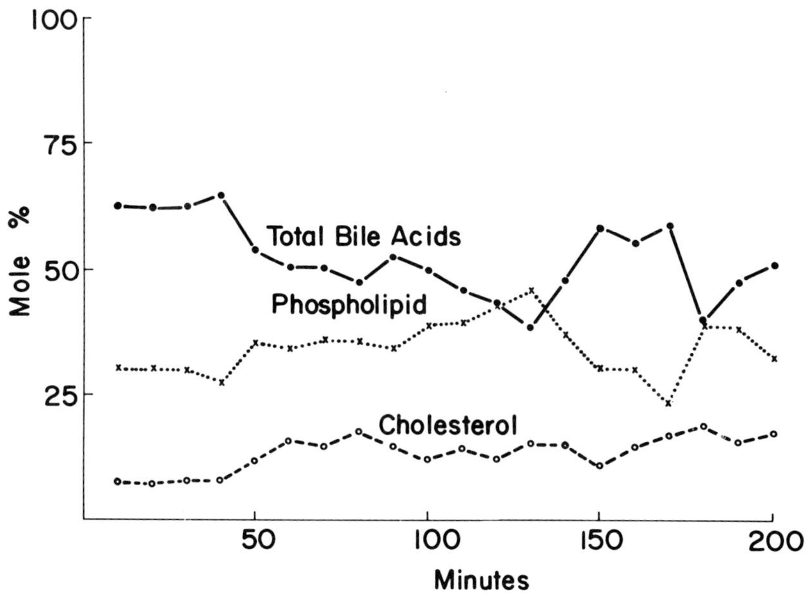

Fig. 15 a,b. The effect of a large pool outside the gallbladder (a) or a small pool (b) on the rate of change in hepatic bile composition. At time 0 which was 12 hours after the last meal the gallbladder was removed and a common duct catheter was inserted. From Portman et al. (40).

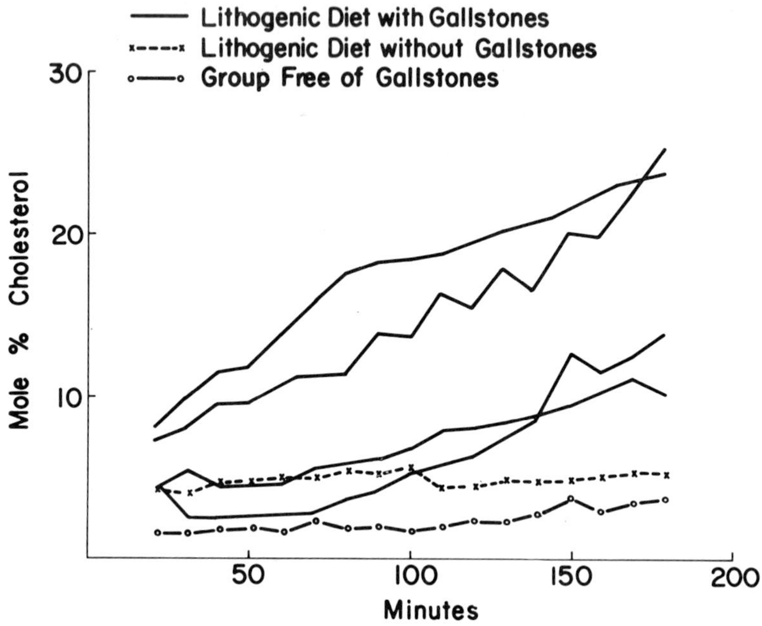

Fig. 16. The relative cholesterol concentration in hepatic bile of different squirrel monkeys in response to insertion of a common duct cannula and removal of the gallbladder. Each line is for a representative monkey from the group indicated. From Osuga et al. (61).

XVI. SEX AND STRAIN FACTORS ON SQUIRREL MONKEY GALLSTONES

We have recently turned our attention to the apparent lack of a sex difference in gallstone incidence for monkeys fed the most lithogenic diet. Because male castrates seemed to be unusually susceptible we are now comparing intact and gonadectomized animals of both sexes. There is a coincidence of high rates of gallstone formation (8) and high plasma HDL concentrations (64) in women. Although the increases in plasma VLDL and triglyceride concentrations that occur with use of estrogenic oral contraceptives have received much attention (65), the HDL fraction is also increased, particularly when no progestins or weakly antiestrogenic progestins are used (66,67). These observations, together with the recent work (21) which indicated that HDL cholesterol is selectively secreted into the bile, formed the basis of a working hypothesis of the mechanism of the difference in gallstone formation between men and women. According to this hypothesis the higher the ratio of plasma $\frac{HDL}{LDL}$ the greater the incidence of gallstones and perhaps the lower the severity of atherosclerosis. One line of support for the role of HDL in lithogenesis came from comparisons of male Bolivian with Brazilian squirrel monkeys. We are testing the Bolivians because additional Brazilians are no longer available to us. Unfortunately, it has become clear that with the diet which is lithogenic for Brazilian monkeys, the Bolivians are completely resistant to cholesterol gallstone formations and their bile has a lower relative concentration of cholesterol. Although there were no differences between the groups in the composition and concentrations of VLDL or LDL nor in the composition of HDL, the Bolivian monkeys had much lower concentrations of HDL than the Brazilians (Table VIII). This was true for studies of monkeys fed semipurified diets that contained butter with or without supplementary cholesterol. There was much less difference in lipoprotein concentrations between male and female and gonadectomized Brazilian squirrel monkeys than was seen between the two strains of male squirrel monkeys.

It was possible that other experimental diets will be more suitable for demonstrating a difference between the sexes in the incidence of cholesterol gallstones. We also hope that we will be able to develop a diet that will induce gallstones in the Bolivian monkeys.

TABLE VIII A Comparison of the Plasma High Density Lipoprotein (HDL) Apoprotein Concentrations in Strains of Squirrel Monkeys which were Susceptible (Brazilian) and Resistant (Bolivian) to gallstone formation.

Diet Determination	Brazilian Males	Bolivian Males
High Butter + Cholesterol		
Incidence of gallstones	9/11	0/12
HDL protein (μg/ml)	3563 ± 231	2374 ± 174
% of HDL as protein	46.1 ± 0.61	45.6 ± 1.15
High Butter		
HDL protein (μg/ml)	3840 ± 179	2360 ± 176
% of HDL as protein	50.7 ± 0.49	53.7 ± 1.08

XVII. FUTURE RESEARCH GOALS

It now appears that most human subjects with or without gallstones that have been studied in the United States or Western Europe have bile that is saturated with cholesterol part of the time. American Indian women who have a very high incidence of gallstones also have very high relative concentrations of biliary cholesterol and small pools of bile salts (68). Certain other populations are reported to be virtually free of gallstones. It is important to confirm the latter observation, and to determine if this resistance is associated with a low concentration of cholesterol in bile. If it is, it would be important to see if changing the diet of those subjects to a Western type would make their bile saturated with cholesterol. If such experiments indicate that nutritional or other environmental factors are responsible for the widespread occurrence of bile saturated with cholesterol and that cholelithiasis like atherosclerosis is in part a disease of the Western diet, therapeutic approaches should continue to emphasize methods of altering the composition of hepatic bile. On the other hand, if further investigations of human bile composition indicate that bile saturated with cholesterol is nearly invariable in all human populations, it should increase the interest in factors that determine the formation of gallstones from the ubiquitous cholesterol crystals and that cause stones to be retained in the gallbladder of particular people. For example, estrogen which increases the incidence of cholesterol gallstones also may reduce the efficiency of gallbladder emptying (69). A number of species of experimental animals will probably be useful in studying such factors as crystal growth, aggregation, and retention by the gallbladder.

Many acute studies on bile kinetics and determinants of bile composition can be carried out on humans. Although there are some limitations on experiments in which triple lumen tubes are used to feed and obtain serial bile samples from people, that technique, together with the use of labelled bile salts to determine pool size and synthesis rate probably makes it desirable to use humans for many of the future kinetics studies. The baboon or rhesus monkey prepared by cholecystopexy and a mechanism for intermittent diversion of common duct bile will also have value.

Since it is difficult to maintain humans on defined diets for long periods of time and to know whether gallstones are present at the beginning and end of the experiment, it would be very desirable to have a suitable test animal for nutritional studies. At this time so little is known about the relationships between nutritional factors and human gallstones that it is impossible to say whether any of the few animal species in which cholesterol gallstones have been induced would be suitable. In humans, obesity has been associated with increased rates of gallstone disease and increased biliary cholesterol secretion (70), cholesterol feeding has induced increased saturation of bile with cholesterol (71) and in one study long-term feeding of a polyunsaturated fat diet increased the incidence of gallstones (72).

Beyond the need for nonhuman primates for toxicity studies, an appropriate model is particularly desirable for the simultaneous testing of potential therapeutic agents on biliary physiology and plasma lipoproteins, on cholelithiasis and atherosclerosis, since, in theory, agents that affect one condition favorably may act unfavorably on the other. This seems to be the case for cholestyramine which in large doses may lower the plasma cholesterol level, but does not reduce and may actually increase the relative concentration of biliary cholesterol (73). It seems theoretically probable that intermediate dosages, perhaps determined from experiments on test animals, will be useful in lowering plasma cholesterol levels without decreasing biliary bile salt concentrations or increasing the lithogenicity of bile. Sex and probably certain sex hormones and clofibrate (74) also represent opposing influences on cholesterol concentrations in human bile and plasma.

There are thus adequate reasons for continuing to search for more desirable animal models of human cholesterol gallstone disease and for continuing studies with existing models. For example, the African green monkey, *Cercopithecus aethiops*, appears to be susceptible to gallstone formation (76) and may deserve more attention. It is probable, however, as has been true of searches for animal models of other human diseases, that the most important insights will come from trying to understand apparent interspecies differences.

ACKNOWLEDGMENTS

This work was supported by a grant-in-aid from the National Institutes of Health, Bethesda, Md. (HL09744), and grant No. RR 00163, to the Oregon Regional Primate Research Center. Publication No. 1011 from the Oregon Regional Primate Research Center.

REFERENCES

1. Danielsson, H. (1963) Present status of research on catabolism and excretion of cholesterol. *Adv. Lipid Res. 1: 335-385.*
2. Bogren, H. (1964) The composition and structure of human gallstones. *Acta radiol. (Stockholm) Supplementum. 226.*
3. Portman, O.W. (1962) Importance of diet, species, and intestinal flora in bile acid metabolism. *Fed. Proc. 21: 896-902.*
4. Heaton, K. (1975) Gallstones and cholecystitis. In: Refined Carbohydrates and Disease. Some implications of dietary fibre. D.P. Burkitt & H.C. Trowell, eds. *Academic Press, New York.*
5. Edington, G.M. (1957) Observations on hepatic disease in the Gold Coast: with special reference to cirrhosis. *Trans. Roy. Soc. Trop. Med. & Hyg. 51: 48-55.*
6. Nakayama, F., and Miyake, H. (1970) Changing state of gallstone disease in Japan. Composition of the stones and treatment of the condition. *Am. J. Surg. 120: 794-799.*
7. Kameda, H. (1964) Gallstone disease in Japan. A report of 812 cases. *Gastroenterol. 46: 109-114.*

8. Sternby, N.H. (1968) Atherosclerosis in a defined population. An autopsy survey in Malmo, Sweden. 13. Atherosclerosis and cholelithiasis, cholesterolosis and arcus senilis. *Acta Pathol. Microbiol. Sand. Suppl. 194: 136-216.*
9. Bennion, L., Ginsberg, R., Garnick, M., and Bennett, P. (1976) Effects of oral contraceptives on the gall-bladder bile of normal women. *New Engl. J. Med. 294: 189-192.*
10. Boston Collaborative Drug Surveillance Programme: Oral contraceptives and venous thromboembolic disease, surgically confirmed gallbladder disease, and breast tumors. *Lancet 1: 1399-1404, 1973.*
11. Boston Collaborative Drug Surveillance Programme: Surgically confirmed gallbladder disease, venous thromboembolism and breast tumors in relation to post-menopausal estrogen therapy. *New Engl. J. Med. 290: 15-19, 1974.*
12. Hofmann, A.F. (1963) The function of bile salts in fat absorption. The solvent properties of dilute micellar solutions of conjugated bile salts. *Biochem. J. 89: 57-68.*
13. Bourges, M., Small, D.M., and Dervichian, D.G. (1967) Biophysics of lipidic associations. III. The quart-ernary systems of bile salt-lecithincholesterol-water. *Biochim. Biophys. Acta 144: 189-201.*
14. Admirand, W.H., Small, D.M. (1968) The physicochemical basis of cholesterol gallstone formation in man. *J. Clin. Invest. 47: 1043-1052.*
15. Hegardt, F.G., and Dam, H. (1971) The solubility of cholesterol in aqueous solutions of bile salts and lecithin. *Z. Ernaehrungswiss. 10: 223-233.*
16. Holzbach, R.T., Marsh, M., Olszewski, M., and Holan, K. (1973) Cholesterol solubility in bile. Evidence that supersaturated bile is frequent in healthy man. *J. Clin. Invest. 52: 1467-1479.*
17. Carey, M.C., and Small, D.M. (1978) The physical chemistry of cholesterol solubility in bile. *J. Clin. Invest. 61: 998-1026.*
18. Thomas, P.J. and Hofmann, A.F. (1973) A simple calculation of the lithogenic index of bile: expressing biliary lipid composition on rectangular coordinates. *Gastroenterol. 65: 698-700.*
19. Metzger, A.L., Heymsfield, S., and Grundy, S.M. (1972) The lithogenic index--a numerical expression for the relative lithogenicity of bile. *Gastroenterol. 62: 499-501.*

20. Schwartz, C.C., Berman, M., Vlahcevic, Z.R., Halloran, L.G., Gregory, D.H., and Swell, L. (1978) Multicompartmental analysis of cholesterol metabolism in man. Characterization of the hepatic bile acid and biliary cholesterol precursor sites. *J. Clin. Invest. 61: 408-423.*
21. Schwartz, C.C., Halloran, L.G., Vlahcevic, Z.R., Gregory, D.H., and Swell, L. (1978) Preferential utilization of free cholesterol from highdensity lipoproteins for biliary cholesterol secretion in man. *Science 200: 62-64.*
22. Tanaka, N., and Portman, O.W. (1977) Effect of type of dietary fat and cholesterol on cholesterol absorption rate in squirrel monkeys. *J. Nutri. 107: 814-821.*
23. Bergstrom, S., Danielsson, H., and Samuelsson, B. (1960) Formation and metabolism of bile acids. In: *Lipid Metabolism, K. Bloch, ed., Wiley, New York, pp. 291-336.*
24. O'Brien, J.J., Shaffer, E.A., Williams, L.F., Jr., Small, D.M., and Wittenberg, J. (1974) A physiological model to study gallbladder function in primates. *Gastroenterol. 67: 119-125.*
25. Lack, L., and Weiner, I.M. (1964) Intestinal bile salt transport: structure activity relationships and other properties. *Am. J. Physiol. 207: 1-2.*
26. Small, D.M., Dowling, R.H., and Redinger, R.N. (1972) The enterohepatic circulation of bile salts. *Arch. Int. Med. 130: 552-573.*
27. Scow, R.O., Stein, Y., and Stein, O. (1967) Incorporation of dietary lecithin and lysolecithin into lymph chylomicrons in the rat. *J. Biol. Chem. 242: 4919-4924.*
28. Illingworth, D.R., and Portman, O.W. Unpublished data.
29. Dowling, R.H., Mack, E., Picott, J., Berger, J., and Small, D.M. (1968) Experimental model for the study of the enterohepatic circulation of bile in rhesus monkeys. *J. Lab. Clin. Med. 72: 169-176.*
30. Dowling, R.H., Mack, E., and Small, D.M. (1971) Biliary lipid secretion and bile composition after acute and chronic interruption of the enterohepatic circulation in the rhesus monkey. IV. Primate biliary physiology. *J. Clin. Invest. 50: 1917-1926.*
31. Dowling, R.H., Mack, E., and Small, D.M. (1970) Effects of controlled interruption of the enterohepatic circulation of bile salts by biliary diversion and by ileal resection on bile salt secretion, synthesis and pool size in the rhesus monkey. *J. Clin. Invest. 49: 232-242.*

32. Kritchevsky, D., and Story, J.A. (1974) Binding of bile salts in vitro by non-nutritive fiber. *J. Nutr. 104: 458-462.*

33. Shaffer, E.A., Brasch, J.W., and Small, D.M. (1972) Bile composition at and after surgery in normal persons and patients with gallstones. Influence of cholecystectomy. *New Eng. J. Med. 287: 1317-1322.*

34. Shaffer, E.A., Small, D.M. (1977) Biliary lipid secretion in cholesterol gallstone disease. The effect of cholecystectomy and obesity. *J. Clin. Invest. 59: 828-840.*

35. Grundy, S.M., Duane, W.C., Adler, R.D., Aron, J.M., and Metzger, A.L. (1974) Biliary lipid outputs in young women with cholesterol gallstones. *Metabolism. 23: 67-73.*

36. Northfield, T.C., Hofmann, A.F. (1973) Biliary lipid secretion in gallstone patients. *Lancet 1: 747-748.*

37. Morrissey, K., Panveliwalla, D., McSherry, C., Deitrick, J., Nieman, W., and Gupta, G. (1977) Effects of contraceptive steroids and pregnancy on bile composition and kinetics in the baboon. *J. Surg. Res. 22: 598-604.*

38. Vlahcevic, Z.R., Bell, C.C., Jr., Buhac, I., Farrah, J. T., and Swell, L. (1970) Diminished bile acid pool size in patients with gallstones. *Gastroenterol. 59: 165-173.*

39. Redinger, R.N., Chow, L., and Grace, D.M. (1978) Cholesterol oxidation in primates by simultaneous sterol balance and breath analysis. *Am. J. Physiol. 235: R55-R63.*

40. Portman, O.W., Osuga, T., and Tanaka, N. (1975) Biliary lipids and cholesterol gallstone formation. *Adv. Lipid Res. 13: 135-194.*

41. Tepperman, J., Caldwell, F.T., and Tepperman, H.M. (1964) Induction of gallstones in mice by feeding a cholesterol-cholic acid containing diet. *Am. J. Physiol. 206: 628-634.*

42. Brenneman, D.E., Connor, W.E., Forker, E.L., and den Besten, L. (1972) The formation of abnormal bile and cholesterol gallstones from dietary cholesterol in the prairie dog. *J. Clin. Invest. 51: 1495-1503.*

43. Dam, H. (1971 Determinants of cholesterol cholelithiasis in man and animals. *Am. J. Med. 51: 596-613.*

44. Danzinger, R.G., Hofmann, A.F., Schoenfield, L.J. (1972) Dissolution of cholesterol gallstones by chenodeoxycholic acid. *New Engl. J. Med. 268: 1-8.*

45. Campbell, C.B., Cowley, D.J., and Dowling, R.H. (1972) Dietary factors affecting biliary lipid secretion in the rhesus monkey. A mechanism for the hypocholesterolaemic action of polyunsaturated fat? *Eur. J. Clin. Invest. 2: 332-341.*
46. Redinger, R.N., Hermann, A.H., and Small, D.M. (1973) Primate biliary physiology. X. Effects of diet and fasting on biliary lipid secretion and relative composition and bile salt metabolism in the rhesus monkey. *Gastroenterol. 64: 610-621.*
47. Martin, D.E., Wolf, R.C., and Houser, W.D. (1973) Naturally occurring cholelithiasis in a rhesus and its affects on plasma and biliary lipid concentrations. *Am. J. Vet. Res. 34: 971-974.*
48. McSherry, C.K., Glenn, F., and Javitt, N.B. (1971) Composition of basal and stimulated hepatic bile in baboons, and the formation of cholesterol gallstones. *Proc. Nat. Acad. Sci. U.S.A. 68: 1564-1568.*
49. Lapin, B.A., and Yakovleva, L.A. (1963) In: Comparative Morphology in Monkeys, Thomas, Springfield, p. 265.
50. McSherry, C.K., Dietrich, J.E., May, P.S., Niemann, W. Morrissey, K.P., Palmer, R.H., and Glenn, F. (1977) Biliary lipid metabolism in the pregnant baboon. *Surg. Gynecol. Obstet. 144: 727-733.*
51. Anver, M.R., Hunt, R.D., and Chalifoux, L.V. (1972) Cholesterol gallstones in *Aotus trivirgatus. J. Med. Primatol. 1: 241-246.*
52. Mann, G.V. (1970) In: Feeding and Nutrition of Nonhuman Primates, R.S. Harris, ed., Academic Press, New York, pp. 159-174.
53. Portman, O.W., and Andrus, S.B. (1965) Comparative evaluation of three species of New World monkeys for studies of dietary factors, tissue lipids, and atherogenesis. *J. Nutr. 87: 429-438.*
54. Middleton, C.C., Clarkson, T.B., Lofland, H.B., and Pritchard, R.W. (1964) Atherosclerosis in the squirrel monkey. Naturally occurring lesions of the aorta and coronary arteries. *Arch. Pathol. 78: 16-23.*
55. Nicolosi, R.J., Hojnacki, J.L., Llansa, N., and Hayes, K. C. (1977) Diet and lipoprotein influence on Primate Atherosclerosis. *Proc. Soc. Exptl. Biol. Med. 156: 1-7.*
56. Osuga, T., and Portman, O.W. (1971) Experimental formation of gallstones in the squirrel monkey. *Proc. Soc. Exp. Biol. Med. 136: 722-726.*

57. Osuga, T., and Portman, O.W. (1972) Relationship between bile composition and gallstone formation in squirrel monkeys. *Gastroenterol. 63: 122-133.*
58. Melchior, G.W., Clarkson, T.B., Bullock, B.C., and Lofland, H.B. (1972) Cholelithiasis in nonhuman primates--effect of species and type of dietary fat. *Circulation. 48: 11-19.*
59. Osuga, T., Portman, O.W., Mitamura, K., and Alexander, M. (1974) A morphologic study of gallstones development in the squirrel monkey. *Lab. Invest. 30: 486-493.*
60. Osuga, T., Mitamura, K., Miyagawa, S., Sato, N. Kintaka, S., and Portman, O.W. (1974) A scanning electron microscopic study of gallstone development in man. *Lab. Invest. 31: 696-704.*
61. Portman, O.W., Osuga, T., and Alexander, M. (1974) Nutritional influences on gallstones, bile lipids, and arterial lipids. In: Atherosclerosis III, G. Schettler and A. Weizel, eds., Springer-Verlag, Berlin, pp. 389-392.
62. Osuga, T., Portman, O.W., Tanaka, N., Alexander, M., and Ochsner, A.J., (1976) III. The effect of diet on hepatic bile formation and bile acid metabolism in squirrel monkeys with and without cholesterol gallstones. *J. Lab. Clin. Med. 88: 649-661.*
63. Tanaka, N., Portman, O.W., and Osuga, T. (1976) Effect of type of dietary fat, cholesterol, and chenodeoxycholic acid on gallstone formation, bile acid kinetics and plasma lipids in squirrel monkeys. *J. Nutr. 106: 1123-1134.*
64. Russ, E.M., Eder, H.A., and Barr, D.P. (1951) Protein-lipid relationships in human plasma. I. In normal individuals. *Am. J. Med. 11: 468-479.*
65. Jackson, R.L., Chan, L., Snow, L.D., and Means, A.R. (1978) Hormonal regulation of lipoprotein synthesis. In: Disturbances in Lipid and Lipoprotein Metabolism. Dietschy, J.B., Gotto, A.M., Jr., and Ontoko, J.A. eds. American Physiological Society Bethesda, Md., pp. 139-154.
66. Krauss, R.M., Lindgren, F.T., Silvers, A., Jatagir, R. and Bradley, D.D. (1977) Changes in serum high density lipoproteins in women on oral contraceptive drugs. Clin. Chim. Acta 80: 465-470.
67. Bradley, D.D., Wingerd, J., Petitti, D.B., Krauss, R.M. and Ramcharan, S. Serum high density lipoprotein cholesterol in women using oral contraceptives, estrogens, and progestins. *New Engl. J. Med. 299: 17-20.*

68. Sampliner, R.E., Bennett, P.H., Comess, L.J., Rose, F.A., and Burch, T.A. (1970) Gallbladder disease in Pima Indians. Demonstrations of high prevalence and early onset by cholecystography. *New Engl. J. Med. 299: 17-20.*

69. Potter, M.G. (1936) Observations of gallbladder and bile during pregnancy at term. *J. Amer. Med. Assoc. 106: 1070-1074.*

70. Grundy, S.M., Metzger, A.L., and Adler, R.D. (1972) Mechanisms of lithogenic bile formation in American Indian women with cholesterol gallstones. *J. Clin. Invest. 51: 3026-3043.*

71. DenBesten, L. Connor, W.E., and Bell, S. (1973) The effect of dietary cholesterol on the composition of human bile. *Surgery. 73: 266-273.*

72. Sturdevant, R.A.L., Pearce, M.L., and Dayton, S. (1973) Increased prevalence of cholelithiasis in men ingesting a serum cholesterol-lowering diet. *New Engl. J. Med. 288: 24-27.*

73. Grundy, S.M. (1972) Treatment of hypercholesterolemia by interference with bile acid metabolism. *Arch. Int. Med. 130: 638-648.*

74. The Coronary Drug Project Research Group. Gallbladder disease as a side effect of drugs influencing lipid metabolism. Experience in the Coronary Drug Project. *New Engl. J. Med. 296: 1185-1190, 1977.*

75. Portman, O.W., and Mann, G.V. (1955) The disposition of taurine-S^{35} and taurocholate-S^{35} in the rat: Dietary influences. *J. Biol. Chem. 213: 733-743.*

76. Lofland, H.B., Clarkson, T.B., and Melchior, G.W. Cholesterol gallstones in nonhuman primates. In: Atherosclerosis III, G. Schettler and A. Weizel, eds., Springer-Verlag, Berlin, 1974, pp. 392-395.

DIET AND ATHEROSCLEROSIS IN NONHUMAN PRIMATES

K. C. Hayes

Department of Nutrition
Harvard School of Public Health
Boston, Massachusetts

I. INTRODUCTION

Historically, nutritional studies utilizing primates in atherosclerosis research were initiated by two considerations of equal importance, one being the strong implication that dietary factors played a significant role in the rising incidence of human coronary heart disease associated with increased affluence and the second being that an animal model representative of human atherosclerosis was needed if substantial progress was to be made in the pathogenesis of the disease. Thus, with the increased availability of imported primates (predominantly rhesus) for use in polio vaccine testing in the late 1940's and 1950's it was not long before nutritionally-induced atherosclerosis in nonhuman primates had been reported (1-3).

ISBN 0-12-334250-3

II. CHARACTER OF INITIAL STUDIES

These were relatively "crude" nutritional studies in the sense that the diet was based on natural ingredients and inadequately defined or the level of suspected atherogenic dietary factors (saturated fat and cholesterol) were supplied in extraordinary concentrations that made physiological interpretation of the data difficult. For example, the dietary fat occasionally exceeded 50% of the calories, and cholesterol was often added at the level of 5% of the diet (for discussion of these diets see page 99). Such a fat load was approximately 300% of the normal intake for monkeys and 25% greater than a high-fat diet for humans, while the cholesterol intake was equivalent to 25-50 times the average per capita consumption of cholesterol in the United States. Not suprisingly, the results of the cardiovascular pathology were extremely positive with production of advanced atherosclerotic plaques, superficial cutaneous and tendinous xanthomas, myocardial infarction, and death resulting from arterial degeneration (3).

III. PRIMATE MODELS OF ATHEROSCLEROSIS

These observations prompted a search for the most appropriate primate species to model the disease process, again generally without particular concern for the diet, eventhough diet was invariably used to effect the change through saturated fat and cholesterol. The search continues today, although somewhat abated, with only 10-20 of the more than 200 primate species tested and examined with any degree of uniformity (4). At the same time the approach has become infinitely more sophisticated and the models appreciably more difficult to obtain.

It is worth reemphasizing that atherosclerosis remains a significant public health problem in the United States and in most developed countries and that dietary factors loom as an increasingly important variable (5) despite the provocative selection of specific epidemiologic data to indicate otherwise (6). The problem is even more complex than originally anticipated due to the numerous dietary factors involved and and their various interactions (7). These factors primarily include cholesterol, fat, carbohydrate and protein in order of importance, but trace minerals, dietary fiber and certain vitamins are influential as well.

Several nonhuman primate species have been studied for their susceptibility to atherosclerosis, and all have responded with arterial lesions if given sufficient dietary challenge (saturated fat and cholesterol) and time to do so. Aging probably enhances susceptibility (8), but age has not been adequately assessed in enough species to elicit meaningful comment. A summary comparison of atherosclerosis in the most important primate models studied to date is contained in Table I.

The specific arteries involved in diet-induced atherosclerosis vary according to species with only a few species ever developing appreciable coronary artery involvement. However, this is probably not a critical issue for study of atherogenesis, *per se*, since the basic process in one artery should generally apply to the next even though the rates of degeneration and specific characteristics may differ. Nonetheless, certain anatomic, hemodynamic, and biochemical differences have an impact on atherogenesis, and diet may pre-empt or reinforce certain aspects of these variables.

IV. DIETARY LIPEMIAS IN PRIMATES

Of considerable importance to this review are the relative responses in character and degree of lipemias between primate species and individuals subjected to various dietary manipulations. Meaningful information is not easily gleaned from the literature because most investigators have not been concerned with physiological tolerances to the various dietary ingredients. Generally, mechanisms underlying the dietary influence of the lipemia have been ignored in favor of rapid induction of overt arterial lesions to satisfy a requirement for morphologic study of arterial degeneration. The morphology of lesions has varied with species and diet, but most lesions are thought to resemble human atheromatous plaques. Specific dietary variables were not carefully scrutinized until recently, but the initial methods of induction have helped to select the several models now utilized (Table I).

Following the initial surge in model development of 15-20 years, which necessarily must continue to some degree, several new approaches involving diet manipulation and lipoprotein metabolism have evolved (9) particularly in conjunction with additional dietary variables. Much of the original investigation in this area was that of Portman and his colleagues who studied cebus and squirrel monkeys for effects of diet on lipid metabolism, *per se* (10). This continues to be the most

TABLE I Comparison of Major Primate Models of Diet-Induced Atherosclerosis

Species	Approximate Number Studied	Estimated Relative Susceptibility[1] Spontaneous	Induced
Humans	2000+	4.0	-
Apes and Old World			
Chimpanzee	30	1.5	2.0
Rhesus	500	0.5	2.5
Baboon	500	0.5	1.5
Cynomolgus	100	0.5	3.0
African Green	50	1.0	4.0
Stumptail	25	1.0	4.5
New World			
Squirrel	400	2.5	4.5
Cebus	100	0.1	0.5
Spider	75	0.5	1.5
Wooly	50	1.5	3.0

[1]Scale 0-5. Since no standard measure has been utilized for such estimations between species, an attempt was made to interpret observations reported in the literature.

intensely investigated association between diet and atherosclerosis, i.e., an altered lipid metabolism. Portman's results demonstrated a relationship between serum lipids, arterial wall free (FC) and esterified (CE) cholesterol and phospholipid (PL) concentrations, and gross aortic sudanophilia in squirrel monkeys. We have extended these observations to include both cebus and squirrel monkeys and have found age-related differences in the CE/PL ratio of spontaneous arterial changes between the two species (Table II). It will be noted that the CE concentration eventually exceeds that for PL in aging squirrel monkeys, but not in cebus, a ratio that is associated with lipid-related degeneration of the human aorta (11).

Using this comparative approach, it has become apparent that the lipid metabolism of different species is variable and, consequently, the responses to dietary challenges differ. This is not only true for species of monkeys, but also applies to individual monkeys within a species. In the category of species differences, New World cebus and squirrel monkeys, like humans, are remarkably prone to saturated fat-induced hypercholesterolemia whereas Old World rhesus and cynomolgus are typically less responsive (Table III). Dietary cholesterol and carbohydrate have evoked additional species differences and striking dietary interactions that are exemplified by data presented in Tables IV and V and Fig. 1.

One of the more important contributions of modeling has been the identification and characterization of the individual animal variation in response to dietary challenge with cholesterol (18). Dubbed "hypo-" or "hyper-responders" these individual monkeys should prove extremely useful models for study of control mechansims associated with specific dietary factors within a given species. Similar individuals have been identified as a function of dietary saturated fat alone and have been studied in at least two species--squirrel monkeys (Table VI) and cynomolgus (19, also see page 206). Thus, the nutritionist has both species and individual comparisons to model in future dietary studies concerning the metabolic control of lipid metabolism and lipemia in primates.

V. DIET AND LIPOPROTEIN METABOLISM

Investigation of control mechansism requires an appreciation of the general scheme of lipoprotein (LP) metabolism and the points at which dietary components may interact. Figure 2 is a summary of basic concepts in lipoprotein metabolism, including their source and catabolic pathway and current

TABLE II Aortic Lipid Composition in Geriatric and Neonatal Cebus and Squirrel Monkeys

Lipid Class	Geriatric		Neonates	
	Cebus n = 4	Squirrel n = 4	Cebus n = 5	Squirrel n = 6
	µg/mg wet wt tissue			
Cholesteryl ester	0.90 ± 0.44[a]	5.91 ± 1.73[b]	0.14 ± 0.03	1.01 ± 0.65
Cholesterol	3.22 ± 0.58	7.85 ± 1.40	2.02 ± 0.49	2.19 ± 0.30
Phospholipid	1.34 ± 0.53	2.72 ± 0.29	2.02 ± 0.31	6.08 ± 1.34
Free fatty acid	0.45 ± 0.04	1.38 ± 0.34	2.53 ± 0.97	2.02 ± 0.66
Tyiglyceride	0.68 ± 0.35	2.22 ± 1.80	2.14 ± 1.09	2.33 ± 1.68
Total lipid	6.59 ± 1.97[a]	20.08 ± 1.91	8.87 ± 1.15	13.33 ± 4.13

Values expressed as Mean ± SE. (data from Hoover and Hayes, unpublished)

[a]Difference between species significant ($P < 0.05$).

[b]Difference between age groups significant ($P < 0.05$).

TABLE III Species Differences in the Plasma Cholesterol Response of Adolescent Primates to Dietary Fats

Species	Plasma Cholesterol, mg/dl Corn Oil		Coconut Oil	
Cebus[1]	154 ± 32	(11)	267 ± 57	(11)
Squirrel[1]	142 ± 27	(11)	215 ± 32	(10)
Cynomolgus[2]	138 ± 27	(5)	174 ± 16	(7)
Rhesus[3]	192 ± 16	(6)	184 ± 16	(6)
Human[4]	181 ± 22	(10)	276 ± 39	(10)

Values are mean ± SD; numbers in parentheses indicate subjects.

[1]Lichtenstein et al. (12)
[2]Corey et al. (13)
[3]Ershow et al. (14)
[4]Hegsted et al. (15) (adult men, safflower oil vs. coconut oil)

TABLE IV Interactions Between Dietary Fats and Dietary Cholesterol Reflected in the Plasma Cholesterol Response of Various Primate Species

Level of Dietary Cholesterol and Species	Dietary Fat Polyunsaturated	Dietary Fat Saturated	Low Fat
	mg/dl		
0.1% Cholesterol			
Cebus[1]	158 ± 26 (6)	300 ± 36 (6)	168 ± 20 (6)
Squirrel[1]	138 ± 19 (6)	232 ± 10 (6)	140 ± 28 (6)
Rhesus[2]	167 ± 14 (6)	276 ± 32 (6)	---
Human[3]	213 ± 34 (6)	202 ± 17 (5)	---
0.2% Cholesterol			
Cebus[1]	196 ± 26 (5)	357 ± 34 (5)	---
Cynomolgus[1]	426 ± 55 (7)	506 ± 137 (7)	---

Values are mean ± SD; number in parentheses indicate subjects.

[1]Corey et al. (13)
[2]Ershow et al. (14)
[3]Connor et al. (16)

TABLE V Dietary Fat and Carbohydrate Interactions Reflected in the Plasma Cholesterol Response of Spider and Squirrel Monkeys

Dietary Fat and Carbohydrate	Primate Species	
	Spider[1]	Squirrel[2]
	mg/dl	
Safflower oil		
Glucose	200 ± 17 (3)	225 ± 43 (7)
Sucrose	203 ± 2 (3)	229 ± 32 (7)
Coconut oil		
Glucose	186 ± 12 (5)	219 ± 40 (7)
Sucrose	286 ± 35 (5)	317 ± 53 (7)
Low fat (1%)		
High glucose	175 ± 14 (5)	---

Values are mean ± SD; number of monkeys in parentheses.

[1]Corey et al. (13)
[2]Westmoreland and Hayes (17)

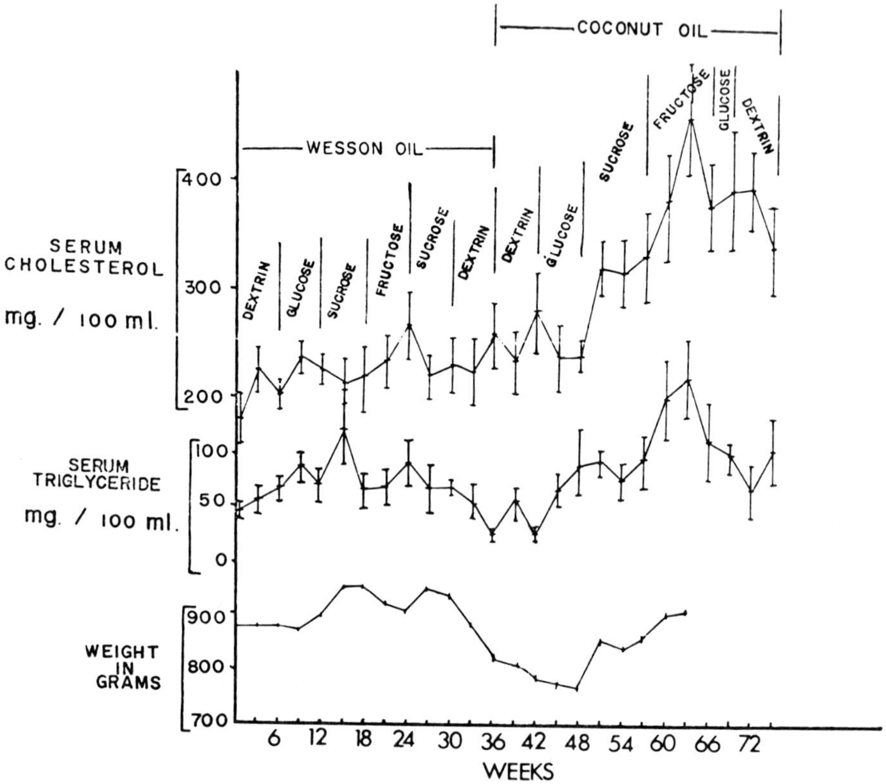

Fig. 1. The interaction between the type of dietary carbohydrate and saturation of the dietary fat on serum lipids is plotted for a group of 7 adult male squirrel monkeys (Letitian) (17). The synergism between fructose-sucrose and saturated fat is noteworthy. The decrease in body weight associated with dextrin reflects the unpalatability of that starch.

TABLE VI Association Between In Vitro Liver Perfusion Data and a High or Low Cholesterolemic Response to Dietary Fats in Squirrel Monkeys

Dietary Fat	Plasma Cholesterol	Triglyceride Secretion Rate	^{3}H-Glucose Into Lipid
	mg/dl	mg/100 g liver	dpm/100 g liver
Coconut oil			
High (1)	312	52	2652
Low (1)	158	27	1627
Lard			
High (2)	250 ± 8	63 ± 36	3051 ± 409
Low (1)	150	36	1999
Corn oil			
High (1)	196	68	2236
Low (2)	105 ± 14	27	1337 ± 37

Values in parentheses indicate number of animals (data from Nicolosi and Hayes, unpublished).

knowledge of several dietary factors that influence their metabolism. There are two main sources of input, the gut with chylomicrons (largely apo B and apo A_1) and the liver with very low density lipoproteins (VLDL) (apo B, apo C, apo E) (20). The liver also probably contributes high density lipoprotein (HDL) and low density lipoprotein (LDL) directly as nascent particles, but the extent of this contribution is uncertain for primates, including man. Other important factors are the two enzymes involved with intravascular metabolism, lipoprotein lipase (LPL) which removes triglyceride (TG) from VLDL and chylomicrons, and lecithin cholesterol acyl transferase (LCAT) which esterifies cholesterol, changing the free sterol from a surface constituent to a stored core component within the lipoprotein (21).

As suggested by Figure 2, diet influences these aspects in various ways. The fatty liver and gut in the infant with kwashiorkor provide clinical evidence that dietary protein is needed for apoprotein synthesis and adequate LP secretion from these organs, and experimental evidence on gut lipoprotein secretion substantiates this observation (22). In a similar vein inositol and choline are required for phospholipid synthesis needed to package and secrete LP particles from gut and liver (23, 24).

Alcohol induces hyperlipemia in man presumably by a combination of mechanisms (25) and also appears to increase HDL cholesterol by increasing LPL activity (26). The type and amount of dietary fat exert a poorly defined effect on TG secretion (27) while possibly increasing both gut VLDL output (28) and total apo B turnover (29). Polyunsaturated fat (PUF) on the other hand, decreases the fasting TG output by the liver (27).

As for key serum enzymes affecting lipid metabolism, LPL activity is enhanced by moderate alcohol intake and PUF feeding (26,30), and LCAT efficiency can be enhanced in cebus and squirrel monkeys by feeding PUF rather than saturated fat (12). Lecithin acyl transferase activity has been impaired by vitamin E deficiency in monkeys (31) and severe alcoholism in man (32). A high carbohydrate diet can increase the fractional rate of intravascular HDL apo A_1 catabolism, whereas nicotinic acid decreases it (33), either of which might alter LCAT activity.

Various dietary factors also alter the size of the serum cholesterol or triglyceride pools, including altered copper, zinc and iron ratios (34) and vitamin A toxicity (35), which result in hyperlipemias of undetermined mechanism that may, in turn, influence atherogenesis. Not to be overlooked are ascorbic acid which is required for hydroxylation of

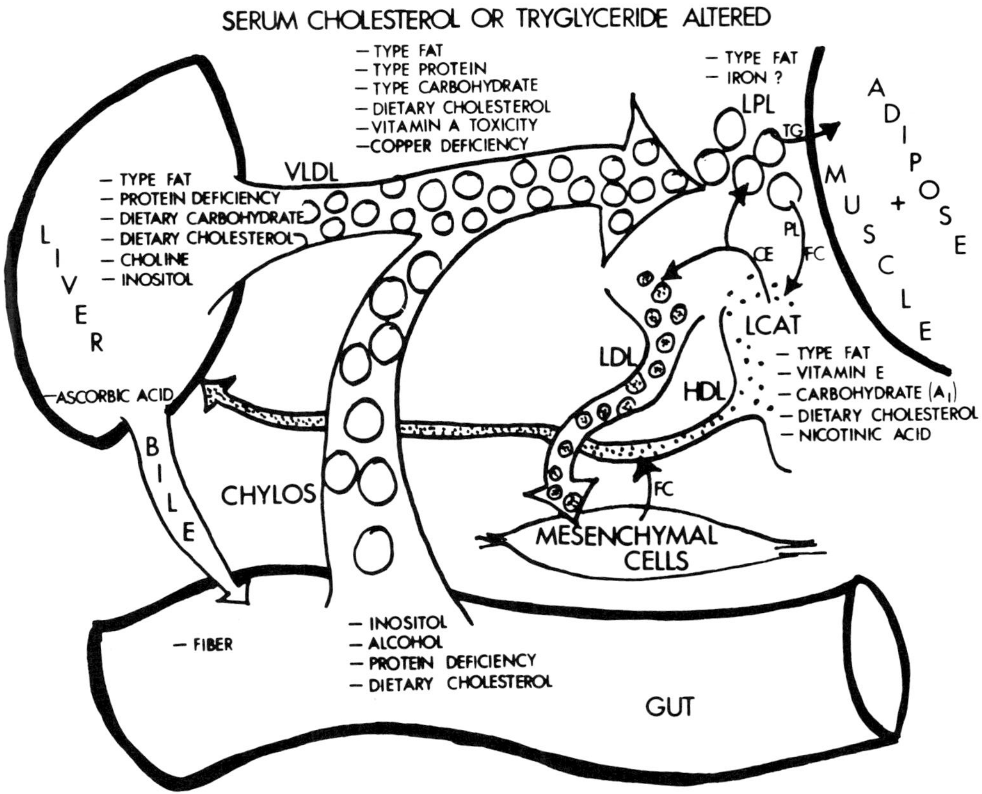

Fig. 2. Dietary factors and lipemia: Scheme depicting the interaction of dietary factors with lipoprotein metabolism and lipemia. See text for interpretation.

steroids including that required for bile acid synthesis (36), and dietary fiber which may selectively bind and remove bile acids and cholesterol to influence cholesterol metabolism (37). Thus, the mechanisms involved in numerous nutritional aspects of lipemias and their effect on atherogenesis remain to be investigated.

The importance of delineating those factors which regulate lipoproteins is emphasized by the epidemologic data implicating the atherogenicity of LDL and the potential protective association provided by HDL (38). This has been reemphasized recently by data on the LP response of two monkey species - the atherosclerosis-susceptible squirrel and the resistant cebus monkey - where saturated fat-induced hypercholesterolemia resulted in appreciable increases in the LDL and HDL cholesterol pools, respectively (39).

VI. FUTURE STUDIES

Thus, at this stage of our knowledge on the comparative aspects of atherogenesis we can refer to the mobile reservoir of body cholesterol circulating predominantly as LDL and HDL in nonhuman primates, but our comprehension of this lipoprotein distribution is inadequate, particularly as it relates to dietary factors and their relationship to the pathogenesis of arterial disease. Several questions that require further nutritional-atherosclerosis investigation are suggested below:

1) How do certain dietary factors influence the LDL-HDL reservoir size?

2) Is it true that this ratio is important to atherogenesis in primates and, if so, how?

3) How does diet affect LPL and LCAT enzymes and their control over triglyceride and cholesterol clearance or mobilization?

4) How are the apoproteins influenced by dietary factors, and how does this affect atherogenesis?

5) Are primates the preferred models for these studies? In fact, for diet and atherogenesis they (and pigs) would appear superior since their LP metabolism seems most closely aligned to that of man.

It will be important in future studies to identify dietary-plasma LP arterial relationships that will facilitate our ability to predict the degree of arterial insult and damage from an altered biochemical or physiological parameter changed by diet. Thus, a salt load might be translated into changes in blood pressure with predictable effects on the arterial wall, whereas dietary fat, carbohydrate, or cholesterol might be evaluated by specific lipoprotein changes and subsequent deterioration of the artery. Experiments should be designed which logically predict an arterial change, i.e., testing a logical hypothesis rather than the generalized descriptive investigations of the past.

It is still apparent that the ultimate test of dietary or other intervention used for altering arterial metabolism and inducing atherosclerosis will be histopathologic evaluation of arteries, but it will be the responsibility of future investigators to more fully develop hypotheses for the mechanism of action of the proposed intervention and to test these by antemortem evaluation of biological phenomena which can later be correlated with arterial change. A simple serum cholesterol determination is no longer adequate. It will be important to learn how lipoprotein metabolism is controlled and where diet impacts this control.

Thus, a well-conceived experiment should consider:

1) Justification of the specific primate model(s) and not simply their availability.

2) Utilization of controlled, semipurified diets with due consideration for physiological tolerances if it is to be a true dietary study of atherogenesis. Other approaches are justified only if dietary considerations are not of primary concern, i.e., when attempting to induce severe atherosclerosis for physiological or pathological studies where diet only represents a means to attain the starting point.

3) Meaningful measures to assess endpoints of induced atherosclerosis including lipoprotein changes and arterial wall degeneration. For instance, enzymatic measures of key regulatory points in cholesterol or lipoprotein metabolism such as LPL, LCAT, 7-α-hydroxylase, or HMG-CoA reductase might be compared to arterial changes including lipid composition, connective tissue assimilation, cell proliferation, or enzymatic profiles.

In summary, to further our understanding of the relationship between atherosclerosis and diet, studies focusing on basic metabolism of primates are needed related to dietary lipid, carbohydrate, protein, and minerals. Such studies should involve a comprehensive understanding of lipoprotein and bile metabolism with particular reference to cholesterol, *per se*, eventually including interaction with hormones such as insulin and thyroid hormone. These studies should strive to integrate the essence of energy metabolism in primates with mechanisms of atherogenesis in order that we might better understand how diet influences the arterial wall.

REFERENCES

1. Mann, G.V., Andrus, S.B., McNally, A., and Stare, F.J. (1953) Experimental atherosclerosis in cebus monkeys. *J. Exp. Med. 98: 195.*
2. Cox, G.E., Taylor, B.C., Cox, L.G., and Counts, M.A. (1958) Hypercholesteremia induced by dietary fat and cholesterol. *Arch. Path. 66: 32.*
3. Mann, G.V., and Andrus, S.B. (1956) Xanthomatosis and atherosclerosis produced by diet in an adult rhesus monkey. *J. Lab. Clin. Med. 48: 533.*
4. Strong, J.P. Ed. (1976) Atherosclerosis in Primates. *Prim. Med. vol. 9.* Karger.
5. Glueck, C.J., and Connor, W.E. Diet-coronary heart disease relationships reconnoitered. *Am. J. Clin. Nutr. 31: 727.*
6. Mann, G. (1977) Diet-Heart: End of an era. *N.E.J. Med. 297: 6441.*
7. Truswell, A.S. (1978) Diet and plasma lipids - a reappraisal. *Am. J. Clin. Nutr. 31: 977.*
8. Bullock, B.C., Clarkson, T.B., Lehner, N.D.M., Lofland, H.B., and St. Clair, R.W. (1969) Atherosclerosis in cebus albifrons monkeys. III. Clinical and pathologic studies. *Exp. Mol. Path. 10: 39.*
9. Rudel, L.L., and Lofland, H.B. (1976) Circulating lipoproteins in nonhuman primates. *Prim. Med. 9: 224.*
10. Portman, O.W., and Illingworth, D.R. (1976) Arterial Metabolism in Primates. *Prim. Med. 9: 145.*
11. Smith, E.B., and Slater, R.S. (1973) Lipids and low-density lipoproteins in intima in relation to its morphological characteristics in Atherogenesis: Initiating factors. Elsevier, New York.
12. Lichtenstein, A.H. (1979) Lecithin cholesterol acyltransferase: effect of diet in New World primates. Doctoral Dissertation. Harvard School of Public Health.
13. Corey, J.E., Hayes, K.C., Dorr, B., and Hegsted, D.M. (1974) Comparative lipid response of four primate species to dietary changes in fat and carbohydrate atherosclerosis. 19: 119.
14. Ershow, A.S., Nicolosi, R.J., and Hayes, K.C. (1978) Response of rhesus monkey lipoproteins to saturated fat. *Fed. Proc. 37: 836.*
15. Hegsted, D.M., McGandy, R.B., Myers, M.L., and Stare, F.J. (1965) Quantitative effects of dietary fat on serum cholesterol in man. *Am. J. Clin. Nutr. 17: 281.*

16. Connor, W.E., Stone, D.B., and Hodges, R.E. (1964) The interrelated effects of dietary cholesterol and fat upon human serum lipid levels. *J. Clin. Invest. 43: 1691.*
17. Westmoreland, N.P., and Hayes, K.C. (1974) Dietary carbohydrate influence on serum lipids in squirrel monkeys. *Fed. Proc. 33: 718.*
18. Clarkson, T.B., Lofland, H.B., Bullock, B.C., and Goodman, H.O. (1971) Genetic control of plasma cholesterol. *Arch. Path. 92: 37.*
19. Malinow, M.R., McLaughlin, P., Papworth, L., Naifo, H. K., Lewis, L., and McNulty, W.P. (1976) A model for therapeutic interventions on established coronary atherosclerosis in a nonhuman primate. In. Atherosclerosis Drug Discovery, C.E. Day, ed., Plenum, New York. p. 3.
20. Schaefer, E.J., Eisenberg, S., and Levy, R.I. (1978) Lipoprotein apoprotein metabolism. *J. Lipid Res. 19: 667.*
21. Glomset, J.A., Verdery, R.B. (1977) Role of lecithin cholesterol acyltransferase in cholesterol metabolism. In. Cholesterol Metabolism and Lipolytic Enzymes. (J. Polonovski, ed., Masson Publ. New York, p. 137.
22. Gangl, A., and Ockner, R.K. (1975) Intestinal metabolism of lipids and lipoproteins. *Gastroenterology 68: 167.*
23. Kuksis, A., and Mookerjea, S. (1978) Choline. *Nutr. Revs. 36: 201.*
24. Kuksis, A., Mookerjea, S. (1978) Inositol. *Nutr. Revs. 36: 233.*
25. Bardona, E., and Lieber, C.S. (1979) Effects of ethanol on lipid metabolism (a review). *J. Lipid Res. 20: 289.*
26. Nikkla, E.Q. (1978) Metabolic and endocrine control of plasma high density lipoprotein concentration. In. High Density Lipoproteins and Atherosclerosis. Gotto, Miller, and Oliver, eds. Elsevier, New York, p. 177.
27. Nicolosi, R.J., Hayes, K.C., elLozy, M., and Herrera, M.G. (1977) Hypercholesterolemia and triglyceride secretion rates in monkeys fed different dietary fats. *Lipids 12: 936.*
28. Portman, O.W., Alexander, M., Tanaka, N., and Illingworth, D.R. (1977) Triacylglycerol and very low density lipoprotein secretion into plasma of squirrel monkeys. *Biochem. Biophys. Acta 486: 470.*
29. Shepherd, J., Packard, C.J., Gotto, A.M., Jr., and Taunton, O.D. (1978) Effects of dietary saturated and polyunsaturated fat on the metabolism of apolipoproteins, A. and B. *Clin. Chem. Acta. 87: 395.*

30. Bagdade, J.D., Hazzard, W.R., and Carlin, J. (1970) Effect of unsaturated dietary fat on plasma lipoprotein lipase activity-normal and hyperlipidemic states. *Metabolism 19: 1020.*
31. Mickel, H.S., Hill, P.L., and Hayes, K.C. (1975) Depression of lecithin-cholesterol acyltransferase esterification in vitamin E-deficient monkeys. *Am. J. Clin. Nutr. 28: 994.*
32. Sabesin, S.M., Hawkins, H.L., Kuiken, L., and Ragland, J.B. (1977) Abnormal plasma lipoproteins and LCAT deficiency in alcoholic liver disease. *Gastroenterology 72: 510.*
33. Blum, C.B., Levy, R.I., Eisenberg, S., Hall, M., Goebel, R.H., and Berman, M. (1977) HDL metabolism in man. *J. Clin. Invest. 60: 795.*
34. Sherman, A.R., Guthrie, H.A., and Wolinsky, I. (1977) Interrelationships between dietary iron and tissue zinc and copper levels and serum lipids in rats. *Proc. Soc. Exp. Biol. Med. 156: 396.*
35. Hayes, K.C. Hypervitaminosis A in monkeys (unpublished data).
36. Bjorkhem, I., and Kallner, A. (1976) Hepatic 7-α hydroxylation of cholesterol in ascorbate-deficient guinea pigs. *J. Lipid Res. 17: 360.*
37. Myant, N.B. (1972) The effect of drugs on the metabolism of bile acids. *Adv. Exptl. Med. Biol. 26: 137.*
38. Gotto, A.M., Jr. (1979) Status report: Plasma lipids, lipoproteins and coronary artery disease. *Athero. Revs. 4-17.*
39. Nicolosi, R.J., Hojnacki, J.L., Llansa, N., and Hayes, K. C. (1977) Diet and lipoprotein influence on primate atherosclerosis. *Proc. Soc. Exp. Biol. Med. 156: 1.*

IMPORTANT CONSIDERATIONS FOR NONHUMAN PRIMATE MODELS OF DIET-INDUCED ATHEROSCLEROSIS

M. R. Malinow

Oregon Regional Primate Research Center
Beaverton, Oregon

University of Oregon Health Sciences Center
Portland, Oregon

I. INTRODUCTION

Despite the achievements of numerous investigators that have greatly advanced our knowledge of atherogenesis, several unsolved questions preclude successful therapy for atherosclerosis in man, and the consequences of this process are still the main cause of death in the adult populations of industrialized nations. This discussion addresses unanswered questions concerning atherosclerosis in nonhuman primates.

ISBN 0-12-334250-3

II. THROMBOSIS IN MONKEYS

The main anatomical features of human atherosclerosis can be observed in atherosclerotic lesions induced in monkeys by cholesterol feeding. This includes, in exceptional cases, arterial occlusion by cell proliferation (1). Notably absent in most experiments are ulceration and thrombotic occlusion and, thus, advanced lesions as observed in humans are seldom present in monkeys fed atherogenic diets (2). Since it is precisely those complicating features that are responsible for the majority of human illness and death from atherosclerosis (3), models to successfully reproduce thrombotic occlusion superimposed on complicated plaques are needed. Because of the possible involvement of platelets in arterial thrombosis, we have studied these cells in monkeys fed regular and high-fat diets and have shown that the reactivity of platelets of five macaque species differs from the reactivity in man. Thus, monkey platelets react poorly to collagen, epinephrine, or thrombin at concentrations that caused maximal aggregation in human platelets. Furthermore, contrary to man, they disaggregate in almost every case after the initial aggregation brought about by ADP, and they fail to release serotonin after stimulation with ADP or epinephrine (4,5). It would be useful to understand the mechanism of these differences and whether the hyporeactivity of platelets is responsible for the lack of arterial thrombosis in monkeys. Research in nonhuman primates is also needed on endothelial function in different arterial territories (functions summarized in Table I). Other factors besides platelet function and interaction with arterial intima are involved in thrombosis. Seldom mentioned, for instance, is the possible effect of dynamics of blood flow. Monkeys have a very short circulation time with a linear flow approximately 10 times faster than humans (6). The rapid blood flow probably affects platelets, the local concentration of activated clotting factors, and the breaking off of thrombi. Studies are needed to determine if the parameters mentioned above are modified by diet-induced atherosclerosis.

III. VARIATION IN ARTERIAL RESPONSE

Many observations on experimental atherosclerosis in monkeys deal with the aorta--mainly an elastic vessel--but the clinical importance of atherosclerosis in humans is due to the involvement of medium-sized muscular arteries. Although

Table I Functions of the Endothelium

Synthesis of surface glycoproteins
Synthesis of prostacyclin
Degradation of Thromboxane
Removal of circulating prostaglandins
Metabolism of vasoactive substances (bradykinin, angiotensin I, serotonin, adenine nucleotides)
Binding of activated coagulation factors (thrombin, Xa)
Release of inhibitors of platelet function
Release of activators of plasminogen

Adapted from R.G. Mason, D. Sharp, H.Y.K. Chuang, and S.F. Mohammad. The endothelium. Arch. Path. Lab. Med. 101: 61, 1977.

certain findings concerning aortic atherosclerosis may be relevant to other arterial territories, experimental lesions observed in cerebral, coronary, carotid, renal, mesenteric, and femoral arteries may have a more direct bearing on human arterial pathology. In this respect, advantage may be taken of the variations in the extent of atherosclerotic lesions induced in different medium-sized muscular arteries in cholesterol-fed monkeys (Table II); such differences have even been observed between the right and left coronary arteries (Fig. 1). Little is known about varying involvement of several arterial beds within the same animal.

It has been repeatedly observed--since the early work of Anitschkow in rabbits (7)--that within a group of animals ingesting similar atherogenic diets, there is large intragroup variability in the extent of the arterial lesions in a given location when plasma cholesterol concentrations are comparable (8). Monkeys are no exceptions in this respect. Although there is a significant positive correlation between plasma cholesterol--or beta-lipoproteins--and arterial involvement, some animals show minimal aortic or coronary

TABLE II Correlation Coefficients (r) between Atherosclerotic Involvement of Various Arteries in 18 Macaca fascicularis Monkeys Maintained on a High-butter, High-cholesterol Semipurified Diet[a] for 6 Months

	Left coronary	Carotid artery	Femoral artery	Mesenteric artery	Renal artery
Right coronary artery (% encroachment)	.926* (17)	.591* (17)	.369 (17)	.720* (17)	.441 (17)
Left coronary artery (% enroachment)		.633*	.189	.694*	.405
Carotid artery**			.085	.703* (17)	.756* (16)
Femoral artery**				.254 (17)	.020 (16)
Mesenteric artery**					.730* (16)

[a]Diet composition per 100 grams: casein, 25; sucrose, 43; butter, 25.5; cholesterol, 0.5; salt mix, 4 (Hegsted IV, Nutritional Biochemical Corp., Cleveland, Ohio); vitamin mix, 2 (OWP, Nutritional Biochemical Corp.) (see ref. 14).

*$P < 0.01$.

**Severity of lesions graded 1 to 7. Number of animals in parentheses.

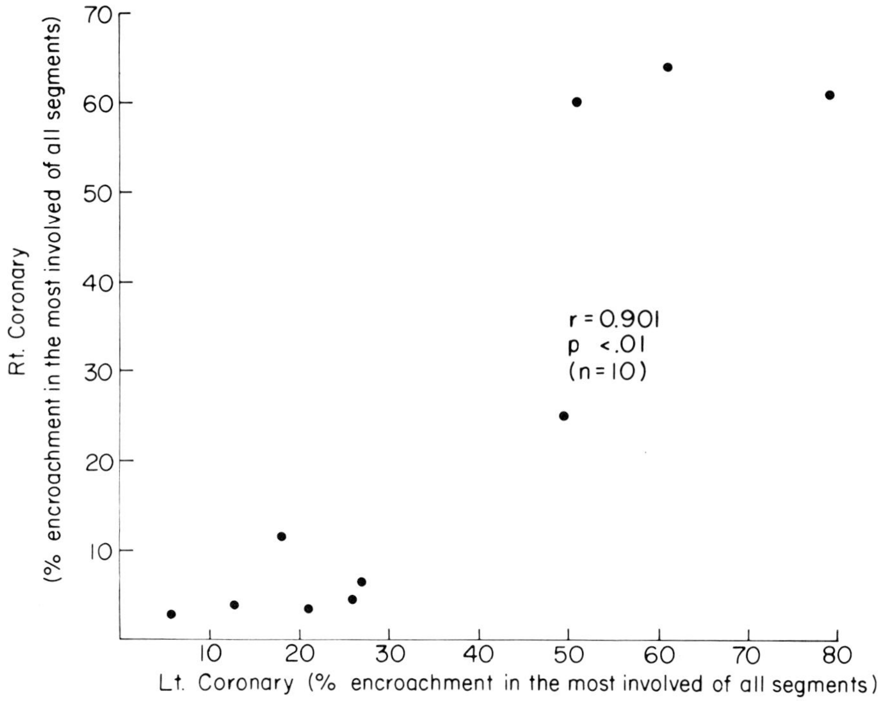

Fig. 1. Atherosclerosis in the left and right coronary arteries of Macaca fascicularis. The monkeys were fed a high-fat, high-cholesterol semipurified diet for 6 months (see Table 2). The extent of atherosclerosis was measured by planimetry in sections obtained from 5 preselected sites (see ref. 14 for details). The extent of involvement in both arteries show a high degree of correlation ($r = 0.901$ $P < .01$) although in most animals the lesions were more extensive in the left coronary arteries. Each point represents a single monkey.

involvement in spite of persistent and marked hypercholesterolemia (Fig. 2). This "hyporeactivity" to elevated plasma cholesterol is not always consistent between the aorta and the coronary arteries of an individual animal (1). Local differences in the amount of arterial cholesterol may also be demonstrated between the abdominal aorta and the carotid and subclavian arteries (1) (Fig. 3). Further information on the factors responsible for the apparent resistance--or susceptibility--to the development of diet-induced atherosclerosis in certain arterial territories of nonhuman primates may be useful in the management of the human disease. Moreover, the anatomical and chemical discrepancies between several arterial territories in monkeys suggest that observations derived from one artery should not be extrapolated arbitrarily to other arteries.

IV. VARIATION IN CHOLESTEROLEMIA

Monkeys, like other animals, are hypo- (HO), normo-, or hyper-responders (HR) to dietary cholesterol (1). We have shown that HO cynomolgus monkeys also have lower levels of plasma cholesterol (1), VLDL, LDL, and HDL_2 than HR, even when ingesting a regular chow diet and these differences persist

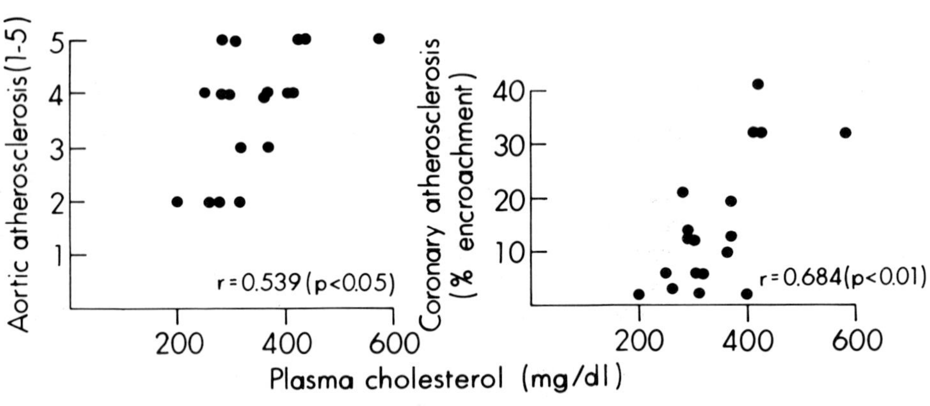

Fig. 2. Atherosclerosis in Macaca fascicularis showing that, although the extent of aortic and coronary atherosclerosis was correlated with the level of plasma cholesterol, a wide range of involvement existed at comparable cholesterolemias. The extent of atherosclerosis was measured as described in Ref. 14. Each point represents a single animal fed a high-fat, high-cholesterol (0.5%) diet for 6 months (see Table 2) followed by a similar diet with 2.5% added corn oil and lower dietary cholesterol (0.0%) for 18 months (see Ref. 14). Plasma cholesterol represents the average of 6 bimonthly determinations during the last 12 months of study only.

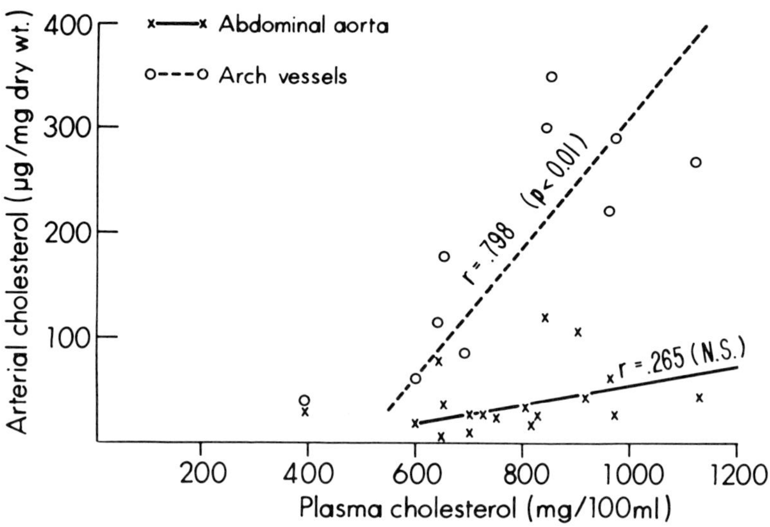

Fig. 3. Correlation of arterial and plasma cholesterol in two arterial territories of Macaca fascicularis. Each symbol represents the abdominal aorta or each vessels (carotid and subclavian arteries) from a single animal, (see Ref. 1 for details). The monkeys were fed the high-fat, high-cholesterol semipurified diet described in Table 2 for 6 months. Plasma cholesterol represents the average of four determinations during that 6 months.

with high-fat, cholesterol-free diets (Table 3). The differences are accentuated by cholesterol in the diet, but, interestingly, HDL levels were lower in HR than in HO monkeys when cholesterol was fed (Table III). In view of the apparent protective effect of HDL with respect to human coronary atherosclerosis, the mechanism of decreased HDL levels in cholesterol-fed HR monkeys needs further evaluation.

V. REGRESSION OF ATHEROSCLEROSIS

Methods free of the limitations of contrast X-ray arteriography (9) are needed for the *in vivo* documentation of the extent of arterial lesions, and it is likely that progress in ultrasound scanning may provide information on the progression and regression of atherosclerosis. In the meantime, investigators will probably continue to study the regression

TABLE III Plasma Lipoproteins in Hyporesponder and Hyperresponder Macaca Fascicularis

Results are the average of two pooled samples of plasma with four monkeys in each pool. Ultracentrifugal separation was effected at density indicated, based on similar ranges in man. Total lipoprotein values were calculated from apolipoprotein concentrations (18) on the assumption that the average protein/lipid ratio in the various density classes was the same as human lipoproteins isolated in the same density range.

Lipoprotein	Density	Dietary regimen*					
		Monkey chow		(A) High-saturated cholesterol-free		(B) High-saturated 0.5% cholesterol	
		HO	HR	HO	HR	HO	HR
		Total lipoprotein (mg/dl)					
VLDL	<1.007	15	65	39	56	24	175
IDL	1.007-1.019	22	30	24	30	65	403
LDL	1.019-1.065	175	258	182	250	307	1152
HDL_2	1.065-1.125	214	344	174	286	192	84
HDL_3	1.125-1.195	125	136	155	175	140	105
$\Sigma\ HDL_2 + HDL_3$		339	480	329	461	336	188

Lipoprotein	*Density*	*Monkey chow*		*(A) High-saturated cholesterol-free*		*(B) High-saturated 0.5% cholesterol*	
		HO	*HR*	*HO*	*HR*	*HO*	*HR*
HDL_2/HDL_3		1.71	2.53	1.12	1.64	1.34	0.82
Total cholesterol, mg/dl (N = 8)		143 ± 9	186 ± 14	144 ± 7	212 ± 7	239 ± 20	952 ± 91

**Diet A (per 100 grams): Casein, 24.5; sucrose, 40; coconut oil, 22: corn oil, 2.5; vitamin mix, 2 (see ref. 1); salt mix, 4 (see ref. 1); alphacel, 5. Diet B is similar to diet A but contains 24% casein and 0.5% cholesterol.*

Adapted from M.R. Malinow, B.W. Shore, V.G. Shore, and P. McLaughlin. The effect of different amounts of exogenous cholesterol on plasma lipoproteins of hypo- and hyper-responder Macaca fascicularis. Hugh Lofland Conference on Arterial Wall Metabolism, University of Chicago Center for Continuing Education, Chicago, Ill., May 24-26, 1976.

of atherosclerosis in nonhuman primates with well-standardized methods, e.g., the comparison of lesions in groups of monkeys killed after an induction period with dietary cholesterol to those in animals maintained for the same period on similar diets but killed after various intervals--months to years--on diets without cholesterol.

Regression* of atherosclerosis documented in rabbits, dogs, chickens, pigeons, and pigs and also demonstrated in monkeys (11,12,13,14) on low-fat, low-cholesterol diets (Fig. 4) suggests the advisability of dietary treatment for humans with atherosclerosis. However, the possibility of "regressing" coronary atherosclerosis during a high-fat, high-cholesterol diet could be more relevant to patients who adhere poorly to restricted diets. We have documented the regression of aortic and coronary atherosclerosis in monkeys after the addition of alfalfa meal or cholestyramine to a high-butter, high-cholesterol diet, and their regression was similar to that observed in animals on low-fat, cholesterol-free diets (14) (Fig 5).

The anatomical features of regression in the coronary arteries in our experiments (14) were similar to those described by other authors (11,12,13), namely, a widened lumen, shrinkage of the intimal lesions, a decrease in intra-and intercellular lipids, and the virtual disappearance of foam cell accumulation in the media and of fat-laden cells in the adventitia. At the same time, there was an apparent increase in intimal cellularity; the intercellular matrix showed an accumulation of material with tinctorial characteristics of acid mucopolysaccharides and collagen, and newly formed elastic fibers were present. Thus, the seemingly favorable changes in lipid removal were associated with a simultaneous increase in fibrosis, and "regression" did not restore the atherosclerotic lesions to normal. Similar observations in other species have been critically reviewed by Prichard (11). Whether fibrotic regressed arteries are less prone to superimposed thrombosis or to progressive occlusion may be of great importance in human pathology and obviously merits detailed investigation.

We have performed experiments in rats as well as in monkeys (15,16) that suggest that alfalfa saponins are involved in the hypocholesterolemic and antiatherogenic effects of

*Although the word "shrinkage" seems preferable to "regression" which implies a decline in the size or severity of a manifestation of disease (10), we have adhered to the term "regression" because of its widespread acceptance (11,12,13).

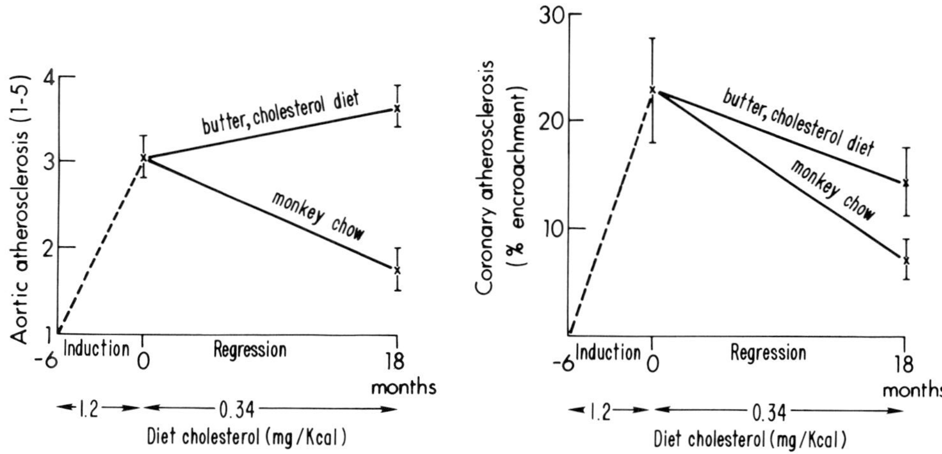

Fig. 4. Aortic and coronary atherosclerosis was induced in Macaca fascicularis fed a high-fat, high-cholesterol (0.5%) diet for 6 months (see Table 2). Other monkeys received that diet followed by either a similar diet with 2.5% added corn oil and low cholesterol (0.1%) or Purina chow for 18 months. Each point represents the average (± S.E.) of 18 monkeys. Regression of atherosclerosis occurred in animals fed monkey chow alone, but those fed the butter-cholesterol diet were not significantly different from lesions present at the end of the induction period. (Details in Ref. 14).

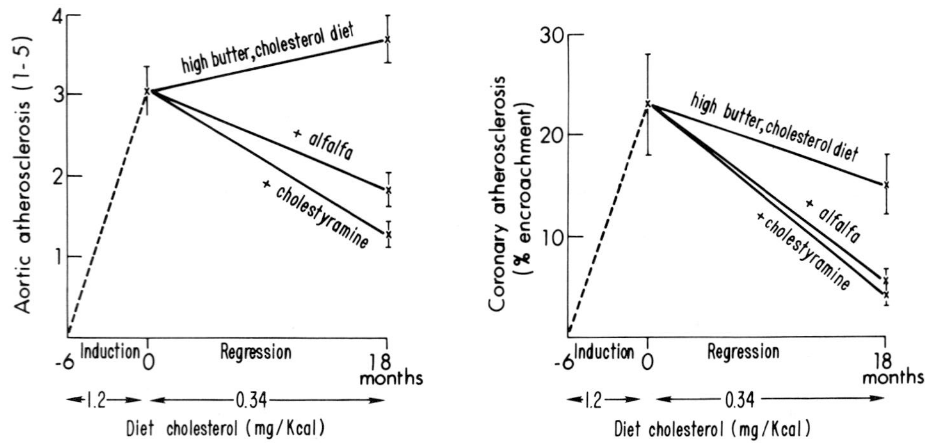

Fig. 5. Aortic and coronary atherosclerosis was induced in Macaca fascicularis as detailed in Fig. 2. During the last 18 month regression period the diet also contained 50% alfalfa meal or 5% cholestyramine. Each point represents the average (± S.E.) of 18 monkeys. The graphs demonstrate the positive influence of alfalfa meal and cholestyramine on regression of atherosclerosis. (Details in Refs. 14 and 17).

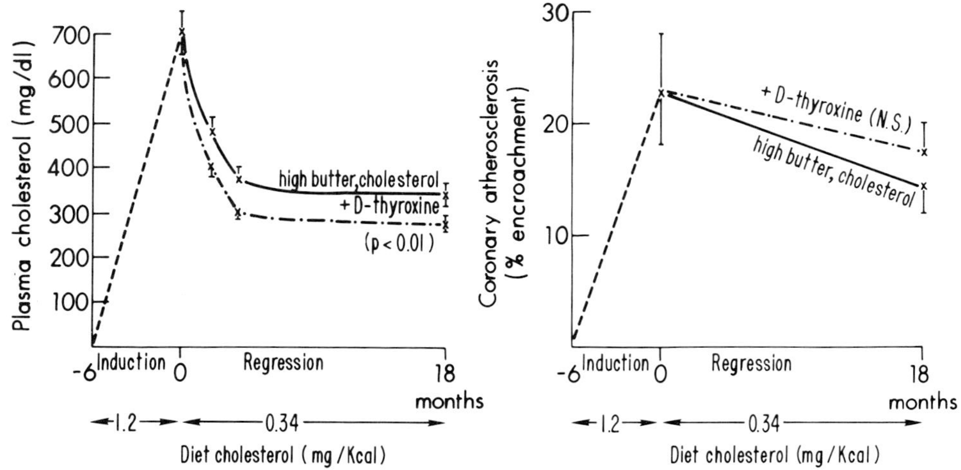

Fig. 6. Plasma cholesterol levels and coronary atherosclerosis are compared in Macaca fascicularis treated as in Fig. 2. During the last 18 months the diet also contained 0.003% dextrothyroxine in the intervention group. Each point--except at month 0 (N = 36)--represents the average (± S.E.) of 18 animals. The average plasma cholesterol in 6 bimonthly determinations performed from months 6 through 18 is indicated. The graph demonstrates lack of regression of coronary atherosclerosis even when cholesterolemia was decreased by 20% in the dextrothyroxine-treated animals. (See ref. 17 for details).

alfalfa meal observed in cholesterol-fed monkeys (14). Diet sapoins are seldom considered in nutritional studies, but the ingestion of saponins from plants may conceivably modify atherogenesis in man. It seems likely that saponins isolated from alfalfa could be useful in the treatment of patients with atherosclerosis through inhibition of absorption of intestinal cholesterol (15). However, studies in nonhuman primates are needed to determine whether alfalfa saponins will influence diet-induced atherosclerosis and whether toxicity might be associated with their prolonged intake.

Because samples of blood are more accessible than arteries, most clinical trials deal with plasma levels of cholesterol or of certain lipoproteins, rather than with the extent of atherosclerosis. That a disparity may indeed exist between plasma cholesterol levels and arterial involvement is suggested by a recent experiment that demonstrated that although dextrothyroxine reduced plasma cholesterol by 20%, it did not affect regression of atherosclerosis in cholesterol-fed cynomolgus monkeys (Fig. 6) (17). Thus, experiments with nonhuman primates, in which the extent of atherosclerosis can be documented through findings in the arteries, may be more indicative of the possible value of therapy than studies on plasma lipid changes.

In conclusion, it seems likely that observations in nonhuman primate models of diet-induced atherosclerosis will provide answers to the problem of atherogenesis and suggest new therapies applicable to man, in whom they must ultimately be tested.

ACKNOWLEDGMENTS

Observations mentioned in this report were performed in association with Phyllis McLaughlin, Carolyn Stafford, Wilbur P. McNulty, and Scott H. Goodnight. Certain plasma lipid studies were carried out by B.W. Shore, and V. G. Shore of Lawrence Livermore Laboratory, Livermore, California.

Publication No. 997 of the Oregon Regional Primate Research Center. Supported with grants No. RR00163 and No. HL16587 of the National Institutes of Health.

REFERENCES

1. Malinow, M.R., McLaughlin, P., Papworth, L., Naito, H.K., Lewis, L., and McNulty, W.P. (1976) A model for therapeutic intervention on established coronary atherosclerosis in a nonhuman primate. *Adv. Exp. Med. Biol. 67: 3.*
2. Taylor, C.B., Estrella, Manalo, and Cox, G.E. (1963) Atherosclerosis in rhesus monkeys. *Arch. Pathol. 76: 239.*
3. Chandler, A.B., Chapman, I., Erhardt, L.R., Roberts, W. C., Schwartz, C.J., Sinapius, D., Spain, D.M., Sherry, S., Ness, P.M., and Simon, T.L. (1974) Special report. Coronary thrombosis in myocardial infarction. Report of a workshop on the role of coronary thrombosis in the pathogenesis of acute myocardial infarction. *Am. J. Cardiol. 34: 823.*
4. Malinow, M.R. (1977) Is the thrombotic process involved in diet-induced atherosclerosis in nonhuman primates? Workship on the Thrombotic Process in Atherogenesis, Reston, Virginia, October 16.
5. Goodnight, S.H., and Malinow, M.R. (1978) Platelet function in nonhuman primates. *Clin. Res. 26: 117A.*
6. Linear flow was estimated here from dimensions of common macaque species and from hemodynamic measurements in: Stahl, W.R., and Malinow, M.R. (1967) A survey of physiological measurements in *Macaca mulatta*. *Folia Primat. 7: 12.*
7. Anitschkow, N. Experimental atherosclerosis in animals. In Cowdry's Arteriosclerosis: A Survey of the Problem, H.T. Blumenthal (Ed.), Charles C. Thomes, Springfield, 1967, pp.21.
8. Malinow, M.R., Pellegrino, A.A., and Ramos, E.H. (1958) Chemical and anatomical correlations in cholesterol-fed rabbits. *Acta Physiol. Latinoamer 8: 37.*
9. Roberts, W.C. (1977) Coronary heart disease: a review of abnormalities observed in the coronary arteries. *Cardiovasc. Med. 2: 29.*
10. Webster's Third New International Dictionary, unabridged, Merriam, Springfield, Mass. 1961).
11. Prichard, R.W. (1974) Recent advances in molecular pathology. Regression of atherosclerosis - a perspective. *Exp. Mol. Pathol. 20: 407.*
12. Armstrong, M.L. (1976) Regression of atherosclerosis. *Atheroscler. Rev. 1: 137.*
13. Wissler, R.W., and Vesselinovitch, D. (1977) Regression of atherosclerosis in experimental animals and man. *Mod. Concepts Cardiovasc. Dis. 46: 27.*

14. Malinow, M.R., McLaughlin, P., Naito, H.K., Lewis, A., and McNulty, W.P. (1978) Effect of alfalfa meal on shrinkage (regression) of atherosclerosis plaques during cholesterol feeding in monkeys. *Atherosclerosis 30: 27.*

15. Malinow, M.R., McLaughlin, P., Papworth, L., Stafford, C. Kohler, G.O., Livingston, L., and Cheeke, P.R. (1977) Effect of alfalfa saponins on intestinal cholesterol absorption in rats. *Am. J. Clin. Nutr. 30: 2061.*

16. Malinow, M.R., McLaughlin, P., Kohler, G.O., and Livingston, A.L. (1977) Prevention of elevated cholesterolemia in monkeys by alfalfa saponins. *Steroids 29: 105.*

17. Malinow, M.R., McLaughlin, P., McNulty, W.P., Naito, H. K., and Lewis, L.A. (1978) Treatment of established atherosclerosis during cholesterol feeding in monkeys. *Atherosclerosis 31: 185.*

18. Shore, V.G., and Shore, B.W. The apolipoproteins: their structure and functional roles in human serum lipoproteins. In: Blood Lipids and Lipoproteins: Quantitation, Composition and Metabolism, G. Nelson (Ed.), Wiley Interscience, 1972 pp. 789.

EFFECTS OF DIETARY FAT ON PLASMA LIPIDS IN NORMAL AND SPONTANEOUSLY DIABETIC *MACACA NIGRA*

Charles F. Howard Jr.

Department of Nutrition and Metabolic Diseases
Oregon Regional Primate Research Center
Beaverton, Oregon

I. INTRODUCTION

Hypotheses concerning diets as exacerbating factors in disease development must take into account genetically controlled susceptibilities. Diabetes mellitus develops spontaneously in a high percentage of Celebes black macaques *(Macaca nigra)* (1). Aortic atherosclerosis is especially prominent in diabetic monkeys and is correlated with the severity of the diabetes (2). Although all monkeys are fed a natural ration, diabetes and atherosclerosis occur only in specific monkeys. This fact suggests a selective genetic predisposition in some monkeys to the disease; diets could exacerbate inherent defects. *Macaca nigra* provide a model for the assessment of dietary effects in normal monkeys, as well as in monkeys with diabetes or the potential for developing it.

Initial dietary studies were designed to measure the acute response to alterations in dietary fat. Quantity and degree of saturation of triglycerides affect circulating

ISBN 0-12-334250-3

lipids and the development of atherosclerosis. Since diabetes and atherosclerosis have developed in *Macaca nigra* consuming natural rations, the chow was kept substantially the same, and only the triglyceride content was changed. The low-fat (LF) diet (Fig. 1) was Purina 15 Monkey Chow [R] with only 2.5% vegetable oil present; the polyunsaturated/saturated fat (P/S) ratio was 1.87. The two high fat diets (13.2% fat, w/w) contained predominantly safflower oil (SFO) with a resultant ratio of 1.18, or predominantly coconut oil (CCO) with a P/S ratio of 0.53 that approached the ratio of typical human diets. When fat was added, carbohydrates (primarily starch) were diminished; thus, the LF diet was higher in carbohydrates than the SFO and CCO diets. Each dietary regimen was fed for 8 weeks, and blood was examined after an overnight fast every 2 weeks. Lipoproteins were isolated by the method of Hatch and Lees (3). All diets contained < 0.01% cholesterol.

Macaca nigra were grouped as either normal or diabetic although their clinical classification was somewhat more extensively characterized. Normal and diabetic values in Table I represent extremes for the range of data which allow differentiation of an intermediate group of borderline diabetics. The K values from an intravenous glucose-tolerance text (IV-GTT) and glucose measures after an overnight fast were quantitatively similar to human values. Insulin decreases in this diabetic syndrome. Increased triglycerides are not quantitatively as great as those in diabetic humans, but are qualitatively similar and correlated with the severity of the disease. A point system (numbers in parentheses and ranges at the bottom) was devised as an aid to

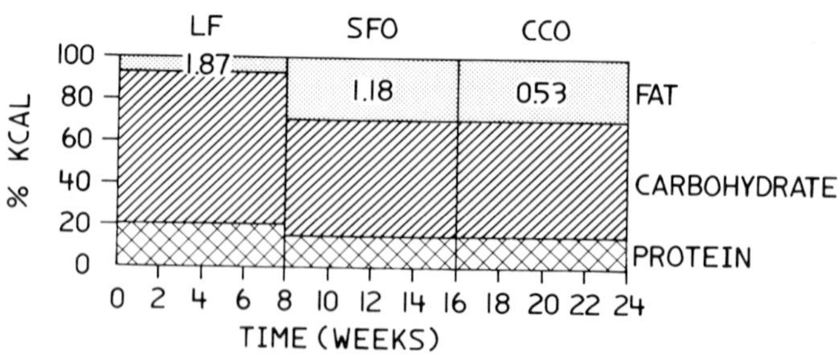

Fig. 1. Comparison of percentages of kilocalories attributable to fat, carbohydrate, and protein in each diet. Diets LF, SFO, and CCO are defined in the text. Numbers in each diet column are the ratios of polyunsaturated/saturated fat.

TABLE 1 Clinical Classification of Mature *Macaca nigra*

Parameter	Normal	Borderline Diabetic	Diabetic	Units
K	>1.8 (1)	1.0-1.8 (3)	<1.0 (5)	%glucose clearance/ minute in an IV-GTT
Glucose	60-105 (1)	106-124 (3)	>125 (5)	mg/dl
Insulin	>20 (1)	18-19 (2)	<17 (3)	μU/ml
Triglycerides	<100 (1)	100-125 (2)	>125 (3)	mg/dl
Average Point Range	(1-1.6)	(1.7-2.7)	(>2.8)	

clinical classification since individual monkeys often had measurements that fit into more than one category. Since there were insufficient overt diabetics, borderline diabetics were included in the diabetic category. Any differences attributable to diabetes would be even more significant since the diabetic group was weighted towards a milder condition.

Diabetes in *Macaca nigra* and other nonhuman primates is attributable to the concurrent appearance of amyloid in the islets of Langerhans and secretory cell degeneration (4). I have proposed the name "insular amyloidotic diabetes mellitus" for this form of maturity-onset diabetes.

II. RESULTS AND DISCUSSION

<u>Triglycerides</u> -- Serum triglycerides were highest in monkeys on the LF diet (Fig. 2); values of diabetics exceeded those for normal monkeys. With the consumption of

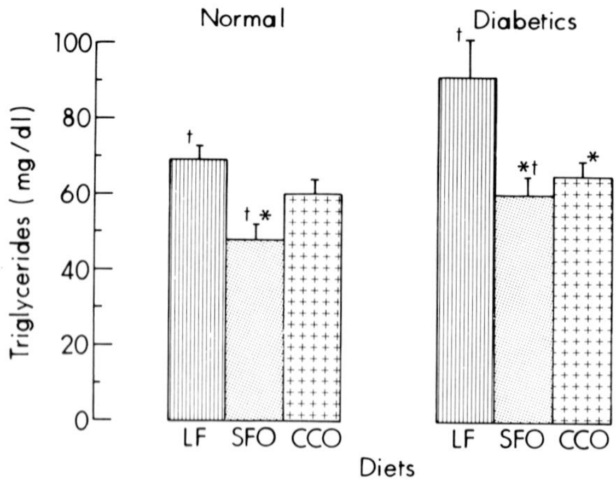

*Fig. 2. Diet induced changes in serum triglycerides. LF, SFO, and CCO are defined in the text. Values are graphed with SEM indicated. * $p < 0.05$ for SFO or CCO diets vs. LF diet within each group. † $p < 0.05$ for normal group vs. diabetic group.*

either of the high-fat diets, serum triglyceride concentrations decreased. The increased saturation in the CCO diet caused a significant ($p < 0.05$) increase in circulating triglycerides in normal macaques, similar to the response in man (5, 6). The relatively small increase in carbohydrates in the diet (18%) over the SFO and CCO diets was sufficient here, as elsewhere (7), to increase circulating triglycerides, especially in diabetics. As with humans, the increased unsaturation of the SFO diet tended to decrease circulating triglycerides (8) with values for diabetics exceeding that for normal.

Cholesterol -- An increased fat concentration caused an increased concentration of serum cholesterol in normals and diabetics, but the groups did not differ significantly from each other (Fig. 3). Normal macaques apparently responded to the quantity of fat, as no differences were attributable to the degrees of unsaturation of the SFO and CCO diets. However, diabetics responded to both the quantity of fat and the degree of unsaturation; the SFO diet increased serum cholesterol and the CCO diet further increased concentrations

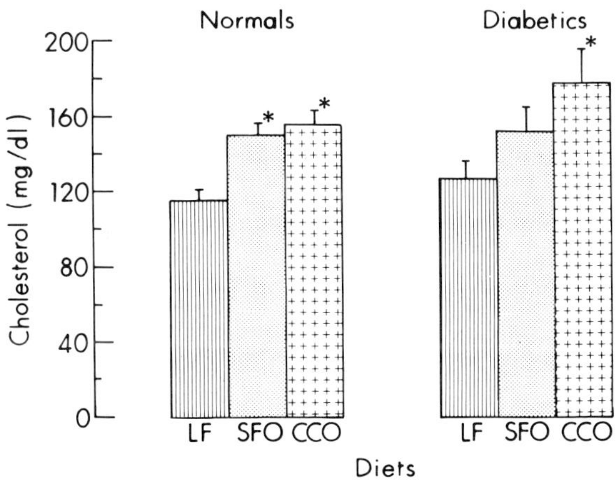

Fig. 3. Diet induced changes in serum cholesterol levels. LF, SFO, and CCO are defined in the text. Values are graphed with SEM indicated. * $p < 0.05$ for SFO or CCO diets vs. LF diet within each group.

to their greatest values. All diets contained < 0.01% cholesterol so that the circulating cholesterol derived mainly from endogenous sources. Dietary saturated fats are associated with metabolic reactions which increase cholesterol in the circulating lipoproteins. Although human studies have usually involved fat alterations in diets that have also contained some cholesterol (9, 10), animal experiments with low-cholesterol diets have verified that unsaturation alone can decrease the circulating cholesterol (11, 12). In the present study, the quantity of fat was more important than the degree of unsaturation.

Lipoproteins -- Very low density lipoprotein (VLDL) protein values correlated with circulating triglyceride concentration ($r = 0.98$; $p < 0.001$), e.g., the greatest concentrations of VLDL protein and triglyceride were in diabetic monkeys on the LF diet and the lowest concentrations were in normal monkeys fed the SFO diet. Similarily, low density lipoprotein (LDL) protein and cholesterol were correlated, though less significantly ($r = 0.82$; $p < 0.05$). The relative sizes of VLDL particles are illustrated in Figure 4. Calculations were based on the ratio of the surface components (protein + cholesterol + phospholipid) to core components (triglycerides + cholesterol ester) (13, 14) and considered as surface area and volume relationships; calculations were empirical and actual space occupied by the particle components was disregarded. Particles from normal monkeys were similar to diabetics. The LF and SFO diets had particles of approximately the same size; the fat content of the SFO diet was five times greater than that of the LF diet, but the P/S ratio of 1.87 and 1.18 indicate greater unsaturation which was associated with larger particle size. The smallest particles were the VLDLs from the CCO diet having a P/S ratio of 0.53. Presumably particles from the CCO diet differed significantly from both the SFO and LF diets because saturated fatty acids would pack more closely than unsaturated fatty acids. Thus, VLDL concentrations were generally greater in diabetic monkeys, but the particle size was related to the degree of dietary fat unsaturation.

III. CONCLUSIONS

Macaca nigra, other experimental animals, and human beings respond to dietary manipulation in comparable ways. On the unaltered Purina 15 Monkey Chow diet, only

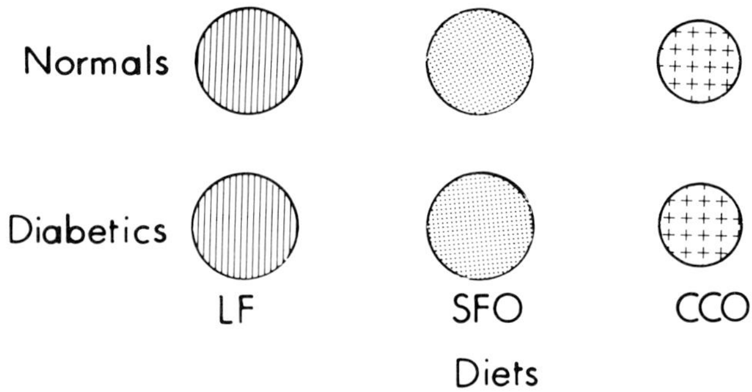

Fig. 4. Relative sizes of VLDL particles. Molar concentrations for the particle components were measured and grouped as surface (protein + phospholipids + cholesterol) or core (triglycerides + cholesterol esters) components. The ratio of surface area ($A = 4\pi r^2$) to core volume ($V = 4/3\pi r^3$) was calculated and average particle diameters were determined from the ratios; sizes of circles are relative to each other and are graphed without regard to the actual volume occupied by each particle component. With the mean CCO particle diameter set at 100 $\pm$6, SFO averaged 132 $\pm$8 and LF particles 143 $\pm$ 6. $p < 0.01$ for CCO vs. LF and SFO.

triglyceride concentrations are related to the severity of the diabetes and the development of atherosclerosis (15); cholesterol is not. Diabetic monkeys have more circulating lipids than nondiabetics. Cholesterol concentrations were increased when dietary fat content of the chow was substantially raised, especially in diabetic monkeys. The rise in circulating cholesterol occurred even though dietary cholesterol was negligible. Thus, since these monkeys will respond to fat manipulation in a manner similar to man, they should be a useful model for elucidating the development and expression of diabetes in human beings as affected by dietary fat.

ACKNOWLEDGMENTS

This is publication No. 1004 from the Oregon Regional Primate Research Center, supported by Animal Resources Branch grant RR 00163 and General Research Support grant RR 05694, both from the Division of Research Resources, N.I.H. This work was supported by PHS grants HL 16661, HL 09744, RR 05694 and AM 21982, and by the Kroc Foundation.

REFERENCES

1. Howard, C.F., Jr. (1972) Spontaneous diabetes in *Macaca nigra*. *Diabetes 21: 1077-1090.*
2. Howard, C.F., Jr. (1975) The relationship of diet and atherosclerosis in diabetic *Macaca nigra*. *Adv. Exp. Med. Biol. 60: 13-31.*
3. Hatch, F.T. & Lees, R.S. (1968) Practical methods for plasma lipid analysis. *Adv. Lipid Res. 6: 1-68.*
4. Howard, C.F., Jr. (1978) Insular amyloidosis and diabetes mellitus in *Macaca nigra*. *Diabetes 27: 357-364.*
5. Ahrens, E.H., Jr., Tsalta, T.T., Hirsch, J., & Insull, W., Jr. (1955) Effects of dietary fats on the serum lipids of human subjects. *J. Clin. Invest. 34: 918.*
6. Beveridge, J.M.R., Jagannathan, S.N., & Connell, W.F. (1964) The effect of the type and amount of dietary fat on the level of plasma triglycerides in human subjects in the postabsorptive state. *Can. J. Biochem. 42: 999-1003.*
7. Ginsberg, H., Olefsky, J.M., Kimmerling, G., Crapo, P., & Reaven, G.M. (1976) Induction of hypertriglyceridemia by a low-fat diet. *J. Clin. Endocrinol. Metab. 42: 729-735.*
8. Nestel, P.J., Carroll, K.F., & Havenstein, N. (1970) Plasma triglyceride response to carbohydrates, fats, and caloric intake. *Metabolism 19: 1-18.*
9. Keys, A., Anderson, J.T., & Grande, F. (1957) Prediction of serum cholesterol responses of man to changes in fats in the diet. *Lancet 2: 959-966.*
10. Hegsted, D.M., McGandy, R.B., Myers, M.L., & Stare, F.J. (1965) Quantitative effects of dietary fat on serum cholesterol in man. *Am. J. Clin. Nutr. 17: 281-295.*

11. Corey, J.E., Hayes, K.C., Dorr, B., & Hegsted, D.M. (1974) Comparative lipid response of four primate species to dietary changes in fat and carbohydrate. *Atherosclerosis 19: 119-134.*

12. Portman, O.W., Alexander, M., Tanaka, N., & Soltys, P. (1976) The effects of dietary fat and cholesterol on the metabolism of plasma low density lipoprotein apoproteins in squirrel monkeys. *Biochim. Biophys. Acta 450: 185-196.*

13. Heimberg, M., & Wilcox, H.G. (1972) The effect of palmitic and oleic acids on the properties and composition of the very low density lipoprotein secreted by the liver. *J. Biol. Chem. 247: 875-880.*

14. Hojnacki, J.L., Nicolosi, R.J., Hoover, G., Llansa, N., el Lozy, M., & Hayes, K.C. (1977) Effects of dietary fat on the composition and size of primate very low density lipoproteins. *Artery 3: 409-420.*

15. Howard, C.F., Jr. (1974) Correlations of serum triglyceride and prebetalipoprotein levels to the severity of spontaneous diabetes in *Macaca nigra*. *J. Clin. Endocrinol. Metab 38: 856-860.*

SELECTED ASPECTS OF THE METABOLIC BEHAVIOR OF THE SQUIRREL MONKEY

Daniel L. Gallina
Lynne M. Ausman

Department of Nutrition
Harvard School of Public Health
Boston, Massachusetts

I. INTRODUCTION

Conventionally, it is possible to divide animal models for human clinical entities into two types--those which present a spontaneous anomaly and are used to focus on the mechanism underlying a specific human disease (for example several strains of mutant mice in diabetes mellitus research) and those that respond comparably to humans as a "whole" in which invasive techniques can be utilized to elicit and study a disease process. *A priori*, primates generally correspond to the second type of model. Nevertheless, species differences expressed by dissimilar characteristics with respect to humans, still do not allow one to classify some primate species strictly in this category of model. The New World monkey, particularly the squirrel monkey, develops a spontaneous or dietary-induced atherosclerosis similar to man, while presenting characteristics of carbohydrate, protein and steroid metabolism which seem dissimilar from the accepted human pattern. The purpose of this review is to discuss some

ISBN 0-12-334250-3

of these characteristics to better appreciate this species as a model for human nutritional and metabolic disorders.

The attainment of an unstressed basal state in the experimental animal is a difficult task due to the obligatory use of some method of restraint or anesthesia. Perhaps the most dramatic effect of these two procedures on monkeys has been observed on the endocrine system and hormone-regulated metabolism (1,2). Anesthesia can markedly influence the response of several hormones such as those from the pituitary (1,3), corticosteroids (4,5) and insulin (6); and its effects appear to persist, at least for some hormones, even with the use of mild depressants of the central nervous system (1,3). In addition, it has been widely documented that both inhalant (7,8) and parenteral anesthetics, particularly barbiturates (6,9), can impair the metabolism of glucose or free fatty acids. Several mechanisms, ranging from the direct or catecholamine-mediated impaired insulin release (10) to a peripheral defect in glucose utilization have been proposed (6,11). Although some varieties of monkeys such as the rhesus can be trained to remain in a restraining chair for prolonged periods, most monkeys are easily alarmed, resulting in the expected stress reaction (12). The few metabolic data available for New World monkeys have been collected almost entirely under anesthesia or hand restraint. A portion of these data will be reviewed in concert with specific aspects of carbohydrate and protein metabolism in the squirrel monkey.

II. CARBOHYDRATE METABOLISM

The response of the squirrel monkey to an oral glucose tolerance test (OGTT) appeared to show a high prevalence of impaired glucose disposal. When a weighted average was chosen to express the blood glucose concentrations in conscious, hand-restrained animals during oral loading, one-half the study population was termed "intolerant" (13,14,15). This bimodal distribution was not confirmed in another laboratory which utilized anesthetized monkeys and defined the entire population intolerant with respect to man (16). Figure 1 depicts the results from these two studies in female monkeys as well as results from a group of conscious hand-restrained males studied in this laboratory (Experiment 2). No bimodal response was found in this last group, and the average blood glucose value fell within the range of "normal" responders (14). We might conclude from this comparative analysis that the greatest effect comes from anesthesia. Nevertheless, several other variables appeared to affect glucose tolerance.

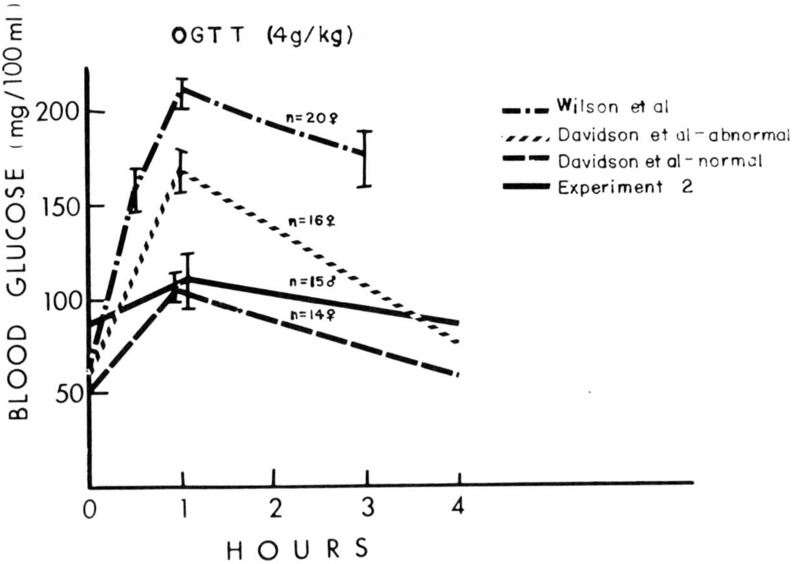

Fig. 1. Results of oral glucose tolerance tests using 4 gm/kg body weight carried out on adult squirrel monkeys in three different laboratories. Blood glucose values for Wilson et al (16) were calculated from plasma glucose values using the formula of Dillon (60). Normal and abnormal responders were derived from Davidson et al (14) and experiment 2 was from this laboratory. Values and vertical bars, when given, represent the mean ± s.d.

Some of these variables have been tested in a group of 16 feral, female squirrel monkeys. The monkeys were removed from a gang breeding colony to evaluate the effects of handling and dehydration. In every case, the oral load was administered after an 18 hour post-absorptive period. The results obtained from three sub-groups are shown in Figure 2. Six animals were removed from the breeding colony, placed in individual cages, in a room with other experimental animals with frequent human exposure, and fed and watered *ad libitum* for seven days before testing (Group A). The other 10 monkeys were also individually caged, but only for the 18 hours before testing and in the same room as the rest of the breeding colony. In six of these animals (Group B), water was available *ad libitum*. In the other 4 (Group C), water was witheld until 2 hours before testing. Water deprivation for a period of 16 hours was associated with a severe glucose intolerance in group C with

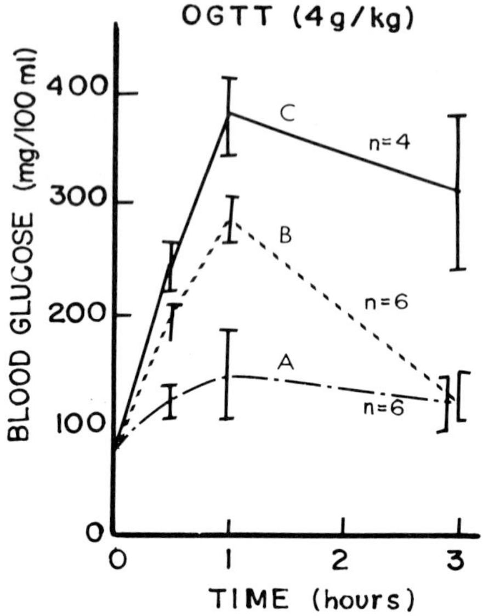

Fig. 2. Response of three groups of feral, chow-fed adult female squirrel monkeys to oral glucose tolerance tests with 4 gm glucose/kg body weight. Monkeys removed from the group 7 days before testing and individually caged for that period with water available ad libitum are Group A; monkeys given water ad libitum and individually caged 18 hours before testing are Group B; and monkeys given no water and caged individually 18 hours before testing are Group C. All monkeys were fasted 18 hours before glucose tolerance testing.

a good degree of this defect persisting in group B despite *ad libitum water*. The response of group A fell into the range of the "normal" responders.

In a different set of experiments with 12 infant squirrel monkeys, the effect of an 8 or 14 hour post-absorptive period on glucose tolerance was evaluated at 8 and 10 weeks of age, respectively. All monkeys were born to dams in the department colony (17), separated from their mothers at birth, and reared in the monkey nursery with a commercial liquid formula (18). Diet was available from 8 a.m. to midnight and water *ad libitum*. Blood glucose responses to 2 and 4 g/kg oral glucose are shown in Figure 3. The curves were evaluated using the

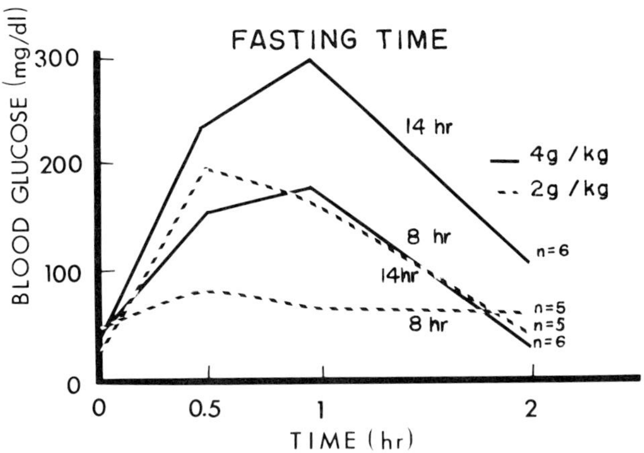

Fig. 3. Response of the infant squirrel monkey to oral glucose loading after fasting periods of both 8 and 14 hours. Two dose levels (2 and 4 gm glucose/kg body weight) were utilized on separate occasions. A significant intolerance was demonstrated at 14 hours relative to that at 8 hours with both doses.

sum of the 1/2 to 1 hour blood glucose levels. A clear intolerance was observed at 14 hours compared to that at 8 hours with both doses.

The response of anesthetized animals to intravenous glucose tolerance testing (IVGTT) shows glucose assimilation rates (K) reported or calculated from the data of others ranging from 1.22 to 2.33 (15,16,19,20). These results appear to correspond with those found in Old World monkeys (21) or man (22) although they were dissimilar to a group of both juvenile and adult animals studied in our lab. Figure 4 includes the results obtained in these two groups (C,D) and are compared with those obtained from two groups of animals from other laboratories in which the same anesthesia was utilized (pentobarbital) (Groups A,B) and two groups of hand restrained animals (with and without prior training) (Groups E,F). Little agreement was found between the values obtained under anesthesia; and the highest degree of intolerance was observed in the group of untrained, conscious animals in which the apparent glucose disposal rate was very low.

From the results of oral and intravenous glucose loading, one may conclude that several other factors affect glucose tolerance in squirrel monkeys besides the well recognized effect of anesthesia. Among these are stress modulated by the degree of prior acclimitization of the monkey to the experimental environment and testing conditions, such as venipuncture and the "pattern" of pre-test food and water consumption. For these reasons, we have adopted a permanent, indwelling, catheterized model that can be studied under strictly controlled, unrestrained conditions.

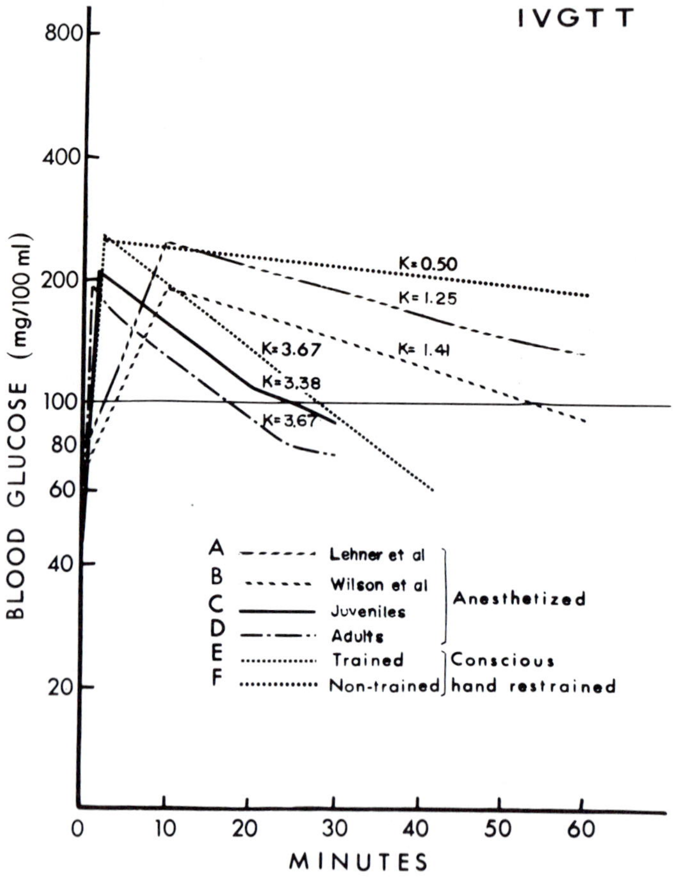

Fig. 4. Comparative results of IVGTT in juvenile and adult squirrel monkeys under various conditions. Four groups were studied under pentobarbital anesthesia: A. Glucose dose 0.75 gm/kg (19). B. Glucose dose 0.4 gm/kg (16); C and D. Dose 0.4 gm/kg in this laboratory. The other two groups, E and F, also from this laboratory, depict the response to both trained and non-trained, conscious animals, respectively, to a glucose dose of 0.5 gm/kg. Plasma glucose values of group B were transformed to blood glucose concentrations as in Fig. 1.

Surgical procedures to implant both arterial and venous catheters, routine maintenance and mechanics of the testing procedures, the training conditions, and the assay techniques for blood glucose (BG), immunoreactive insulin (IRI), and free fatty acids (FFA) have been published elsewhere (23). Radioimmunoassays of growth hormone (GH) and glucagon (IRG) were performed by micromodifications of a double antibody (24) and a single antibody technique using a charcoal-dextran separation (25), respectively.

The response to OGTT (4 g/kg) utilizing this technique was studied in seven animals trained to drink a 50% glucose solution in less than 2 minutes from a syringe inserted through the cage wall. The results obtained with this dose fell into the range previously reported as normal (14) but the additional sampling times in our hands indicated that the peak circulating glucose level occurred between 15-30 minutes after dosing, remaining significantly elevated above fasting until the 90 minute sample. This sustained glucose level, which could be ascribed to the high glucose load (26,27), was also associated with a prolonged response in both insulin release and free fatty acid depression. Figure 5 reproduces the response of the aforementioned three parameters plus growth hormone to an oral load of 4 g/kg body weight as well as the response in blood glucose, free fatty acids, IRI, IRG and GH to an oral load of 2 g/kg. With 2 g/kg, blood glucose levels reached a mean maximum concentration of 96 mg/dl at 15-30 minutes but returned to basal levels by 60 minutes. This trend is followed by the behavior of insulin and FFA in which a symmetrical pattern of release and rapid recovery from antilipolysis was observed, respectively. IRG decreased significantly at 15-45 minutes after loading in close agreement with the results found in man (28). GH shows an earlier release as a consequence of an early rapid glucose fall with 2 g/kg glucose as compared to 4 g/kg. The initial decrease from basal values for growth hormone associated with high blood glucose concentrations (29) was only observed with 4 g/kg. These results suggest that the initial conclusion that 4/g/kg glucose was "more adequate" in this species (14) was erroneous. The temptation to adapt the testing procedures utilized in humans to an animal model to obtain values similar to those obtained in humans rather than utilize a dose to elicit a standard physiological response must be guarded against. The squirrel monkey presents a glucose homeostasis in which the basal blood glucose concentration of about 60 mg/dl after 8-10 hours of postabsorptive period increases 60-65% within 15-30 minutes after oral loading independent of the dose; this is similar to the situation reported for man (27). The sustained glucose levels and parallel prolonged insulin release found with an

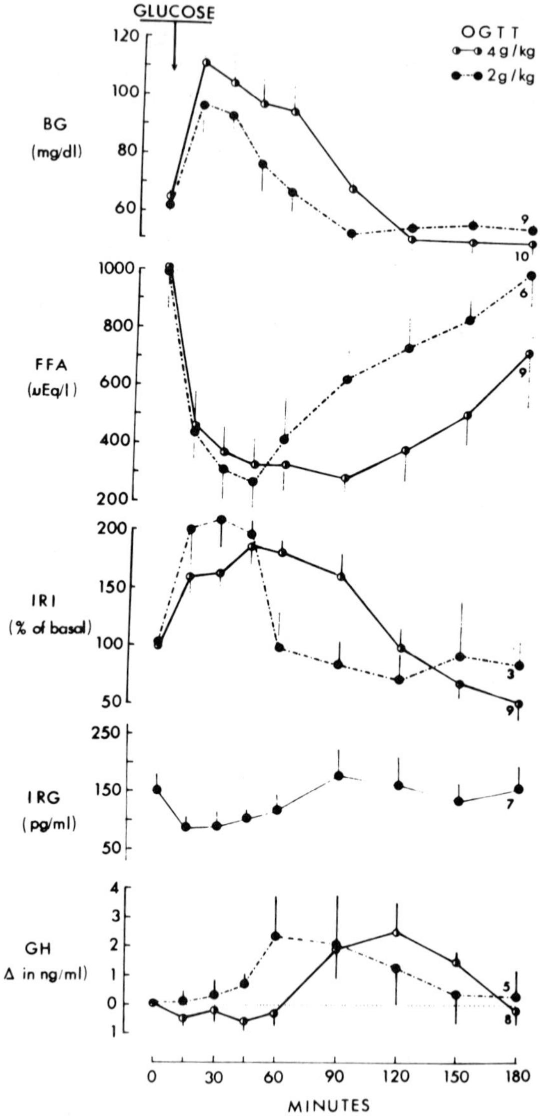

Fig. 5. Response to oral glucose tolerance testing of catheterized squirrel monkeys at two dose levels. Numbers of 180 minutes represent monkeys studied and the vertical bars represent s.e.m.

oral glucose load of 4 g/kg in comparison to 2 g/kg undoubtedly involves a disproportionate share of de novo, with respect to preformed, release of insulin, which is in effect undesirable for the interpretation of the test (27). On the other hand, the fact that the shape of the curves for glucose and insulin are similar to those found with the lowest standard dose for humans (26) and the loss of some physiological responses, such as the initial transient fall in GH (29), appear to suggest that a dose lower than 2 g/kg would be insufficient. For this reason, we believe that 2 g/kg is probably the best dose to be used in standardized tests in this species.

In order to eliminate the absorptive phase from glucose tolerance testing and to isolate the rate of glucose disposal, glucose at 0.5 g/kg body weight was intravenously infused in less than 30 seconds to 16 animals after a basal blood sample was obtained via the arterial catheter. Subsequent samples were withdrawn thereafter until 90 minutes. Normal saline (pH 7.4) was injected in 5 animals at a rate and volume (2 ml/kg) identical for IVGTT. The response to the intravenous injection of glucose and salt solution for the same parameters measured during OGTT are shown in Figure 6 along with the results from a spontaneously diabetic monkey that will be discussed later. The mean glucose disposal rate (K) obtained in this condition was 8.18 ± 2.53. The large standard deviation was due to interanimal variation and not to the testing procedure (23). FFA immediately decreased significantly by 55% from basal values of 1074 ± 352, associated with roughly a one-fold increase in insulin with both parameters returning to basal values more than 60 minutes later. Although a transitory decrease was noted in the FFA concentrations after the saline injection, none of these values or those for blood glucose or insulin were significantly different from basal at any testing point. The rapid fall in the IRG obtained as a consequence of the sharp hyperglycemia was significant up to the 11 minute sample after loading. The growth hormone response was similar to that for OGTT, increasing significantly after the rapid glucose fall (30). This response was not observed when saline solution was injected.

The glucose disposal rates (K) obtained in these conditions were markedly higher than both those previously reported and those obtained in the same monkeys under various conditions prior to catheterization (Table I). With the exception of one animal, both trained and non-trained animals yielded a uniformly depressed response prior to catheterization suggesting that the effects of anesthesia and most probably stress, triggering catecholamine release, were present even in apparently acclimitized monkeys.

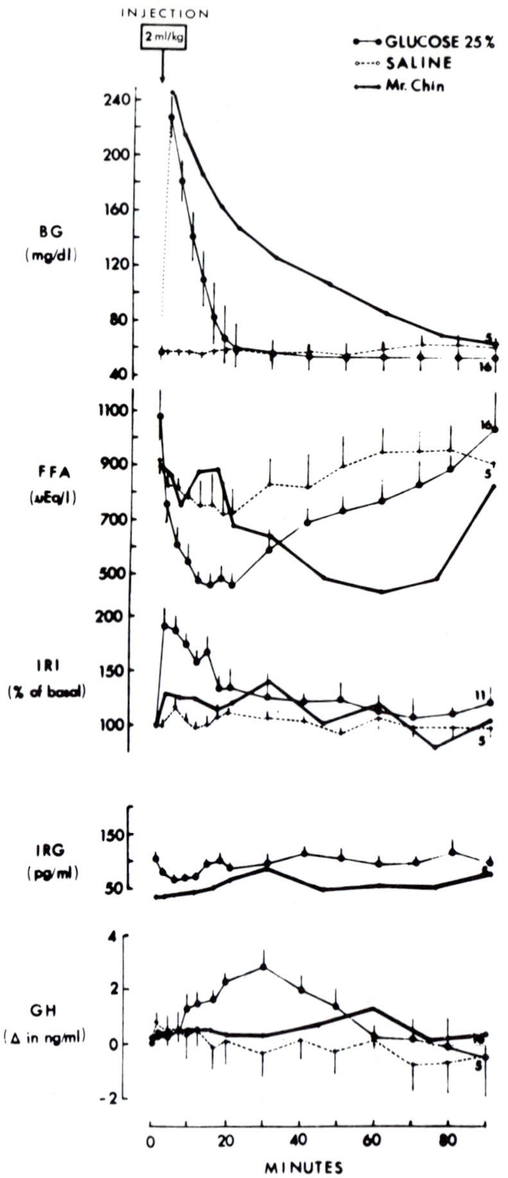

Fig. 6. Response to IVGTT (0.5 g/kg) in catheterized squirrel monkeys. The results of saline loading (using the same volume as for IVGTT) and glucose loading of a single "diabetic" monkey (Mr. Chin) are also included. Blood glucose values are the antilog of a logarithmic mean ± s.d. For all other parameters, the averages are arithmetic ± s.e.m. Small numerals represent the number of animals tested.

TABLE I. Results of IVGTT Before and After Catheterization

Monkey	Before Catheterization		After Catheterization
	Conditions of test	K value	K value
SB-99	Pentobarbital	2.78	5.47
SB-440	Trained, conscious	4.51	11.53 11.56
SB-290	Trained, conscious	2.65	11.22 10.74
SB-262	Untrained, conscious	-0.09	7.61
SFT	Untrained, conscious	1.22 -0.98	6.45
SFR	Untrained, conscious	-0.64	7.45
SFY	Untrained, conscious	5.13	5.54

Source: L. Ausman and D. Gallina (23). Reproduced by permission of the American Physiological Society.

It has been reported that the squirrel monkey has a mean basal metabolic rate of about 107 kcal/kg/24 hours (31,32), a figure which is roughly 4 times higher than that of a 70 kg normal man (33). If we accept that the rate of glucose disposal estimated by the oxygen consumption is inversely proportional to the body size (34), the K after a single postabsorptive pulse would be proportional to the body size. If this assumption is valid, it would be possible to extrapolate the K value of 6-12 found in these 700-1000 gram monkeys into the spectrum of the glucose disposal rate found in mammals of different size (Figure 7). This high rate of glucose disposal, as well as the net decrease in the FFA concentration (antilipolytic effect), appeared to be mediated by a rather small insulin release. As squirrel monkey insulin may be less reactive to anti-human, beef or pork insulin antibodies, the data have been expressed as percent change to partially alleviate the problems involved in interspecies comparisons. Nevertheless, similar amounts of "crude" insulin have been extracted from both the squirrel and rhesus monkey pancreas (16) despite the fact that the rhesus responds like humans

(35) with a 5 to 10-fold insulin increment after glucose loading. The relative high efficiency of the insulin observed in our experiments was confirmed through the use of arginine (see below), a different B-cell stimulus. Glucagon and growth hormone, on the other hand, classically described as "anti-insulinic" hormones, responded in a manner similar to that observed in man (36) and, at least for GH, at levels that would not account for the insulin hypersensitivity.

The low carcass fat content found in these animals (6% of the total extractable fat) (23) demonstrates the extremely small adipose tissue depot of the average adult female or non-fatted male (37). By extrapolation of the data on basal insulin and post-stimulation insulin concentrations in underweight and obese men (38-40), it is possible that these low levels of squirrel monkey insulin correspond to the minimal triglyceride depot in this species.

From the results collected from almost 50 catheterized monkeys, we can conclude that spontaneous glucose intolerence

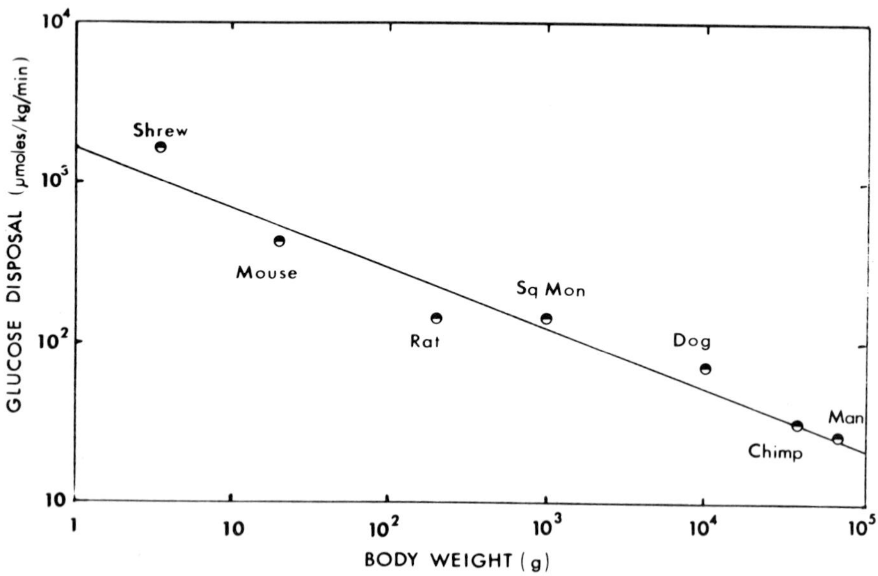

Fig. 7. Glucose disposal rates (oxygen consumption/ 6) are plotted vs. body weights according to Coulson et al (34). Data from the squirrel monkey (31,32) and chimpanzee (33) were calculated in the same manner and are included for comparison. The regression equation, $Y = -0.378 X + 3.234$, was calculated using log values for both X and Y.

is rarely found in this species and that a high rate of glucose disposal and antilipolysis mediated by an efficient insulin action prevails. Only one exception has been found to date. Figure 6 includes the results of the IVGTT performed on a catheterized old breeding male (Mr. Chin). The basal blood glucose of 66 mg/dl was associated with a glucose disposal rate of 2.2 which is reminiscent of values previously reported as normal in non-catheterized monkeys. When this response is compared, however, with the norm for catheterized animals, a high degree of impaired glucose removal was apparent, associated with a delayed decrease in the FFA concentration as a consequence of a delayed and inadequate insulin release. The usual deflections observed in IRG and GH were essentially inverted and non-significant, respectively. These results appear to correspond to a classical case of adult onset diabetes mellitus. Although the small population of monkeys studied does not allow one to draw epidemiological conclusions, a high spontaneous incidence of impaired carbohydrate metabolism in this species seems unlikely. It seems more probable that the crude prevalence of diabetes in squirrel monkeys is not unlike that found in humans. On the other hand, the high incidence of this defect in another primate species (21) may reflect the rather ingrown breeding pattern in restricted geographic areas, similar to that observed in certain Indian cultures such as the Pimas (41).

III. PROTEIN METABOLISM

As shown in a preceding chapter (Ausman and Gallina), the low efficiency of protein utilization found in the young squirrel monkey was not improved by extending the feeding schedule from 16 to 24 hours per day nor by enhancing protein quality with supplemental sulfur amino acids. Aside from the eventuality that these monkeys may present a peculiar amino acid requirement, one might speculate that this low efficiency may be due to poor metabolic utilization of dietary amino acids.

Although the accuracy for quantitation of whole body protein turnover is questionable, the results reported for a normal 70 kg man (42) appear to demonstrate that daily protein synthesis far exceeds the accepted protein requirements (43). This fact indicates the extensive reutilization of amino acids liberated from the endogenous protein breakdown, the degree of which is correlated with the nitrogen intake to yield a rather stable amino acid pool (free plus bound). The urinary nitrogen is the "buffer" of this constancy.

It has been demonstrated that the total protein turnover in different mammals is related to the basal metabolic rate (44). In the squirrel monkey, this parameter has not been measured, but there is no reason to believe that this species will not fit into the range for its body size. Protein requirements, however, were shown to be higher than for other species (45)--even when corrected for metabolic size. This poses the question of whether the fine adjustment between amino acid reutilization and the dietary input into the amino acid pool is set in a different manner. Or, in other words, are the mechanisms of adaptation of protein metabolism in this species altered slightly to the inefficient side? Our initial efforts to answer this question have focused on the post-prandial utilization of the dietary amino acids.

Since muscle, the main protein reservoir, is in negative nitrogen balance in the fasting state, the repletion of this storage depends on the net uptake of amino acids in response to protein ingestion. A large variation in the postprandial vs. postabsorptive blood concentration of the amino acids has been largely recognized, with the branched chain amino acids elevated the highest (46,47). Table 2 shows the venous amino acid concentration obtained from 6 young squirrel monkeys after a mixed meal that represented approximately 30% of their daily caloric intake. These results reflect the usual amino acid pattern found in man (46) but with an earlier drop (already evident at 4 hours after the meal), particularly in the branched amino acids. With extended fasting, the highest concentrations in branched chain amino acids were found at 24 hours post-prandially which probably represents an early mobilization from muscle depots as compared to man where this elevation is found after the 3rd day of starvation (48). These preliminary data do not allow further speculation and the quantitation of the amino acid balance between different organs (particularly liver and muscle) must be studied with the use of regional preparations.

Meanwhile a parallel effort has been initiated by evaluating the same 5 parameters measured in the glucose load test, but in response to a "protein-like" meal: arginine infusion (49). Arginine HCl was infused during a 30 minute period via the venous catheter at a final dose of 0.5 g/kg body weight. Figure 8 depicts the response of 7 animals to this stimulus as well as the response of 5 animals to saline solution alone under the same criteria for volume of infusion (8.3 ml/kg/30 minutes). Most striking was the 5-fold glucagon increase over basal values associated with a 1-fold insulin increase similar to that obtained in the glucose loading test.

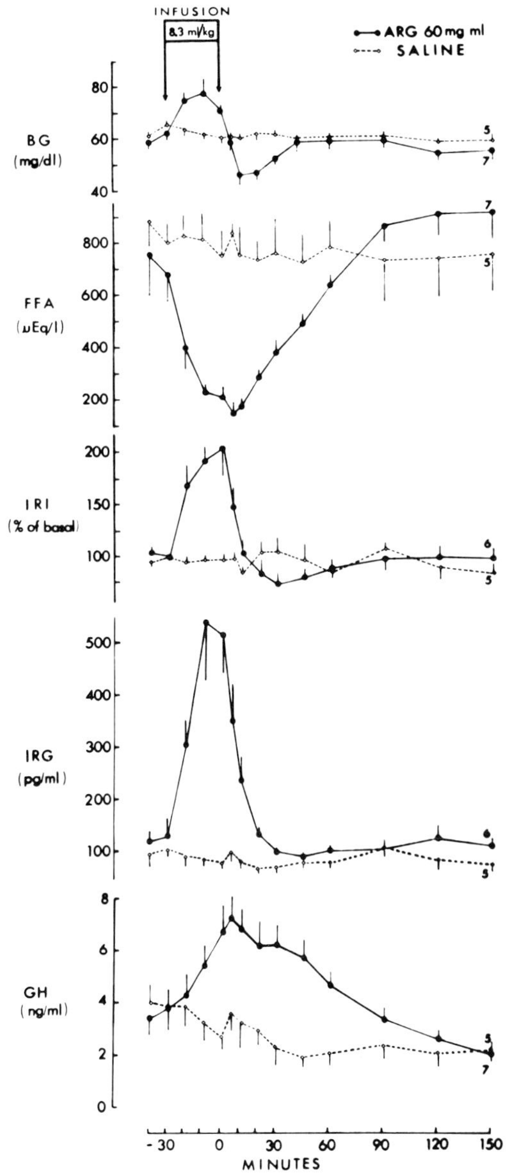

Fig. 8. Response to an arginine HCl infusion at 0.5 g/kg (pH 7.4; 60 mg/ml; infusion period of -30 to 0 minutes). Results of saline loading at the same pH, rate, and volume (8.3 ml/kg/30 minutes) are included for comparison. Vertical bars represent S.E.M. and small numerals represent the animals tested.

TABLE II Venous Plasma Amino Acid Concentrations (μmole/l) in the Squirrel Monkey 2 to 24 Hours After Consuming a Meal

Amino Acid	Hours post prandial				
	2	4	8	12	24
Aspartic Acid	45 ± 4	24 ± 3	31 ± 8	31 ± 10	23 ± 4
Threonine	300 ± 56	134 ± 34	108 ± 28	80 ± 28	104 ± 27
Serine	253 ± 17	131 ± 18	114 ± 29	102 ± 26	82 ± 17
Proline	602 ± 84	276 ± 61	225 ± 60	146 ± 53	98 ± 27
Glycine	384 ± 41	309 ± 50	305 ± 73	275 ± 56	200 ± 32
Alanine	553 ± 116	234 ± 38	251 ± 56	132 ± 15	88 ± 20
Valine	348 ± 71	145 ± 30	108 ± 20	104 ± 40	438 ± 129
Cysteine	35 ± 9	42 ± 9	59 ± 3	48 ± 8	67 ± 4
Methionine	167 ± 20	90 ± 22	64 ± 21	52 ± 11	39 ± 5
Isoleucine	128 ± 33	49 ± 7	46 ± 11	53 ± 24	195 ± 57
Leucine	154 ± 43	60 ± 17	62 ± 18	79 ± 34	256 ± 75
Tyrosine	98 ± 24	37 ± 11	70 ± 28	54 ± 16	44 ± 5

Amino Acid	Hours post prandial				
	2	4	8	12	24
Phenylalanine	109 ± 26	59 ± 12	62 ± 10	65 ± 11	61 ± 4
Ornithine	76 ± 21	56 ± 24	32 ± 5	39 ± 13	64 ± 17
Lysine	263 ± 76	60 ± 23	80 ± 24	72 ± 30	160 ± 42
Histidine	196 ± 32	119 ± 18	112 ± 26	100 ± 23	74 ± 20
Arginine	81 ± 23	36 ± 13	57 ± 21	39 ± 11	70 ± 17

Mean ± SD of 6 animals 28.5 ± 1.5 weeks of age. Animals were studied during the rehabilitation phase 4 weeks after changing from a low protein diet to a control diet containing 13% kcal as protein.

(The collaboration of Dr. Sue Chu in the Amino Acid Analysis is acknowledged).

This bi-hormonal response may explain the fact that the significant rise in blood glucose concentration is of a larger magnitude than can be expected by the equimolar conversion of arginine as a gluconeogenic precursor for glucose (50) and could be ascribed to the enhanced hepatic glucose output following glucagon mediation (51). The decay of blood glucose significantly below basal values following the end of the arginine infusion as well as the decrease in FFA confirms the efficiency of the insulin mediation. No statistically different responses from basal values were found following salt infusion alone for glucose, insulin, FFA, glucagon or growth hormone. The growth hormone response disagrees with the non-specific GH stimulation noted with the same rate of salt infusion in rhesus monkeys (52). The response to arginine shows the typical GH pattern observed in humans (53) although of a lower magnitude at the point of maximal stimulation.

Although most insulin values have been expressed as percent of basal values, levels of 7-10 uU/ml are usually obtained. Coupled with basal glucagon values of 120-130 pg/ml, the insulin:glucagon molar ratio is about 2.3, very similar to a value of 3 found in fasting man (54). By contrast, after arginine infusion, this ratio was 0.9 as compared with the 3-6 found in humans. Although the validity of the insulin: glucagon ratio and its effect on hepatic glucose production is still controversial (55,56), the low ratio found in these animals after arginine infusion would explain the glucose response mentioned previously. Unfortunately, arterial amino acids were not measured in this group of monkeys, but such a low insulin: glucagon ratio would tend to favor hypoaminoacidemia (57) probably as essentially an hepatic effect (58). Whether or not this mechanism is causal in the post-prandial utilization of amino acids in squirrel monkeys remains to be established.

In summary, the squirrel monkey appears to show a high rate of glucose disposal and anti-lipolysis associated with a low release of insulin. This pattern is not surprising in light of its high BMR and relative lack in adipose tissue. On the other hand, independent of the mechanism involved, the low efficiency of dietary protein utilization appears to classify the squirrel monkey as an "hypo-anabolic" mammal similar to another inefficient species, such as the cat (59).

REFERENCES

1. Knobil, E., and Meyer, V. (1968) Observations on the secretion of growth hormone, and its blockade, in the rhesus monkey. *Ann. N.Y. Aca. Sci. 148: 459-470.*
2. Mason, J.W. (1972) Corticosteroid response to chair restraint in the monkey. *Am. J. Phys. 222: 1291-1294.*
3. Ferin, M., Carmel, P.W., Warren, M.P., Himsworth, R.L., and Frantz, A.G. (1976) Phencyclidine sedation as a technique for handling rhesus monkeys: Effects on LH, GH and Prolactin Secretion. *Proc. Soc. Exp. Biol. Med. 151: 428-433.*
4. Oyama, T., and Takiguchi, M. (1970) Plasma levels of ACTH and cortisol in man during halothane anesthesia and surgery. *Anesth. Analg. 49: 363-366.*
5. Oyama, T., Saito, T., Isomatsu, T., Samejima, N., Uemura, T., and Arimura, A. (1968) Plasma levels of ACTH and cortisol in man during diethyl ether anesthesia and surgery. *Anesthesiology 29: 559-564.*
6. Aynsley-Green, A., Biebuyck, J.F., and Alberti, K.G.M.M. (1973) Anaesthesia and insulin secretion: the effects of diethyl ether, halothane, pentobarbitone sodium and ketamine hydrochloride on intravenous glucose tolerance and insulin secretion in the rat. *Diabetologia 9: 274-281.*
7. Cervenko, F.L., and Greene, N.M. (1967) Effect of cyclopropane anesthesia on glucose assimilation coefficient in man. *Anesthesiology 28: 914-919.*
8. Merin, R.G., Samuelson, P.N., and Schalch, D.S. (1971) Major inhalation anesthetics and carbohydrate metabolism. *Anesth. Analg. Curr. Res. 50: 625-632.*
9. Goldsmith, M.W., and Holmes, F. (1957) Thiopentone and carbohydrate metabolism. *Anesthesia 12: 321-325.*
10. Kris, A.O., Miller, R.E., Wherry, F.E., and Mason, J.W. (1966) Inhibition of insulin secretion by infused epinephrine in rhesus monkeys. *Endocrinology 78: 87-97.*
11. Merin, R.G. (1970) The relationship between myocardial function and glucose metabolism in the halothane depressed heart. I. The effect of hypoglycemia. *Anesthesiology 33: 391-395.*
12. Natelson, B.H., Smith, G.P., Stokes, P.E., and Root, A.W. (1973) Changes of plasma glucose and insulin during defense reactions in monkeys. *Am. J. Phys. 224: 1454-1462.*
13. Lang, C.M. (1966) Impaired glucose tolerance in the squirrel monkey (*Saimiri sciureus*) Proc. Soc. Exptl. Biol. Med. 122: 84-85.

14. Davidson, I., Lang, C.M., and Blackwell, W.L. (1967) Impairment of carbohydrate metabolism of the squirrel monkey. *Diabetes 16: 395-401.*
15. Davidson, I.W.F., and Blackwell, W.L. (1968) Changes in carbohydrate metabolism of squirrel monkeys with chromium dietary supplementation. *Proc. Soc. Exptl. Biol. Med 127: 66-70.*
16. Wilson, R.B., Martin, J.M., Kelly, H., and Newberne, P.M. (1971) Plasma and pancreatic insulin concentrations in adult squirrel and rhesus monkeys. *Diabetes 151-155.*
17. Hayes, K.C., Fay, G., Roach, A., and Stare, F.J. (1972) Breeding New World monkeys in a laboratory environment. In: Breeding Primates, ed. W.I.B. Beveridge, *Basel: Karger, 122-129.*
18. Ausman, L.M., Hayes, K.C., Lage, A., and Hegsted, D.M. (1970) Nursery care and growth of old and new world infant monkeys. *Lab. Animal Care 20: 907-913.*
19. Lehner, L.D.M., Clarkson, T.B., and Lofland, H.B. (1971) The effect of insulin deficiency, hypothyroidism, and hypertension on atherosclerosis in the squirrel monkey. *Exptl. Mol. Pathol. 15: 230-244.*
20. Mann, G.V., and Crofford, O.B. (1970) Insulin levels in primates by immunoassay. *Science 169: 1312-1313.*
21. Howard, C.F. (1975) Diabetes and lipid metabolism in nonhuman primates. *Advances in Lipid Research. 13: 91-134.*
22. Andres. R. (1971) Effect of age in interpretation of glucose and tolbutamide tolerance tests. In: Diabetes Mellitus: Diagnosis and Treatment, ed. S.S. Fajans and D.E. Sussman, N.Y. *Am. Diabetes Assoc. 115-120.*
23. Ausman, L.M., and Gallina, D.L. (1978) Response to glucose loading of the lean squirrel monkey in unrestricted conditions. *Am. J. Physiol. 234: R20-R24.*
24. Boden, G., and Soeldner, J.S. (1967) A sensitive double antibody radioimmunoassay for human growth hormone (HGH): Levels of serum HGH following rapid tolbutamide infusion. *Diabetologia 3: 413-421.*
25. Unger, R., and Eisentraut, A. (1967) Glucagon. in: *Hormones in Blood,* ed. C. Grey, and A. Bacharach. Academic Press, Inc., N.Y., 2nd Edition 1: 83-128.
26. Jourdan, M.H. (1972) The influence of different amounts and concentrations of glucose on the oral glucose tolerance test. *Guy's Hospital Rep. 121: 155-162.*
27. Castro, A., Scott, J.P., Grettie, D.P., Macfarlane, D., and Bailey, R.E. (1970) Plasma insulin and glucose responses of healthy subjects to varying glucose loads during three-hour oral glucose tolerance tests. *Diabetes. 19: 842-851.*

28. Müller, W.A., Faloona, G.R., Aguilar-Parada, E., and Unger, R.H. (1970) Abnormal alpha cell function in diabetes: Response to carbohydrates and protein ingestion. *New Engl. J. Med. 283: 109-115.*
29. Pimstone, B., Barbezat, G., Hansen, J.D.L., and Murray, P. (1967) Growth hormone and protein-calorie malnutrition. Impaired suppression during induced hyper-glycaemia. *Lancet 2: 1333-1334.*
30. Glick, S.M., Roth, J., Yallow, R.S., and Berson, S.A. (1965) The regulation of growth hormone secretion. *Recent Progr. Hormone Res. 21: 241-283.*
31. Malinow, M.R., and Wagner, R. (1966) Oxygen uptake in squirrel monkeys *(Saimiri sciurea) Lab. Animal Care 16: 105-108.*
32. Smoake, J.A., Mulvey, P.F., Gerber, M., and Jones, L.G. (1977) Oxygen consumption and thyroid function in the squirrel monkey *(Saimiri sciureus) Laboratory Animal Science 27: 655-659.*
33. Brody, S. (1945) ed. Basal energy and protein metabolism in relation to body weight in mature animals of different species. in: *Bioenergetcis and Growth, N.Y.* Reinhold, 352-403.
34. Coulson, R.A., Hernandez, T., and Herbert, J.D. (1977) Metabolic rate, enzyme kinetics *in vivo*. *Comp. Biochem. Physiol. 56A: 251-262.*
35. Perley, M., Kipnis, D.M. (1966) Plasma insulin responses to glucose and tolbutamide of normal weight and obese diabetic and nondiabetic subjects. *Diabetes 15: 867-874.*
36. Knopf, R.F., Conn, J.W., Floyd, J.C., Fajans, S.S., Rull, J.A., Guntsche, E.M., and Thiffault, C.A. (1966) The normal endocrine response to ingestion of protein and infusions of amino acids. Sequential secretion of insulin and growth hormone. *Trans. Ass. Amer. Phys. 79: 312-321.*
37. Coe, C.L., and Rosenblum, L.A. (1978) Annual reproductive strategy of the squirrel monkey *(Saimiri sciureus). Folia primatol. 29: 19-42.*
38. Bagdade, J.D., Bierman, E.L., and Porte, D. (1967) The significance of basal insulin levels in the evaluation of the insulin response to glucose in diabetic and nondiabetic subjects. *J. Clin. Invest. 46: 1549-1557.*
39. Bjorntorp, P., Berchtold, P., and Tibblin, G. (1971) Insulin secretion in relation to adipose tissue in men. *Diabetes 20: 65-70.*
40. Shreeve, W.W., Hoshi, M., Oji, N., Shigeta, Y., and Abe, H. (1968) Insulin and the utilization of carbohydrates in obesity. *Am. J. Clin. Nutr. 21: 1404-1418.*

41. Bennett, P.H., Burch, T.A., and Miller, M. (1970) Diabetes mellitus in American *(Pima)* Indians. *Lancet 2: 125-128.*
42. Waterlow, J.C. (1969) The assessment of protein nutrition and metabolism in the whole animal, with special reference to man. in: *Mammalian Protein Metabolism.* Vol. III, Academic Press, N.Y., ed. H.N. Munro, 325-390.
43. Energy and protein requirements. World Health Organization. WHO Tech. Rep. Ser. No. 522, Geneva, 1973.
44. Waterlow, J.C. (1968) Observations on the mechanism of adaptation to low protein intakes. *Lancet 2: 1091-1097.*
45. Ausman, L.M., Gallina, D.L., Samonds, K.W., and Hegsted, D.M. (1979) Assessment of the efficiency of protein utilization in young squirrel and macaque monkeys. *Am. J. Clin. Nutr.* In Press.
46. Frame, E.G. (1958) The levels of individual free amino acids in the plasma of normal man at various intervals after a high-protein meal. *J. Clin. Inves. 37: 1710-1723.*
47. Armstrong, M.F., and Stave, U. (1973) A study of plasma free amino acid levels. I. Study of factors affecting validity of amino acid analysis. *Metab. 22: 549-560.*
48. Felig, P., Owen, O.E., Wahren, J., and Cahill, G.F. (1969) Amino acid metabolism during prolonged starvation. *J. Clin. Inves. 48: 584-594.*
49. Fajans, S.S., and Floyd, J.C. (1972) Stimulation of islet cell secretion by nutrients and by gastrointestinal hormones released during digestion. Handbook of Physiology Section 7: *Endocrinology.* American Physiological Society, Vol. I. *Endocrine Pancreas. Wash., D.C. Chapter 30, 473-493.*
50. Felig, P., and Marliss, E. (1972) The glycemic response to arginine in man. *Diabetes 21: 308-310.*
51. Felig, P., Wahren, J., Sherwin, R., and Palaiologos, G. (1977) Amino acid and protein metabolism in diabetes mellitus. *Arch. Intern. Med. 137: 507-513.*
52. Chambers, J.W., and Brown, G.M. (1976) Stimulation of rhesus monkey GH release: Arginine vs plasma volume expansion. *J. Clin. Endocrinol. Metab. 42: 169-172.*
53. Merimee, T.J., Rabinowitz, D., Fineberg, S.E. (1969) Arginine-initiated release of human growth hormone: factors modifying the response in normal man. *New Eng. J. Med.* 280: 1434-1438.
54. Unger, R.H. (1971) Glucagon and the insulin: glucagon ratio in diabetes and other catabolic illnesses. *Diabetes 20: 834-838.*

55. Vranic, M., Ross, G., Doi, K., and Lickley, L. (1976) The role of glucagon-insulin interactions in control of glucose turnover and its significance in diabetes. *Metabolism 25: 1375-1380.*
56. Unger, R.H. (1978) Glucagon in the pathogenesis of diabetes: The status of controversy. *Metabolism 27: 1691-1709.*
57. Marliss, E.B., Aoki, T.T., Unger, R.H., Soeldner, J.S., and Cahill, G.F. (1970) Glucagon levels and metabolic effects in fasting man. *J. Clin. Invest. 49: 2256-2270.*
58. Boden, G., Wilson, R.M., Owen, O.E. (1978) Effects of chronic glucagon excess on hepatic metabolism. *Diabetes 27: 643-648.*
59. Morris, J.G., Rogers, Q.R. (1978) Ammonia intoxication in the near-adult cat as a result of a dietary deficiency of arginine. *Science 199: 431-432.*
60. Dillon, R.S. (1965) Importance of the hematocrit in interpretation of blood sugar. *Diabetes 14: 672-674.*

ACUTE AND CHRONIC EFFECTS OF ETHANOL IN NONHUMAN PRIMATES

Adrianne E. Rogers
James G. Fox
Katherine Whitney
Gregory Lenhart
Albert Wallstrom
Leonard S. Gottlieb

Department of Nutrition and Food Science
Massachusetts Institute of Technology
Cambridge, Massachusetts

Division of Laboratory Animal Medicine
Mallory Institute of Pathology
Department of Pathology
Boston University School of Medicine
Boston, Massachusetts

ISBN 0-12-334250-3

I. ACUTE AND CHRONIC EFFECTS OF ALCOHOL

Ethanol has been ingested for mood alteration and ceremonial purposes since antiquity, and the history of the development of methods of production of alcoholic beverages is of great interest (76). Psychological, physiological, biochemical and pathological results of ethanol ingestion have been studied in humans and several species of experimental animals and are at present an active focus of research in nonhuman primates. Studies in nonhuman primates and other species can be divided into general categories of ethanol effects on: 1) learning, behavior, and CNS development, 2) hepatic structure and function, 3) organs other than liver and brain, and 4) nutritional status. Primates offer obvious advantages over rodents for studies of behavior and learning and have proved receptive to long-term ingestion of large amounts of ethanol. Baboons and several species of *macaques* have been adapted to alcohol ingestion as part of the diet or in response to external stimuli. Evidence of addiction has been obtained in rhesus monkeys, chimpanzees and baboons. Baboons are the only animal species in which severe, probably irreversible hepatic damage, biochemically and morphologically similar to damage found in alcoholic humans, has been induced by alcohol alone, i.e. without the intentional superimposition of malnutrition.

This review includes primarily studies in monkeys of ethanol effects on tissue structure and function which are or may be related to interactions between nutritional status and ethanol toxicity. There are other aspects of alcoholism in humans which are the subject of active research and controversy which are not included, eg. the influence of heredity on development and tissue effects of alcoholism, the role of metabolism of ethanol in the brain, and its effects on neurotransmitters. As primate models of alcoholism are developed, they will undoubtedly be of use in studies of these and other areas.

Definitions of alcoholism in humans are generally based on quantity and duration of ethanol intake and may include behavioral criteria (74,75). Criteria for animal models of alcoholism vary with the interest of investigators and include oral or intragastric administration of ethanol, development of physical dependence and withdrawal and, sometimes, development of psychological dependence and intoxication. Prolonged, voluntary intake which results in significant blood ethanol content has been achieved in several animal species (12,15,16, 74).

Physical dependence on ethanol has been induced in chimpanzees (*Pan troglodytes*) and baboons (*Hamadryas*) by feeding increasing amounts of ethanol in liquid diets and in rhesus monkeys (*Macaca mulatta*) by the same procedure or by repeated intragastric or continuous intravenous administration (33,55, 57). Pieper & Skeen (57) concluded that physical dependence was induced by administration of ethanol in quantities sufficient to maintain blood ethanol above zero for at least 4 days. Chair-restrained rhesus given ethanol by continuous iv infusion for 96 hours to maintain blood ethanol concentration (BEC at 50, 100, 200 or 300 mg/dl) showed increased frequency of extremity tremor over the 33 hours after withdrawal compared to measurements made during periods of restraint without ethanol infusion. The magnitude of the increase in tremor frequency was positively correlated with BEC, in general, but variation was great. Voluntary oral intake of ethanol in drinking water, combined with simultaneous intravenous infusion in chair-restrained rhesus, induced voluntary oral intakes of 3-10.5 g/kg/d. Increased intake persisted for 11 days to 9 weeks in 3 of 5 monkeys and withdrawal symptoms were reported in 1 of the 3 (55,57). Four of 9 baboons, fed ethanol in a liquid diet to supply 50% of calories for an average of 15 months after consumption of smaller amounts of ethanol in drinking water for 17-34 months, developed seizures or other signs of withdrawal when they decreased their ethanol consumption because of respiratory tract infections. Mean ethanol intake was 5.75 g/kg/d, and BEC in clinically inebriated animals as 258-358 mg/dl (35).

Rhesus monkeys given ethanol in formula from birth and then in drinking water from 40 until approximately 1000 days of age did not show signs of withdrawal and did not choose ethanol solutions at the end of that period although they had ingested 4-12 g/kg/d of ethanol and had BEC between 0 and 200 mg/dl with occasional measurements of 300-400 mg/dl (41). Infant chimpanzees given ethanol in liquid diet for 6-10 weeks to supply 45% of calories had seizures or other signs of withdrawal at the end of the time period, and mild withdrawal signs were noted between feedings. Their BEC was greater than achieved in the infant rhesus with peak BEC of 500 mg/dl (58).

II. EFFECTS OF ETHANOL ON REARING, BEHAVIOR AND CNS

Development and retention of tolerance to the effects of ethanol on learning have been demonstrated in rhesus monkeys. Four monkeys, trained to perform a two-choice discrimination-reversal task, were tested daily for 36 days 90 minutes after an ig dose of ethanol, 3 g/kg, which yielded an average BEC of 235 mg/dl. Performance decreased significantly on initiation of ethanol treatment but improved during the testing period and was essentially normal after about 3 weeks. Tolerance was not related to changes in BEC. Subsequent testing after 6-day periods of lactose administration or after 1 year, during which no testing or administration of psychoactive drugs had been performed, showed insignificant decrements in performance at initiation of ethanol treatment which disappeared on subsequent test days (57).

Interactions between behavior patterns and ethanol consumption have been studied in several species of monkeys under varying social or experimental conditions (10,28). For example, *Macaca nemestrina* given free choice of water or 10% (v/v) ethanol in orange juice and living in groups had variable ethanol intakes which ranged between zero and 8.7 g/kg/d and apparently depended on personality and social factors (10). Details of the studies are outside the scope of this review, but they may prove useful for developing primate models in this area.

Developmental abnormalities in children born to alcoholic mothers have been identified and designated as the "fetal alcoholic syndrome". The abnormalities include microcephaly, mental retardation, fine motor dysfunction and craniofacial, musculoskeletal, cardiac and cutaneous anomalies (20,25,51). Fetal death or developmental abnormalities have been induced in guinea pigs, rats and mice exposed to ethanol *in utero* (5,52,60,61); and in rats delayed cerebral cortical and cerebellar differentiation was observed (Jacobson, S. personal communication).

The complexity of studies needed to define the pathogenesis of CNS and other abnormalities related to *in utero* alcohol exposure requires a large animal which closely resembles the human in the time course of CNS development. A primate model for the fetal alcohol syndrome would permit pathogenetic studies, studies of time- and dose-response relationships between ethanol intake and fetal damage, and studies to define the role of malnutrition and intake of drugs other than ethanol in production of the abnormal fetus. There is one report of reproduction in 4 alcohol-fed *Macaca nemestrina*. The monkeys, given free access to 10% ethanol in an orange drink, consumed an average of 3.68 g/kg/d of ethanol before conception and 0-6 g/kg/d during gestation. The monkey which had the highest average intake (4.6 g/kg/d) during gestation produced a tremulous, hyperactive offspring but no other abnormalities were reported (11). Placental transfer of ethanol was demonstrated in *Macaca fasicularis* monkeys given 0.5 g/kg iv in the last month of pregnancy. Unchanged ethanol was detected in fetal blood and other tissues 15 and

90 minutes after injection; the highest concentrations were present in brain and liver (23).

Structural effects of ethanol intake on brain of adult animals have been studied in rodents and recently reported also in rhesus monkeys. Mice given 35-42% of calories as ethanol in a nutritionally adequate diet for 4 months had significant abnormalities of hippocampal neurons by light microscopy (65). Rhesus monkeys, given 5-8 g/kg ethanol per day ig in 2 doses for 16 weeks and fed a natural ingredient diet, had cellular abnormalities in the hippocampus (43). In both cases, as in all studies of the effects of ethanol consumption, the question of nutritional effects arises. The control mice were pair-fed to the alcohol-treated mice; both groups gained weight normally, and previous studies cited had demonstrated normal blood vitamin content in alcohol-fed mice. Nutritional data is not given in the preliminary report of the study in monkeys. The critical relationships between supply of choline and tryptophan, and probably other nutrients, to brain and metabolism synthesis of neurotransmitters (6,50,89) make careful control and analysis of nutrient content important in interpretation of such studies.

III. EFFECTS OF ETHANOL ON HEPATIC STRUCTURE AND FUNCTION

Models used for study of acute and chronic toxicity of alcohol have only partially reproduced the biochemical and histologic pathology associated with alcoholism in humans. The relationships between toxicity of alcohol and altered morphology or function of the liver, kidney, heart, skeletal muscle, spleen, endocrine and exocrine glands and hematologic and immune systems have been studied and disputed throughout this century with the development of experimental pathology. A major question has been and continues to be the relative importance of toxicity of ethanol itself and of other environmental factors associated with alcoholism, of which the most intensively studied has been malnutrition. Nutritional problems in alcoholics include deficiencies of protein, vitamin A, folate, vitamin B_6, zinc and other nutrients as a result of decreased intake, poor absorption and possible increased metabolic demands.

Experimental results obtained in the early 1960s in rats and humans indicated that ethanol itself induced fatty liver and suggested that alcoholic liver disease might be the result of ethanol toxicity rather than, or in addition to, malnutrition (36,62). The observations reported at that time stimulated a great deal of research which has added to and revised theories of the pathogenesis of alcoholic liver disease and provided information on fat metabolism and other aspects of hepatic metabolism. A major contribution has been made by studies in nonhuman primates. The development of primate models of alcoholism, undertaken primarily to permit studies of ethanol effects on the liver, is now being expanded for study of the pathogenesis of lesions in other organs and of psychological responses to ethanol.

Early experimental studies of the roles of ethanol and malnutrition in induction of cirrhosis have been reviewed (21, 59). Male rats fed diets deficient in lipotropes (choline, methionine, vitamin B_{12} and folate) developed fatty liver and cirrhosis which were morphologically and physiologically good models for alcoholic cirrhosis. Superimposition of ethanol ingestion increased the hepatic damage but was not required for development of cirrhosis, and the effects of ethanol could be blocked by suitable choline or protein supplementation of the diet (31,32,77,78,79).

Ethanol feeding alone induced a moderate increase in hepatic triglycerides but did not induce cirrhosis in rats fed a high-fat diet, minimally adequate in choline and low in vitamin B_6 and zinc; hepatic lipid was reduced to or near normal by choline supplementation (3,34). Ethanol or its metabolites apparently enhanced the choline requirement (31, 32) possibly by increasing choline oxidation in the mitochondria (3,82). Patek *etal* (54) reviewed several long-term studies in rats fed adequate diets which provided 18-36% of calories as ethanol and presented their own data from rats fed 36% of calories as ethanol for 19 months. Hepatic lipids increased in some but not all studies, and fibrosis or cirrhosis were not induced. No significant ultrastructural abnormalities related to alcohol feeding were observed.

Primate models for hepatic cirrhosis were sought first by feeding lipotrope deficient diets. The results were not satisfactory, possibly because mature monkeys were used, the number studied was small, or the investigation was too brief (5 months to 1 year) (24,87). Recently Patek *etal* (53) induced cirrhosis in 1 of 5 young rhesus monkeys fed a lipotrope and protein deficient diet for 2 years. A second monkey fed the deficient diet for 14 months and then the same diet with 2% cholesterol added developed cirrhosis more rapidly. In all other studies in which lipotrope deficiency has induced cirrhosis in monkeys, 2% cholesterol has been included in the diet. Wilgram (86) induced fatty liver and cirrhosis in 2 out of 2 cebus monkeys fed for 17 months a diet low in protein and choline with cholesterol added. Rutherford *etal*(70) induced cirrhosis in 4 of 4 rhesus monkeys fed a similar diet for 2 years.

Rhesus monkeys fed a similar diet, with or without administration of ethanol, 5 g/kg ig 3 times a week, developed portal hypertension and abnormal liver function tests after 4 to 8 months. Two animals died after 8 months; 1 fed the diet alone had fatty liver and cirrhosis and the other fed the diet and given ethanol had severe fatty liver. Three monkeys given the same amount of ethanol but fed a natural ingredient diet supplemented with choline and methionine had no portal hypertension or abnormality of liver function. In both dietary groups ethanol administration was associated with gastritis and upper gastrointestinal bleeding, but in neither group was there evidence that ethanol caused structural liver damage (7).

Nath *etal* (49) attempted to demonstrate hepatotoxicity of ethanol in combination with other agents by administering ethanol ig, 5 g/kg/d divided into 2 doses, to rhesus monkeys fed an apparently nutritionally adequate natural ingredient diet or an 8% casein diet and given iron dextran, 50 mg iv, 3 times a week until 0.5 g/kg had been given. Monkeys survived 11-110 weeks and at autopsy had extensive hepatic siderosis which was increased by ethanol treatment. Ethanol induced a moderately fatty liver, which was more marked in monkeys fed 8% casein, but no monkeys had fibrosis or cirrhosis.

The studies discussed above suggested that fatty liver and cirrhosis, analagous to human alcoholic liver disease could be induced in nonhuman primates by combined treatment with lipotrope-deficient, high cholesterol diets and that ethanol might enhance the damage but could be blocked by choline and methionine.

The first report of induction of cirrhosis by ethanol in a nonhuman primate (baboon) was published in 1974 by Rubin and Lieber (69), and their laboratories have continued to publish interesting and valuable data from the model as they have refined it. Other investigators have induced fatty liver, alteration of serum lipids and myocardial abnormalities in rhesus monkeys by ethanol treatment, but there have been no other reports of cirrhosis.

To induce cirrhosis Rubin and Lieber fed adolescent baboons (12 female, 1 male) a purified diet for 9-18 months in which 50% of calories was supplied by ethanol. Six of the animals had been given ethanol in drinking water to supply 36% of total calories and fed a natural ingredient diet for 2-3 years before introduction of the purified diet and had not developed cirrhosis. Livers were fatty after consumption of the liquid diet and ethanol for 1-2 months. Subsequently lesions resembling alcoholic hepatitis were found in 4 monkeys,

3 of which had been exposed to ethanol in liquid diet for 9 months; and fibrosis or cirrhosis were present in 1 animal each, exposed for a total of 4 years. Ethanol consumption was 4.5-8.3 g/kg/day. Serum glutamic-oxalacetic transaminase (SGOT) was elevated in some baboons with fatty liver and in all with cirrhosis or hepatitis. The authors concluded that hepatic damage was the result solely of ethanol toxicity and was not related to nutritional status of the monkeys because they had maintained their weights and normal peripheral blood counts, serum proteins and electrolytes (69). In another publication, 16 baboons were fed 50% of calories as ethanol for 8-22 months and failed to gain weight although pair-fed controls gained significantly. Ethanol-fed baboons had reduced hemoglobin and hematocrit, increased serum cholesterol, clinical inebriation with BEC 258-358 mg/dl, elevated SGOT and developed fatty liver (9 baboons), hepatitis, and fibrosis (7 baboons) (33,35). In other studies the investigators described development of increased collagen and peptidylproline hydroxylase in baboons with a histologic diagnosis of fatty liver and a progressive increase in the width of the connective tissue surrounding terminal hepatic venules which correlated with later development of fibrosis or cirrhosis (13,84). Enlarged abnormal mitochondria and increased, vesicular SER were seen in hepatocytes in EM studies (69).

The composition of the liquid diet fed in these studies meets most nutrient recommendations for nonhuman primates (48). However, it is low in manganese and does not contain chromium. The vitamin mix is low in pantothenate and niacin and does not contain vitamin D, but these vitamins were supplied in a vitamin tablet. The vitamin D is present as D_2 which probably is satisfactory for baboons although D_3 is recommended for all nonhuman primates. It is unlikely that the dietary deficiencies noted contributed to the development of liver disease. However, questions which remain to be answered are: 1) the adequacy of the lipotrope content of the diet; 2) the effect of ethanol on nutrient requirements; 3) the effect of the extremely low dietary carbohydrate content on fat metabolism and other aspects of metabolism, and 4) the effect of added L-cystine.

Lipotropes: Data in rats indicate that ethanol feeding markedly increases the choline requirement, as previously discussed; the other lipotropic nutrients which interact with choline may be affected similarly. The primate requirement for choline is not established although the other major lipotropes, methionine and folate, are, of course, required; 0.2-0.35% choline is usually added to primate diets (48).

The baboon diet contains only 0.04% choline chloride, which is well below the rat requirement. The metabolic interactions between the lipotropes and between them and other nutrients are extensive and have been studied in many species including humans (17,50). Requirements are influenced most notably by dietary fat but also by the total lipotrope, protein, and caloric content of diets, as well as by content of several vitamins. The choline requirement may increase if methionine or folate is limited in the diet or poorly absorbed. Primates, like rats, may have an increased requirement with ethanol ingestion.

Other Nutrients: The contribution of malnutrition on cirrhosis in humans has been questioned, but evidence in patients given jejunoileal bypass for treatment of obesity supports the concept that malnutrition can induce cirrhosis even in the absence of alcoholism. In a review of 88 patients given bypass surgery, Maribbio *etal* (39) reported that fatty change was present in livers of 68% of patients at the time of surgery and in 94% 1 year later associated with an increased amount of fat. Centrolobular, pericellular fibrosis was present in 8.6% before surgery and in 48% 1 year later. The fibrosis persisted and extended to form bands joining central and portal areas in 6.8% and to development of cirrhosis in 3.4% at 1 year. Centrolobular fibrosis appears to be a precursor of cirrhosis in alcoholics as well (84). Maribbio *etal* postulated that biochemical abnormalities related to morbid obesity and exacerbated by post-bypass malabsorption, fatty acid mobilization, and protein deficiency were the cause of the structural damage to the liver. Protein malabsorption and malnutrition, deterioration of liver function, and development of hepatic steatosis and fibrosis have been demonstrated in other studies following bypass surgery (44). Similarly interactions between fat metabolism and intake and absorption of protein and other nutrients may be the basis for development of cirrhosis in alcoholic patients.

Ethanol, like fat, is metabolized to acetate, and influences metabolism of fatty acids and triglyceride synthesis (27,34). Effects of ethanol on lipid metabolism include mobilization of fatty acids from lipid stores, increased hepatic synthesis of fatty acids and tyiglycerides, and possibly decreased hepatic triglyceride release. The effects vary with alcohol dosage and duration of administration (1). Plasma lipids and lipoprotein composition and distribution are markedly altered in alcoholics; this may be at least partially the result of the reduced lecithin-cholesterol acyl transferase (LCAT) activity in the serum associated with alcoholic hepatitis (4,71,72).

Monkeys given ethanol iv, 2.5 g/kg in 30-45 minutes, which yielded a BEC of 300-390 mg/dl, had elevated plasma free fatty acids and slightly decreased plasma triglycerides up to 6 hours (47). Administration of Triton WR-1339 to block lipoprotein lipase (14) indicated that ethanol reduced hepatic triglyceride release (46).

Rhesus monkeys fed 36% of calories as ethanol for 3 months in a liquid diet, which was stated to be nutritionally adequate, developed fatty liver but no evidence of fibrosis (85). In contrast to the results of Nair *etal* (47) serum triglycerides were increased. Hepatic triglyceride synthesis, determined *in vivo* by incorporation of labeled fatty acid, appeared increased, but release was not measured.

Dietary Carbohydrate: The possible contribution of the extremely low carbohydrate content of the diet fed to the baboons has been discussed with respect to earlier studies in ethanol-fed rats (21). Carbohydrate supplied 11% of calories given to ethanol-fed baboons and 61% of calories given to controls. Purified diets fed to laboratory primates generally supply 50-73% of calories as carbohydrate (48). The effect of the low carbohydrate intake on the liver is unknown.

L-Cystine: A fourth question which arises with respect to the liquid diet fed to the baboons is the contribution of L-cystine to lipotrope requirements. This amino acid is added to many lipotrope-deficient diets used to induce cirrhosis in rats because it enhances the deficiency, possibly by increasing growth (17).

IV. EFFECTS OF ETHANOL IN OTHER ORGANS

In humans the cardiovascular, gastrointestinal and central nervous systems all may be severely affected by excessive intake of ethanol, and other organ systems also may show its effects (74,75). Cardiovascular disease associated with alcoholism was recognized initially in patients with thiamin and other clinically apparent nutritional deficiencies which responded to thiamin treatment. In the U.S. today it appears that only 10% or less of heart disease related to alcoholism is associated with clinical nutritional deficiency; the majority of cases may result from the myocardial toxicity of ethanol, the myocardium possibly rendered more susceptible to toxicity by subclinical nutritional deficiency (63,80,88).

Alcoholic cardiomyopathy is characterized clinically by congestive failure and arrhythmias and morphologically by cardiomegaly, interstitial fibrosis and occasionally myocytolysis (18,63).

Acute toxic effects of ethanol on the myocardium demonstrated in humans and dogs include decreased contractility, loss of cations and triglyceride accumulation. Chronic toxicity in dogs fed 36% of calories as ethanol for 18 months was manifest by abnormal left ventricular pressure-volume relationships, conduction delay, accumulation of glycoprotein-like material, dilatation of the sarcoplasmic reticulum and triglyceride accumulation (63,64). Studies in rats gave similar morphologic and functional results, but were less extensive (63). Adult male rhesus monkeys, fed by gastric intubation a purified liquid diet stated to be nutritionally adequate which provided 40% of calories as ethanol for 3 months, gained weight and showed no clinically detectable abnormalities of cardiovascular or other system function. At autopsy the hearts of ethanol-fed monkeys were flacid and slightly increased in weight compared to control monkeys fed the diet alone. Myocardial fibers contained lipid droplets and showed variations in size and staining, focal fragmentation and fibrosis. Livers were fatty but had no fibrosis. Increased triglyceride content was measured in plasma, liver and heart (85). In our experiments described below, the hearts of the 2 monkeys that died showed no significant abnormalities.

Morphologic and biochemical effects of ethanol in the CNS and other organ systems have begun to be characterized in primates (see below) and psychological and pharmacological interactions with other drugs have been studied. Research into the interactions of alcohol, narcotics and other drugs with biogenic amines and acetylcholine, which may be expressed through alterations in neurotransmitter content, has been reviewed recently (45,74). Most studies have been performed in rats, mice and hamsters, but primates are being used and offer greater potential for studies relating morphology, biochemistry and behavior. Behavioral effects of ethanol are complex, and the extension of studies into interaction with drugs has yielded complex results which require detailed pharmacologic and biochemical studies for resolution (2,29, 30,45). Two observations which may be of interest on further study are moderate reduction of CSF pressure in baboons following iv infusion of ethanol(1.6 mg/kg in 30 minutes) (8) and retarded dark adaptation in rhesus monkeys given 3.4 g/kg ethanol iv over 90 minutes (83).

The pathogenesis of renal failure in alcoholics is not established but appears to be the result of cardiovascular, hormonal, mechanical and other changes elicited by end-stage liver disease (40,73). Primate models which can be induced to develop severe cirrhosis with its attendant fluid and electrolyte abnormalities will be of great value in elucidating the pathogenesis and treatment of the disorders. The hyponatremia which has appeared in alcohol-fed rhesus monkeys (see below) may be dilutional and indicate expansion of plasma volume which precedes more severe changes in humans.

Similarly, primate models will be useful in exploration of pancreatic, immunological, hematological and other abnormalities associated with alcoholic cirrhosis in humans.

V. EFFECTS OF ETHANOL ON NUTRITIONAL STATUS

This point focuses again on the interaction of nutrition and alcohol in the production of disease. Alcoholism may be associated with malnutrition through decreased intake of nutrients because of economic and organizational factors in obtaining food, because it is supplying calories, or through toxicity to the gastrointestinal tract which decreases digestion and absorption of food ingested. There are studies in humans which demonstrate inhibition of nutrient absorption of D-xylose, thiamin, folate, vitamin A, vitamin B_{12}, fat and methionine under some, but not necessarily all, conditions of testing (42,38). Nutrient utilization also may be decreased by ethanol. *Macaca radiata* given 50% of calories as ethanol in the diet for 1 year wasted a greater fraction of parenterally administered folate and incorporated a smaller fraction into hepatic polyglutamates than control monkeys (81). In our studies (see below) serum and red cell folate content have generally been higher in alcohol-fed than in control monkeys, an observation which does not support inhibition of folate absorption and utilization by ethanol feeding. The differences in dietary composition may be responsible. Ethanol induced reduction of pyridoxine phosphorylation in human alcoholics (22). The decreases in absorption and utilization of these nutrients in primate models of alcoholism may be significant in studies of the pathogenesis of cirrhosis.

VI. CURRENT STUDIES

Experiments in rhesus monkeys to examine the question of interactions between dietary content of lipotropes and ethanol-induced liver damage are in progress.

Twelve male rhesus monkeys initially weighing 6.1 ± 0.4 kg (mean ± S.E.) have been fed a liquid, supplemented (S) diet or a diet marginally deficient (D) in lipotropes for 4 years (Table 1). Within each group, 2 control monkeys have been fed diet alone and 4 were given 40% of calories as ethanol (substituted isocalorically for sucrose) for 9-14 months followed by 50% of calories as ethanol for 2 years, with one exception: one deficient monkey died after approximately 1 year of ethanol feeding (40% of calories), and its replacement was given 40% of calories as ethanol for only 2 months and then increased to 50% at the same time as the other monkeys. Increasing ethanol to 50% of calories required reduction of both Wesson Oil and sucrose (Table I). The total exposure to ethanol for all monkeys except the last has been approximately 40 months including the introductory period before an intake of 40% of calories was reached.

The deficient diet is similar to a lipotrope-deficient diet previously used in rats but contains vitamins and minerals in amounts recommended for monkeys and vegetable oil in place of tallow (66). Amino acids are added to the supplemented diet to make it equivalent in essential amino acids to a diet containing 20% casein; folate content of the S diet has been 4-6 mg/kg, and 0.5% choline chloride is added compared to 0.1% in the deficient diet.

Nutritional and hepatic status of the monkeys has been monitored by body weight, serum chemistries, blood counts, serum and red cell folate (Schwartz-Mann, folate radioassay) serum vitamin A (measured fluorimetrically (19)) and needle biopsies of liver for examination by light and electron microscopy. Food intake has been measured daily, but animals have not been pair-fed. Other measurements have been made at approximately 6 month intervals and representative results are given in tabulated data. Control monkeys fed supplemented diet (S monkeys) have gained weight while 3 of 4 supplemented monkeys given ethanol (SA monkeys) have maintained their weight or gained slowly and one has apparently stabilized at 80% of his initial weight (Figure 1). Mean daily caloris intake has been slightly greater in SA than in S monkeys for the past 20 months, but was lower prior to that time (Table II).

Monkeys fed the deficient diet alone (D monkeys) have gained weight and appear healthy. Deficient monkeys fed

TABLE I Composition of Supplemented or Deficient Diets Fed During Alcohol Studies in Rhesus Monkeys

Component	Diete (g/100g) Supplemented (S)	Deficient (D)
Methanol Extracted Peanut Meal	11.9	11.9
Gelatin	5.9	5.9
Vitamin Free Casein	3.0	3.0
Fibrin	1.0	1.0
Amino Acids[1]	9.0	----
L-Cystine	0.5	0.5
Wesson Oil[2]	14.8	14.8
Corn Oil	3.0	3.0
Sucrose[2]	41.4	50.8
Non-Nutritive Fiber	2.0	2.0
Modified Hegsted Salts[3]	5.0	5.0
Vitamins[4]	2.0	2.0
Choline Chloride	0.5	0.1
Vitamin B_{12}	0.000005	0.000005
Folic Acid	0.00005	0.00005

[1]Percent by Wt. in diet: L-Arginine, 0.63; l-Glutamic Acid, 2.28; L-Isoleucine, 0.60; L-Leucine, 1.17; L-Lysine, 0.92; DL-Methionine, 0.92; L-Phenylalanine, 1.22; L-Threonine, 0.33; L-Tryptophan, 0.14; L-Valine, 0.79.

[2]When ethanol is added to supply 50% of calories, Wesson Oil is reduced to 80%, and sucorse is reduced to zero in the supplemented diet and to 9% in the deficient diet. Peanut meal supplies approximately 3% carbohydrates in the diet.

[3]Hegsted IV Mineral Mix (Teklad test diets, Madison Wisconsin) supplemented per kg with: Zinc Chloride ($ZnCl_2$), 0.27 g; Chromium acetate hydrated $Cr(C_2H_3O_2)_3 \cdot H_2O$,o.178 g; Sodium selenite anhydrous (Na_2SeO_3) 0.00436 g; Magnesium sulfate ($MgSo_4 \cdot 7H_2O$), 249.55 g. Minerals supplied by diet: Ca, 0.55%; Cl, 0.41%; Mg, 0.14%, P, 0.28%; K, 0.58% Na, 0.26%; S, 0.06%; per kg: Cr, 1.5 mg; Cu, 3.0 mg; I, 24.4 mg; Fe, 183 mg; Mn, 49 mg; Se, 0.08 mg; Zn, 10.0 mg.

[4]Supplies per kg diet: A, 30,000 IU; D_2, 3000 IU; D_3, 1600 IU: E, 100 IU; K_3, 1.0 mg; Biotin, 0.6 mg; Niacin, 127 mg; Pantothenate (Ca): 30 mg; B_6, 10 mg; Riboflavin, 10 mg; Thiamin, 10 mg; C, 1267 mg.

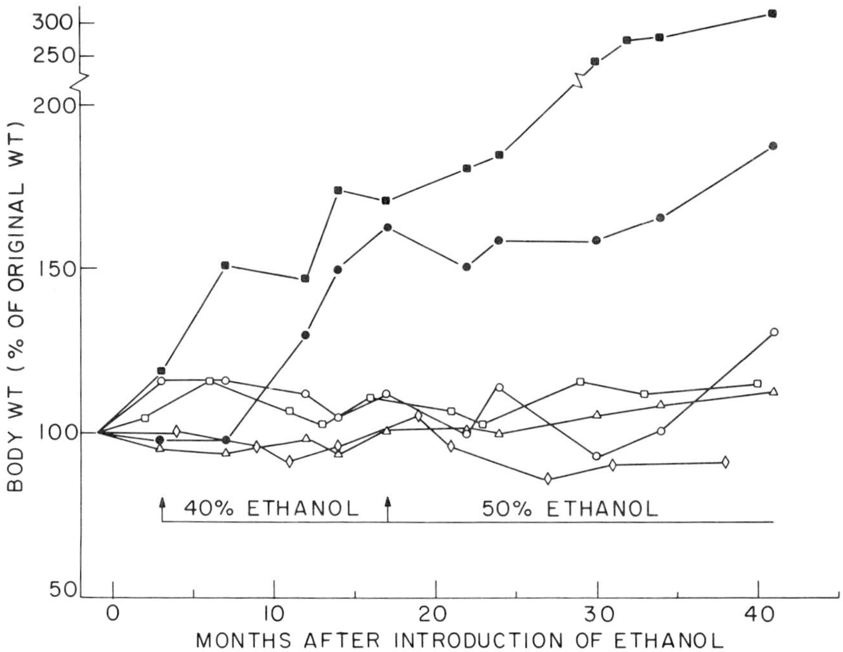

Fig. 1. Body weight of S (closed dots) and SA (open dots) monkeys expressed as percent of initial weight which was 6 ± 0.4 kg.

TABLE II Daily Caloric Intake of Supplemented (S) or Deficient (D) monkeys with or without Dietary Ethanol

Treatment N	Months after introduction of ethanol 1-8	9-16*	17-24	25-32+	33-40
			kcal/kg body wt.		
S (2)	108	90	60	59	56
SA (4)	82	60	59	64	64
D (2)	71	68	56	48	58
DA (4)	77	66	64	70	82

*Ethanol increased from 40% to 50% of calories at this time
+Coconut oil replaced the Wesson Oil (soybean-cottonseed oils) at 30 months in all diets; D and DA diets contained no choline but included 0.1% cholesterol between 30-34 months.

alcohol (DA) monkeys have not gained consistently (Figure 2). Two DA monkeys have died in addition to the one that died earlier. Both deaths followed a dietary alteration introduced to increase liver damage. Thirty months after introduction to ethanol, choline was omitted from the deficient diets (D + DA); 0.1% cholesterol was added, and coconut oil was substituted for Wesson Oil. The new diet was fed for approximately 5 months during which time 2 of the 4 DA monkeys developed marked edema of legs and scrotum; 3 lost weight, and 2 died. One of the 2 had been losing weight before the diet alteration; the other was one which had become edematous. Caloric intake by DA monkeys has been greater than by D monkeys for the past 2 years (Table II).

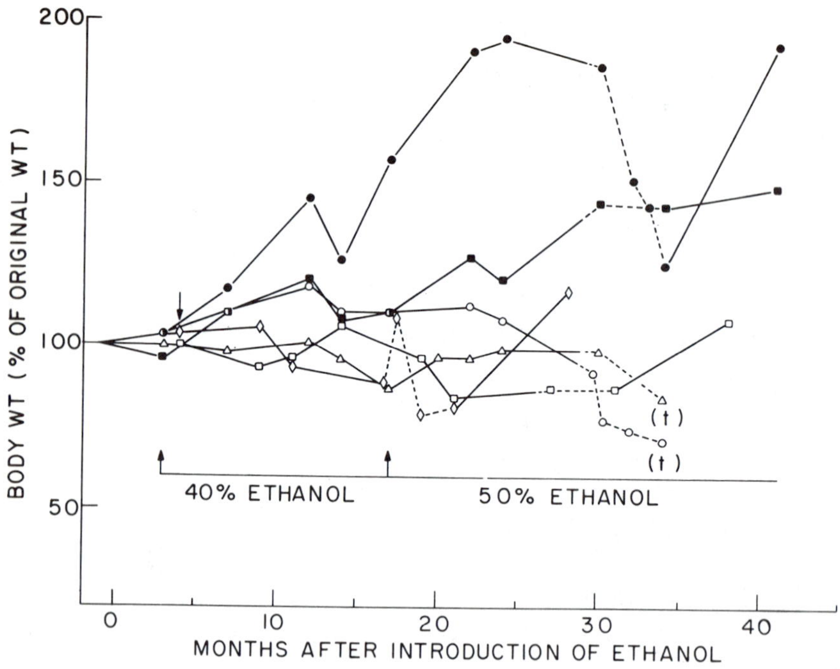

Fig. 2. Body weight of D (closed dots) and DA (open dots) monkeys expressed as percent of initial weight which was 6.2 ± 0.4 kg. Dotted line indicates period when diet altered by deletion of choline, addition of 0.1% cholesterol and substitution of coconut oil for Wesson Oil.

Blood ethanol was measured on one occasion, 9 months after ethanol has been raised to provide 50% of calories, and was 60-250 mg/dl 2 hours after the monkeys had ingested 1.8-4.2 g ethanol/kg in one hour of diet consumption. Monkeys ingesting ethanol are less active following feeding and lose some muscular coordination, but do not lose consciousness. Ethanol and diet intake have been decreased for 24-48 hours following liver biopsy without participation of tremor or other signs of withdrawal. During part of the period the monkeys are, of course, under the influence of ketamine hydrochloride used to tranquilize them for the procedure.

A. Serum Chemistries and Blood Counts

These values have been normal with the following exceptions. Hemoglobin and hematocrit have been consistently lower in SA than in S animals and in DA than in D animals since approximately 12 months after they began to ingest ethanol. MCV was slightly increased before the blood counts were first noted to be low; serum and red cell folate content was decreased; (Table III) serum iron was normal. Dietary folate was increased to 6 mg/kg in S diet and 3 mg/kg in D diet at that time. Serum and red cell folate rose in all groups, but the mild anemia in SA & DA monkeys has persisted. D monkeys became mildly anemic when the diet was made more severely deficient.

Serum transaminases (SGOT, SGPT) have been highly variable but show no consistent abnormality in S or SA. Gamma glutamyl transpeptidase (GGTP) has been elevated in 2 of the 4 SA monkeys at one or more times (Figure 3).

In 2 of the 4 DA monkeys these enzymes became elevated approximately 1 year after introduction to ethanol and increased in D and DA monkeys when the diet was made more severely deficient at 30 months (Figure 4).

Serum albumin dropped in DA monkeys after the dietary change at 30 months but has returned toward normal with refeeding the original D diet at 35 months. Bilirubin and alkaline phosphatase have remained within the normal range in all groups; serum sodium decreased in SA & DA monkeys, while other electrolytes and BUN remained normal (Table IV).

Serum content of vitamin A was normal in all animals (49-86 g/dl) 1 year after introduction of ethanol and again 7 months after ethanol intake was raised to 50% of calories. It has not been measured subsequently. Hair zinc content, measured by atomic absorption, was 143-257 mg/kg with an average of 180 mg/kg 10 months after ethanol intake was raised to 50% of calories. Hair zinc content of 6 monkeys fed a natural ingredient diet in another study and measured at the same time was 241-734 mg/kg with an average of 363 mg/kg. The zinc

TABLE III Hematology of Supplemented (S) or Deficient (D) Monkeys with or without Ethanol (A)

Assay & treatment	Months after introduction of ethanol				
	3	12	24*	34*	39
Hematocrit, % **					
S	36 ± 2	40 ± 1	41 ± 1	42 ± 1	42 ± 1
SA	35 ± 0	38 ± 1	36 ± 1	36 ± 2	37 ± 1
D	35 ± 1	40 ± 1	41 ± 1	34 ± 4	38 ± 2
DA	34 ± 1	39 ± 1	36 ± 1	32 ± 2	35 ± 0
Hemoglobin, gm; dl **					
S	13.0 ± 0.8	13.6 ± 0.1	11.1 ± 0.9	12.2 ± 0.2	12.8 ± 0.1
SA	13.1 ± 0.3	11.1 ± 0.2	9.7 ± 0.4	10.8 ± 1.0	11.1 ± 0.4
D	13.4 ± 0.2	14.0 ± 0.1	11.2 ± 0.2	10.5 ± 1.0	11.8 ± 0.4
DA	13.5 ± 0.7	13.3 ± 0.3	10.2 ± 0.2	9.5 ± 0.7	10.0 ± 0.2
Mean cell volume,^{3}					
S	70 ± 0	68 ± 0	67 ± 1	74 ± 2	71 ± 0
SA	70 ± 1	71 ± 2	72 ± 2	76 ± 1	72 ± 0
D	69 ± 2	69 ± 1	72 ± 2	70 ± 1	68 ± 0
DA	66 ± 1	73 ± 1	71 ± 3	77 ± 2	69 ± 3

Assay & treatment	Months after introduction of ethanol				
	3	12	24*	34*	39
Serum folate, ng/ml					
S	7.0 ± 2.0	3.2	6.8 ± 2.0		
SA	13.7 ± 1.7	6.2 ± 1.4	12.5 ± 4.1		
D	0	< 1.0	6.6 ± 1.3		
DA	2.3 ± 2.3	1.3 ± 0.1	8.3 ± 2.0		
RBC folate, ng/ml					
S	87 ± 79	79 ± 4	337 ± 16		
SA	411 ± 76	149 ± 29	437 ± 21		
D	205 ± 23	66 ± 4	416 ± 24		
DA	157 ± 66	68 ± 3	414 ± 14		

*Ethanol increased from 40% to 50% calories by 14 months.

+Coconut oil replaced Wesson oil at 30 months in all diets; D and DA diets contained no choline but included 0.1% cholesterol between 30-34 months.

**By analysis of variance the decreased hematocrit in SA and DA monkeys at 24, 34, and 39 months is significant ($P < 0.05$); the hemoglobin is significantly lower in SA and DA monkeys at 24 and 39 months and in D monkeys at 39 months.

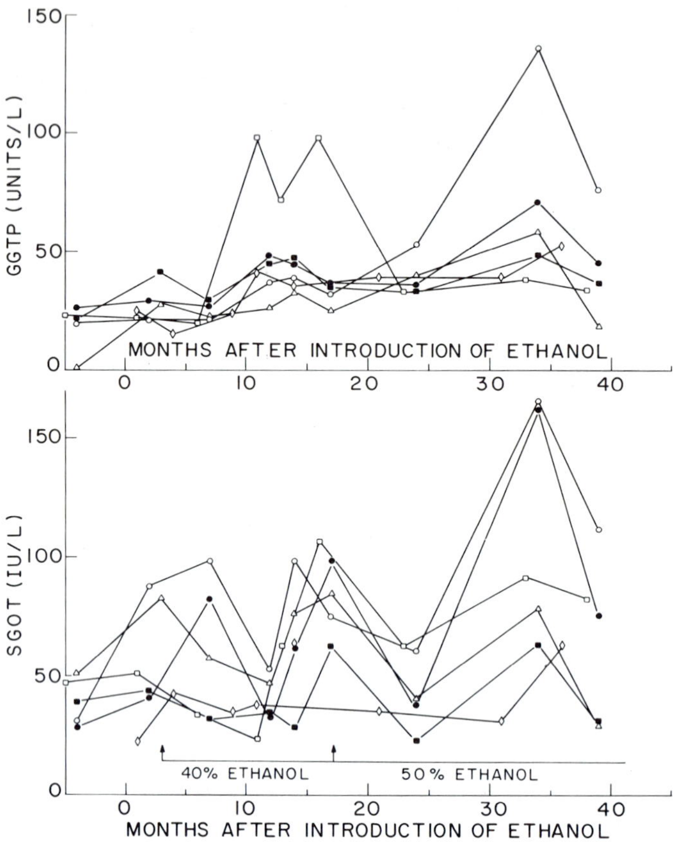

Fig. 3. Serum GGTP and SGOT in S (closed dots) and SA (open dots) monkeys.

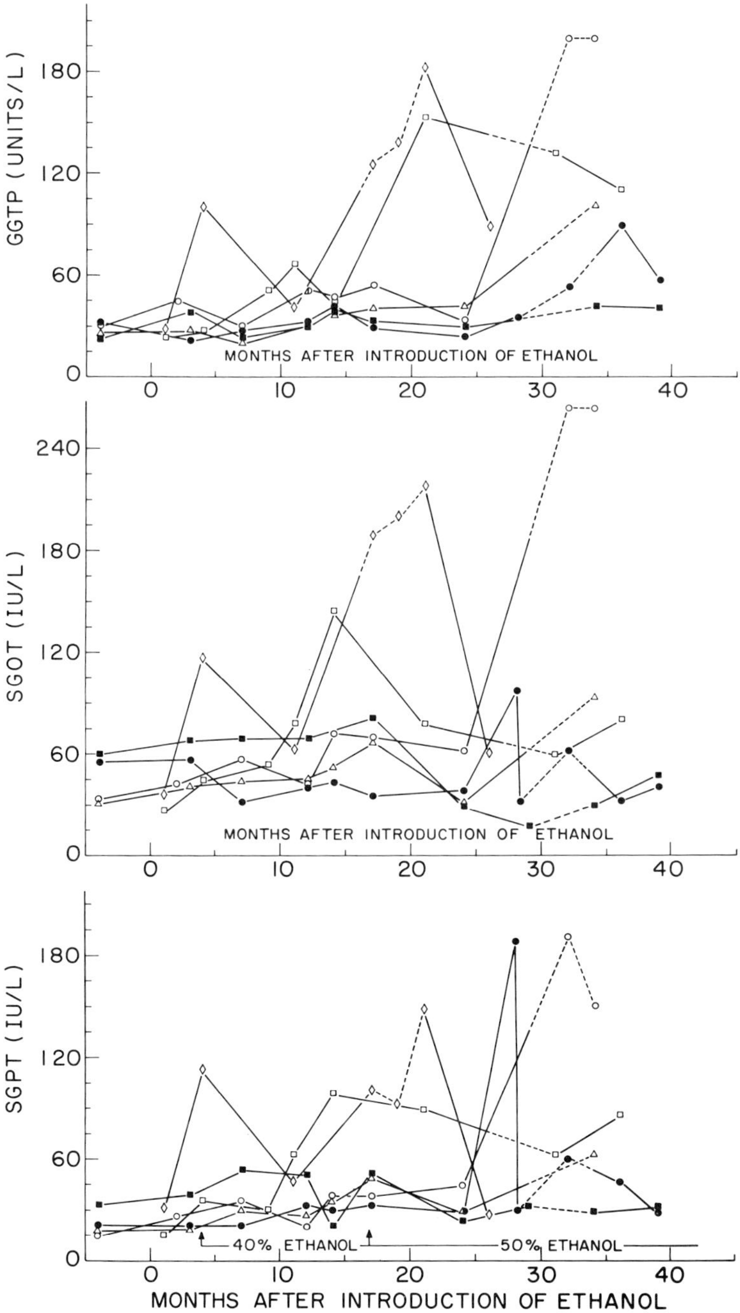

Fig. 4. Serum GGTP, SGOT and SGPT in D (closed dots) and DA (open dots) monkeys.

content of both SD diets was subsequently raised from 5 to 10 mg/kg which is the content recommended in the recent NAS-NRC publication (48).

B. Serum Lipids

Fasting serum triglycerides have been higher in SA monkeys than in S monkeys on all but one measurement, but differences have not been statistically significant. Serum cholesterol has shown no consistent differences in the 2 groups (Table V). On one occasion postprandial serum triglyceride content was measured and again was higher in SA (193 ± 54 mg/dl) than in S (60 ± 4 mg/dl) monkeys. Cholesterol distribution in lipoprotein fractions showed no effect of ethanol. (Assay kindly performed by K.C. Hayes and R. Nicolosi).

Thirty months after introduction of ethanol, dietary lipid was changed from Wesson Oil to coconut oil, to determine whether saturated fat would enhance the effect of ethanol on serum or hepatic lipids. Since that time, serum cholesterol has risen in all groups (Table V). Serum cholesterol has been lower in D monkeys than in S monkeys throughout the experiment, but DA monkeys have not shown this dietary effect; triglycerides were not affected by diet or ethanol in D or DA groups (Table V).

C. Hepatic Morphology

Needle biopsies of liver have been taken at intervals of 5-12 months. Occasional fatty hepatocytes have been present in SA monkeys since 1 month after inclusion of ethanol in the diet to supply 40% of calories, but there has been no evidence of fibrosis or cirrhosis. S monkeys have had no visible fat in the liver until the most recent biopsies which were taken at month 34, approximately 4 months after substitution of coconut oil for Wesson Oil. Both S and SA monkeys had small clusters of fatty hepatocytes (Figure 5) and 1 of the 4 SA monkeys had more marked fat accumulation (Figure 6). There was no evidence of fibrosis or cirrhosis in any monkey fed the supplemented diet. Ultrastructural changes are described below.

Liver biopsies of D monkeys contained no visible fat until approximately 4 months after alteration of the diet by deletion of choline, addition of 0.1% cholesterol and substitution of coconut oil for Wesson Oil. Liver tissue obtained in 1 of the 2 D monkeys was moderately fatty similar to that demonstrated previously in Figure 6. DA monkeys have had scattered

TABLE IV Serum Sodium, Potassium and Blood Urea Nitrogen from Supplemented (S) or Deficient (D) Monkeys with or without Ethanol (A)

Assay & treatment	Months after introduction of ethanol				
	3	12*	24	34+	39
			(mean ± SE)		
Sodium, mmol/l					
S	145 ± 1	146 ± 1	144 ± 8	144 ± 4	143 ± 0
SA	145 ± 3	146 ± 1	140 ± 1	133 ± 6	140 ± 1
D	145 ± 4	146 ± 1	146 ± 2	142 ± 10	148 ± 2
DA	145 ± 3	145 ± 2	138 ± 3	134 ± 3	135 ± 0
Potassium, mmol/l					
S	3.6 ± 0.2	4.0 ± 0.1	3.2 ± 0.2	4.4 ± 0.8	4.6 ± 0.2
SA	3.6 ± 0.6	4.2 ± 0.2	3.5 ± 0.2	3.7 ± 0.5	5.1 ± 1.1
D	3.7 ± 0.1	3.6 ± 0.2	3.4 ± 0.0	4.0 ± 0.2	4.0 ± 0.2
DA	4.2 ± 0.2	4.0 ± 0.1	3.8 ± 0.4	4.3 ± 0.5	3.9 ± 0.0
Blood urea nitrogen, mg/dl					
S	14 ± 2	22 ± 3	20 ± 4	13 ± 0	14 ± 1
SA	23 ± 3	12 ± 1	17 ± 2	13 ± 1	17 ± 1
D	13 ± 1	12 ± 2	12 ± 2	10 ± 2	12 ± 2
DA	18 ± 1	11 ± 0	13 ± 2	12 ± 1	14 ± 0

*Ethanol increased from 40% to 50% calories by 14 months.

+Coconut oil replaced Wesson oil at 30 months. In all diets; D and DA diets contained no choline but included 0.1% cholesterol between 30-34 months.

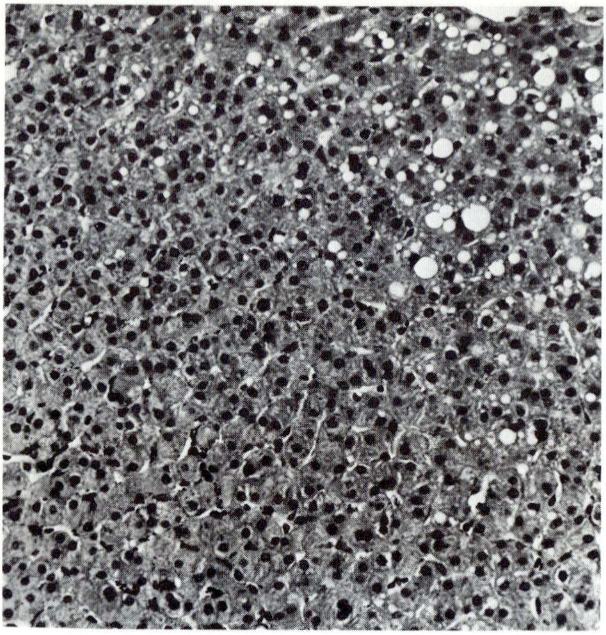

Fig. 5. Liver biopsy from monkey fed S diet without ethanol for approximately 3 1/2 years, with substitution of coconut oil for Wesson Oil 4 months prior to biopsy, reveals clusters of fatty hepatocytes (H + E, x 100).

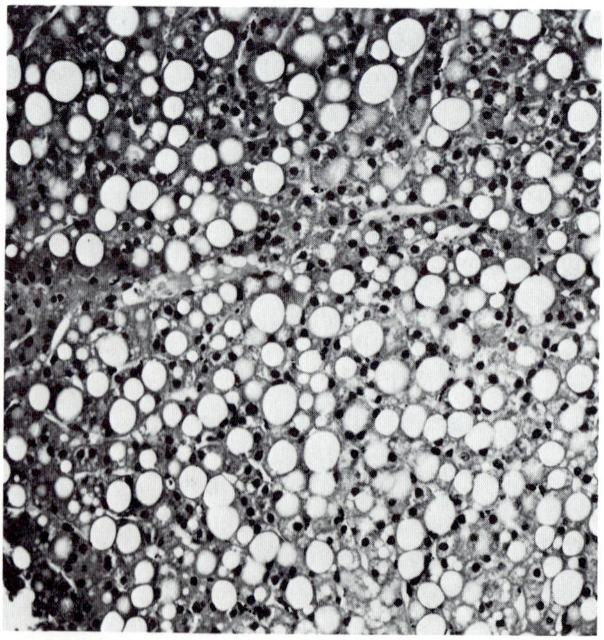

Fig. 6.

fatty hepatocytes, comparable to the findings in SA monkeys, until the last biopsy in which all had moderately fatty livers. Two DA monkeys died, 1 immediately after the biopsy and the other 2 months later, despite return to the original D diet. Both had moderately fatty livers, bile duct proliferation and prominent centrolobular, perivenular and sinusoidal fibrosis with focal nodule formation (pseudolobules) (Figures 7,8). Neither had grossly visible cirrhosis. Of the 2 surviving DA monkeys, 1 had fatty liver and the other had fibrosis and bile duct proliferation but no definite evidence of cirrhosis in the biopsy.

D. Electron Microscopy

In S monkeys the ultrastructure of hepatic cells remained normal for 30 months (Figure 9) until the dietary fat was changed to coconut oil. In the three remaining groups (Groups SA, D and DA) there was an increase in the amount of smooth endoplasmic reticulum during the same period (Figure 10). In Groups SA and D, the increase was variable from cell to cell, while in Group DA nearly every cell showed a marked increase in smooth endoplasmic reticulum.

Mitochondrial changes were noted in all groups except S monkeys. The mitochondria of both alcohol-fed groups, SA and DA, were somewhat pleomorphic and swollen with reduced numbers of cristae and occasional crystalline material within the mitochondria (Figure 11). Megamitochondria also were found in these groups with the most marked changes in Group DA. In Groups SA, D and DA there was a tendency toward vesiculation of rough endoplasmic reticulum and loss of ribosomes with disruption of their usual parallel array.

In biopsies taken 4 months after the dietary fat has been changed to coconut oil (month 34), livers of SA monkeys showed an accumulation of small myelin-like figures in membrane-bound cytoplasmic vesicles. The mitochondria appeared somewhat enlarged with loss of cristae. There appeared to be basement membrane formation about some of the sinusoids with an increased amount of collagenous material in the space of Disse (Figure 12). In the earlier biopsies there was no evidence of basement membrane or collagen formation.

In biopsies and autopsy material from DA animals, fed 0.1% cholesterol, coconut oil and no choline in the diet, there were marked accumulations of myelin-like figures with smaller amounts in the D monkeys (Figure 13). The 3 DA monkeys fed

---

Fig. 6. Liver biopsy from monkey fed SA diet with ethanol for 2 years and 8 months, with substitution of coconut oil for Wesson Oil 4 months prior to biopsy, reveals moderately fatty liver (H + E, x 150).

TABLE V Serum Lipids in Supplemented (S) or Deficient (D) Monkeys with or without Ethanol (A)

Serum Lipid	Treatment	Months after introduction of ethanol				
		3	12*	24	34+	39
Triglycerides** mg/dl	S	104 ± 8	62 ± 2	49	92 ± 8	129 ± 13
	SA	171 ± 41	103 ± 15	138 ± 25	83 ± 12	185 ± 55
	D	186 ± 20	104 ± 4	71	83 ± 1	123 ± 4
	DA	142 ± 13	132 ± 16	95 ± 9	82 ± 11	120 ± 2
Cholesterol** mg/dl	S	178 ± 10	206 ± 24	167 ± 33	224 ± 40	230 ± 13
	SA	195 ± 16	178 ± 23	192 ± 28	233 ± 29	252 ± 24
	D	106 ± 26	110 ± 23	120 ± 14	156 ± 18	168 ± 3
	DA	166 ± 1	137 ± 9	156 ± 9	248 ± 53	207 ± 8

*Ethanol increased from 40% to 50% calories by 14 months.
+Coconut oil replaced Wesson Oil at 30 months in all diets; D and DA diets contained no choline but included 0.1% cholesterol between 30-34 months.
**By analysis of variance there was no significant effect of ethanol on triglycerides; the D diet lowered cholesterol at 3 and 12 months ($P < 0.03$) and at 39 months ($P < 0.06$).

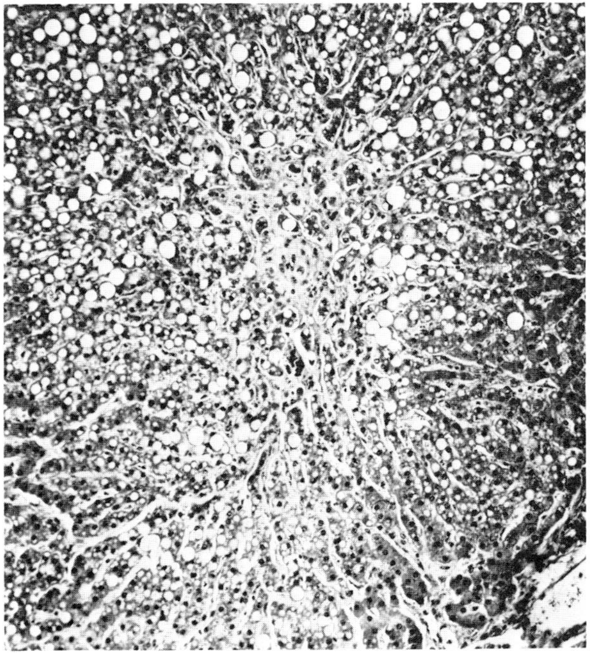

Fig. 7. Postmortem liver section from DA monkey fed ethanol for 3 years followed by deletion of choline, addition of 0.1% cholesterol, and substitution of coconut oil for Wesson Oil for 6 months prior to death and return to DA diet 2 months prior to death. The liver is moderately fatty with centrolobular fibrosis (CAB, x 100).

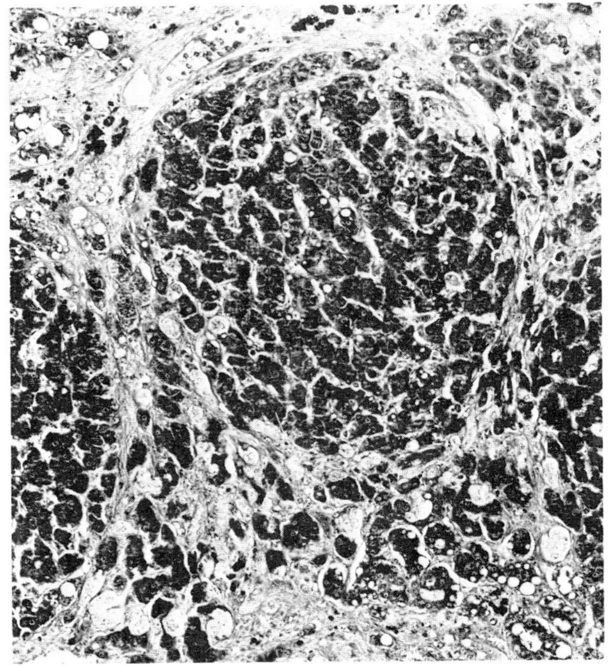

Fig. 8. Postmortem liver section from DA monkey fed ethanol for 2 years and 9 months followed by deletion of choline, addition of 0.1% cholesterol and substitution of coconut Oil for Wesson oil 4 months prior to death. The liver is fibrotic with focal micronodules (CAB, x 100).

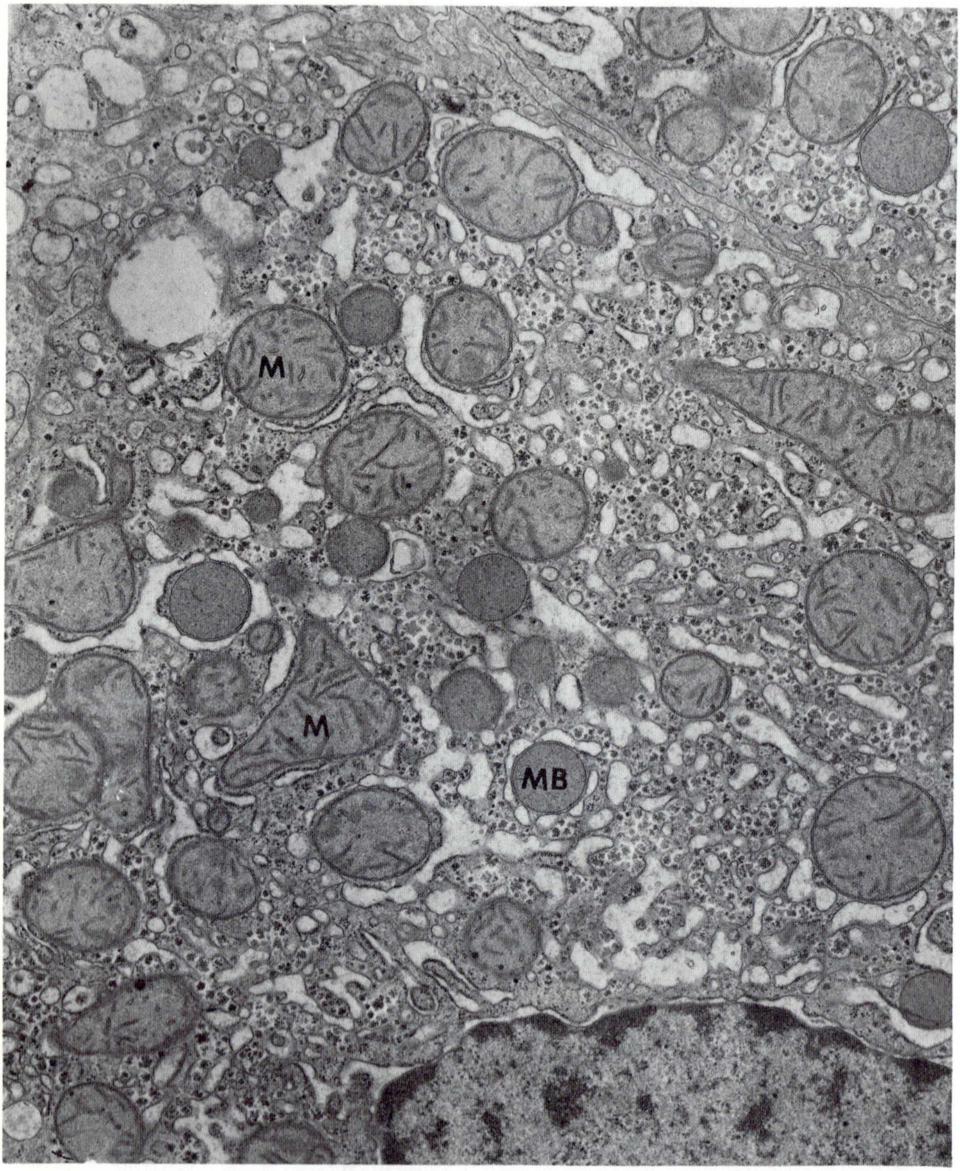

Fig. 9. x20,000 A portion of a typical hepatocyte from a monkey fed S diet before addition of coconut oil. M = mitochondria; MB = microbodies.

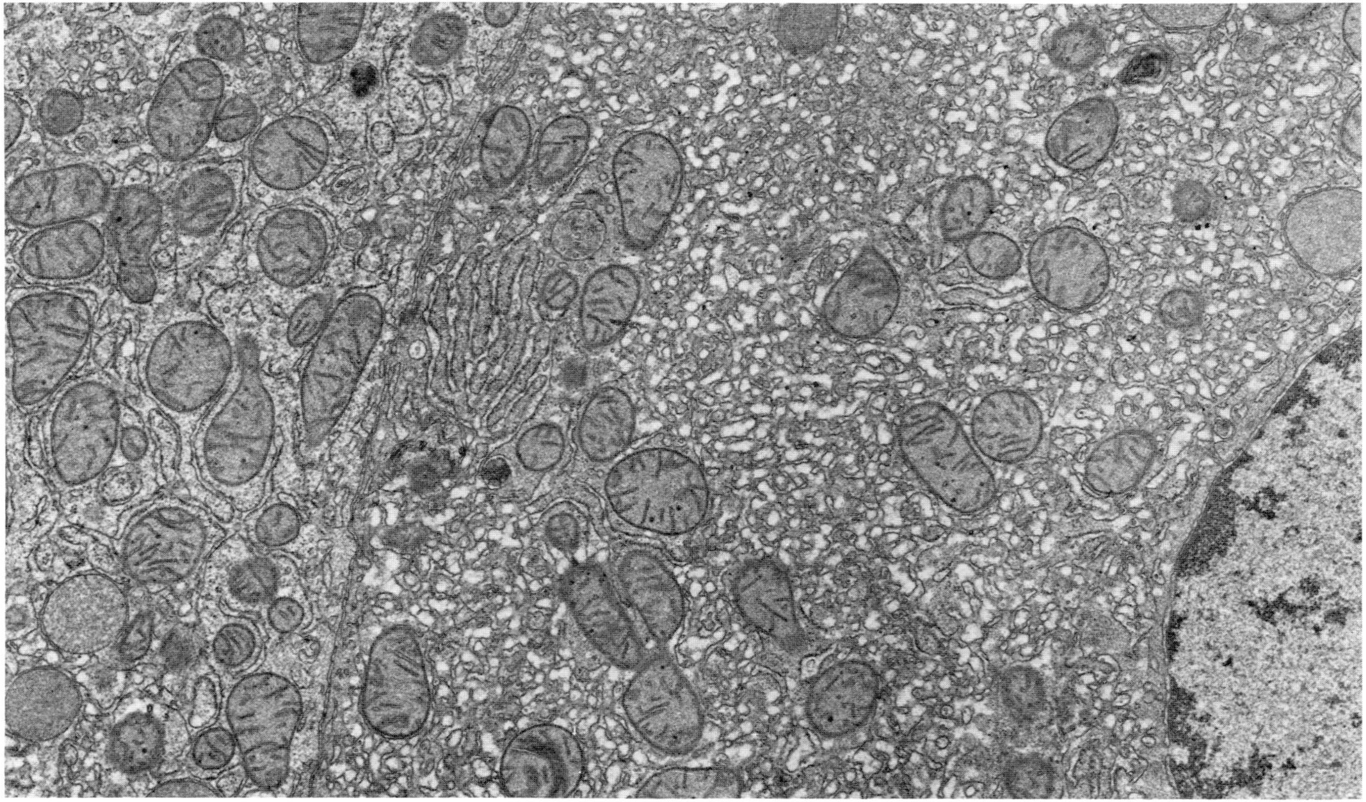

Fig. 10. *x12,000 Liver biopsy from SA monkey. The hepatocyte on the right contains more SER than the cell on the left. Mitochondrial cristae are not altered.*

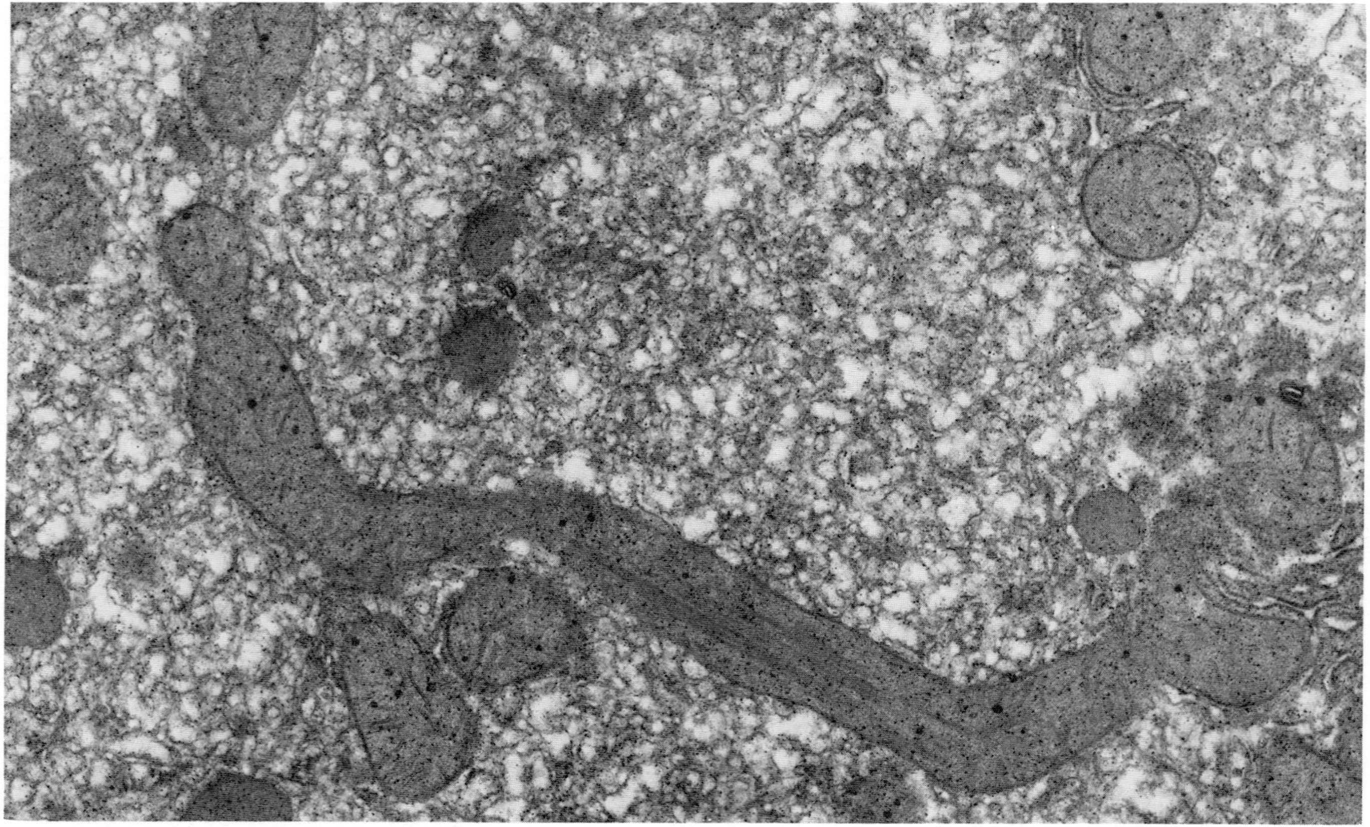

Fig. 11. x19,000 Liver biopsy from DA monkey. A portion of an hepatocyte containing proliferated SER and a giant mitochondion bearing a paracrystalline body.

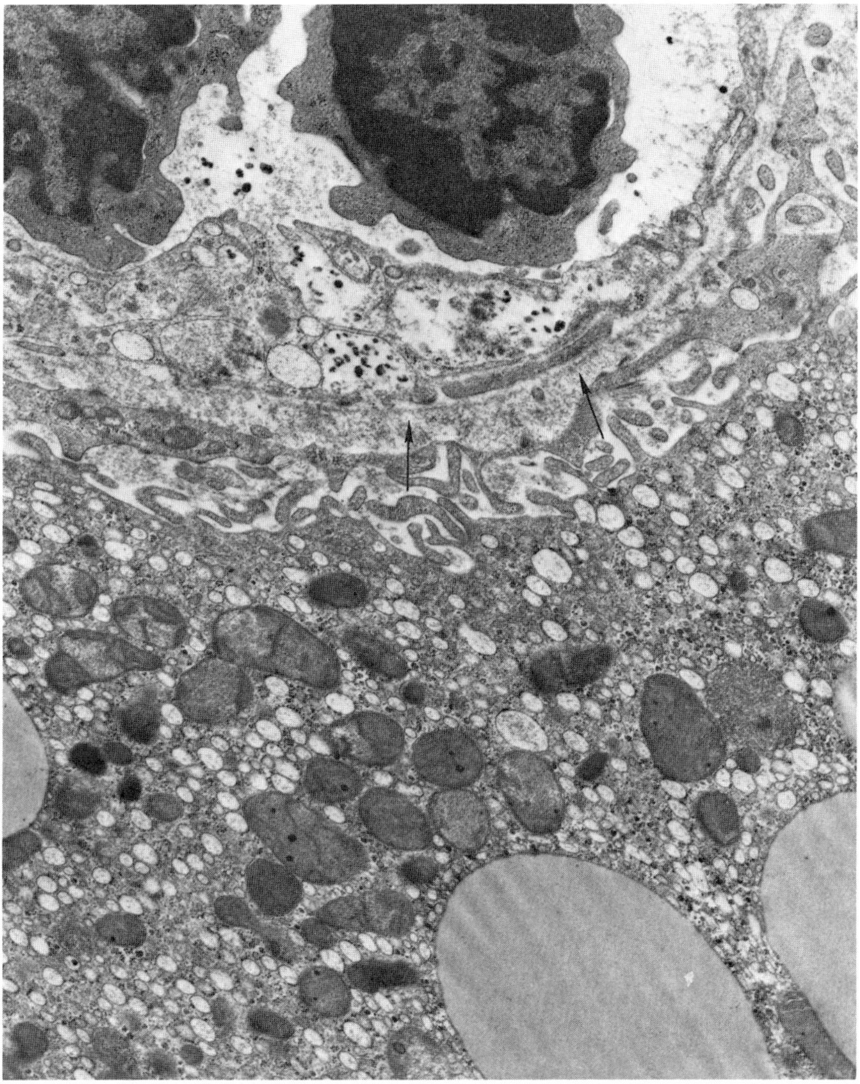

Fig. 12. x16,000 Liver biopsy from SA monkey 4 months after alteration of diet (see text). Portion of sinusoid and adjacent hepatocyte showing capillarization. The basement membrane is visible as a thin layer in the Disse space closely applied to the sinusoidal lining cell (arrows).

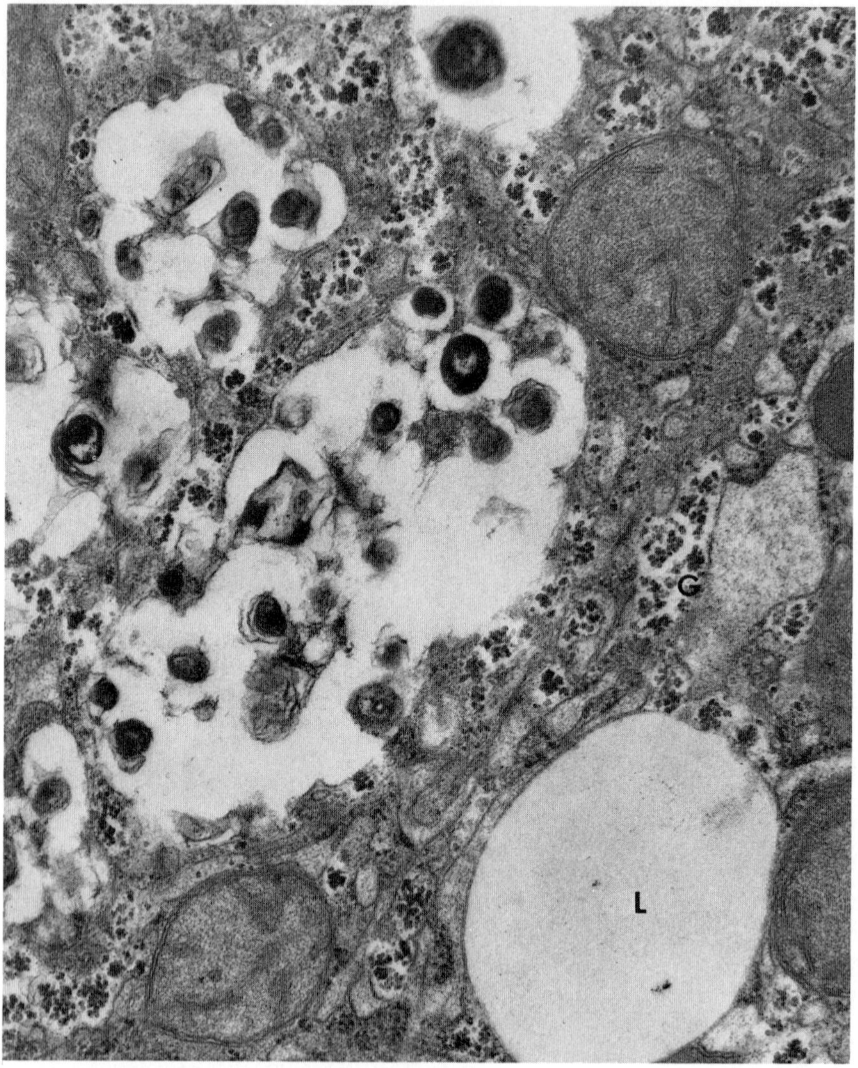

Fig. 13. x35,000 Liver biopsy from DA monkey fed altered diet. A portion of hepatocyte cytoplasm contains myelin-like granules. G = glycogen; L = Lipid.

ethanol for the longer period showed marked increases in collagen deposition in the space of Disse, basement membrane formation, and capillarization of the sinusoids. Fibroblast-like cells, presumably altered lipocytes, were noted in the space of Disse adjacent to the collagen deposition. These changes are consistent with earlier reports of ultrastructural effects of ethanol in humans and nonhuman primates (26,67,68).

E. Discussion

Comparison of the results of our studies in rhesus monkeys and the studies of Lieber and his colleagues in baboons demonstrates that the question of etiology and pathogenesis of alcoholic cirrhosis remains unanswered. Major methodological differences between the 2 models being used include differences in diet, species and sex. Most of the baboons reported were females (69). They developed fibrosis or cirrhosis after 9-18 months of ingesting a purified diet low in choline, but otherwise essentially nutritionally complete, containing 50% of calories as ethanol.

We have induced hepatic fibrosis with focal nodule formation, but not cirrhosis, in monkeys fed 40-50% of calories as ethanol for approximately 3 years in a diet deficient in lipotropes and, for a short period, containing saturated fat and cholesterol. We have not induced fibrosis or cirrhosis in monkeys given the same amount of ethanol in a fully nutritious diet. In both Lieber's and our experiments the diets considered fully nutritious did not support weight gain or hemoglobin levels comparable to controls not given ethanol. Therefore the questions must be raised of what constitutes nutritional adequacy in the face of ethanol ingestion and what dietary manipulations are needed to induce liver damage with ethanol. The SA monkeys show no significant liver damage and in the deficient monkeys we had to decrease dietary choline from 0.1% to zero, add 0.1% cholesterol and change the fat to coconut oil to induce significant liver damage with ethanol. The baboon diet is very low in choline, as pointed out above, but baboons appear to be less susceptible than rhesus monkeys to abnormalities of lipid metabolism induced by diets high in fat and cholesterol, and their response to choline deficiency is unknown except for the early study in 2 animals (9). Other questions raised with respect to the baboon diet, i.e. it's low content of carbohydrate and its cystine content, apply also to the diet we have fed and therefore do not account for the different results. Reduction of serum cholesterol in D monkeys is supporting evidence that lipotrope deficiency was induced. Failure of ethanol to induce hypertriglyceridemia in DA monkeys, as it has done in SA monkeys, while increasing serum cholesterol to the level of S and SA monkeys indicates interactions in lipid metabolism which should be explored using pair-fed animals. The marked deterioration of DA monkeys precipitated by removing dietary choline and increasing the saturated fat and cholesterol load on hepatic lipid metabolism compared to the mild effects in D monkeys also indicates that ethanol and lipotrope deficiency interacted to induce severe hepatic damage in rhesus monkeys whereas either alone did not.

Studies in these and other primate models are needed to resolve the questions raised. Detailed assessment of nutritional status in ethanol-fed animals during development of liver disease and examination of effects of nutrient deficiencies on lipid metabolism and hepatic toxicity of ethanol will answer the basic question: does ethanol induce irreversible liver disease directly or indirectly by inducing nutrient deficiencies, or are dietary nutrient deficiencies or excesses required in conjunction with ethanol?

Subsequent studies of other tissue or biochemical lesions associated with alcoholism can be performed in the models developed. The basic nutritional data obtained in studies of hepatic effects will be available and can be performed in the models developed. The basic nutritional data obtained in studies of hepatic effects will be available and can be utilized to answer the same question of nutrient and ethanol interactions in induction of renal, CNS, cardiac, and other alcohol related diseases. Other results of the studies will be definition of primate requirements for the nutrients of interest, eg. choline, methionine, and others, and elucidation of the effects of ethanol or other toxins on the requirements.

ACKNOWLEDGMENTS

Supported by a grant from the Distilled Spirits Council of The United States and by U.S.P.H.S., Grants 5-S05-PRO7047 and RR01046, Institute Funds, Mallory Institute of Pathology, Dept. of Pathology, Boston University School of Medicine.

REFERENCES

1. Alpers, D.H., and Isselbacher, K. (1975) Fatty Liver: Biochemical and Clinical Aspects, in: *Disease and Liver*, Edition 4; ed. L. Schiff, J.B. Lippincott, Philadelphia, p. 815-832.

2. Barrett, J.E. (1976) Effects of Alcohol, Chlordiazepoxide, Cocaine and Pentobarbital on Responding Maintained Under Fixed-Interval Schedules of Food or Shock Presentation. *J. Pharmacol. & Exper. Therap. 196: 605-615.*
3. Carter, E.A., McCarron, M.J., and Isselbacher, K.J. (1978) Chronic Ethanol Feeding Increases Choline Oxidation in the Rat. Submitted for Publication.
4. Castelli, W.P., Gordon, T., Hyjortland, M.C., Kagan, A., Doyle, T., Halley, S.B., and Zukel, W.J. (1977) Alcohol and Blood Lipids. *Lancet ii: 153-155.*
5. Chernoff, G.F. (1975) A Mouse Model of The Fetal Alcoholic Syndrome. *Teratology 11: 14A.*
6. Cohen, E.L., and Wurtman, R.J. (1975) Brain Acetylcholine Increases After Systemic Choline Administration. *Life Sci., 16: 1095-1102.*
7. Cueto, J., Tajen, N., Gilbert, E., and Curne, R.A. (1967) Experimental Liver Injury in the Rhesus Monkey: 1. Effects of Cirrhogenic Diet and Ethanol. *Ann. Surg. 166: 19-28.*
8. Duthie, A.M., Adams, J.S., and Whik, J.F. (1976) The Influence of Ethyl Alcohol on Cerebro-Spinal Fluid Pressure. *Anesthesia 31: 893-896.*
9. Eggen, D.A. (1974) Cholesterol Metabolism in Rhesus Monkeys, Squirrel Monkeys and Baboons. *J. Lipid Res. 15: 139-145.*
10. Elton, R.H., Greaves, D.A., Bunger, D.R., and Pyle, T.W. (1974) Drinking Patterns of Pigtailed Macaques. *J. Studies on Alcohol 37: 1548-1555.*
11. Elton, R.H., and Wilson, M.E. (1977) Changes in Ethanol Consumption by Pregnant Pigtailed Macaques. *J. Studies on Alcohol 37: 1548-1555.*
12. Falk, J.L., and Tang, M. (1977) Animal Model of Alcoholism: Critique and Progress, in: *Alcohol Intoxication and Withdrawal,* Vol. 3B, ed. M. Gross, Plenum, N.Y., p.465-493.
13. Feinman, L., and Lieber, C.S. (1972) Hepatic Collagen Metabolism: Effect of Alcohol Consumption in Rats and Baboons. *Science 176: 795.*
14. Fiser, R.H., Denniston, J.C., Rindsig, R.B., and Beisel, W.R. (1974) Triglyceride Secretion Rates: Use of Triton WR 1339 in the Rhesus Monkey. *J. Nutr. 104: 223-226.*
15. Freund, G. (1975) Induction of Physical Dependence on Alcohol in Rodents, in: *Biochemical Pharmacology of Ethanol,* ed. F. Majchrowicz, Plenum, N.Y. pp. 311-325.

16. Friedman, H.J., and Lester, D., (1977) A Critical Review of Progress Towards An Animal Model of Alcoholism, in: *Alcohol and Opiates,* ed. Blum, K., Academic Press, N.Y., pp.1-19.
17. Griffith, W.H., and Nye, J.F. (1971) Choline. Effects of Deficiency, in: *The Vitamins,* Vol. III, eds. W.H. Sebrell, and R.S. Harris, Academic Press, pp.81-122.
18. Gunnor, R.M., Demakis, J., Rahimtoola, S.H., Sinno, M.Z., and Tobin, J.R., Jr. (1976) Clinical signs and Natural History of Alcoholic Heart Disease. *Ann. N.Y. Acad. Sci. 252: 264-272.*
19. Hansen, L.G., and Warwick, W.J. (1968) A Flourometric Micromethod for Serum Vitamin A. *Am. J. Clin. Pathol. 50: 515-529.*
20. Hanson, J.W., Jones, K.L., and Smith, D.W., (1976) Fetal Alcohol Syndrome. Experience with 41 Patients. *J.A.M.A. 235: 1458-1460.*
21. Hartroft, W.S., and Porta, E.A. (1971) Choline, Man, in: *The Vitamins,* Vol. III, eds. W.H. Sebrell, and R.S. Harris, Academic Press, N.Y., pp. 123-154.
22. Hines, J.D. (1975) Hematologic Abnormalities Involving Vitamin B_6 and Folate Metabolism in Alcoholic Subjects. *Ann. N.Y. Acad. Sci. 252: 316-327.*
23. Ho, B.T., Fritchie, A.E., Indanpaan-Heikkila, J.E., and McIsacc, W.M. (1972) Placental Transfer and Tissue Distribution of Ethanol-1-^{14}C. Quart. *J. Studies Alcohol. 33: 485-493.*
24. Hoffbauer, F.W., and Zaki, F.G. (1965) Choline Deficiency in Baboon and Rat Compared. *Arch. Path. 79: 364-369.*
25. Hollstedt, C., Olsson, O., and Rydberg, U. (1977) The Effect of Alcohol on the Developing Organism. *Medical Biology 55: 1-14.*
26. Iseri, O.A., Lieber, C.S., Gottlieb, L.S. (1966) The Ultrastructure of Fatty Liver Induced by Prolonged Ethanol Ingestion. *Am. J. Pathol. 48: 535-555.*
27. Isselbacher, K.J. (1977) Metabolic and Hepatic Effects of Alcohol. *N. E. J. Med. 296: 612-616.*
28. Kamback, M.C. (1973) Drinking as an Avoidance Response by the Pigtail Monkey (*Macca Nemestrina*). Quart. *J. Alcohol. 34: 943-946.*
29. Katz, J., and Barrett, J.E. (1978) Effects of d-Amphetamine and Ethanol on Responding of Squirrel Monkeys Maintained under Fixedration Schedules of Food Presentation and Stimulus-shock Termination. *Pharmacol., Biochem., Behavior 8: 35-39.*

30. Katz, J., and Barrett, J.E. (1978) Ethanol, Pentobarbital, and Chloridaizapoxide Effects in Squirrel Monkeys Responding Under Fixed-ration. Food Presentation and Stimulus-Shock Termination Schedules. *Psychopharmacol. 56: 153-155.*

31. Klatskin, G., Krehl, W.A., and Conn, H.O. (1954) The Effect of Alcohol on the choline Requirement I. Changes in the Rat's Liver Following Prolonged Ingestion of Alcohol. *J. Exp. Med. 100: 605-614.*

32. Klatskin, G., and Krehl, W.A. (1954) The Effect of Alcohol on the Choline Requirement II. Incidence of Renal Necrosis in Weanling Rats Following Short Term Ingestion of Alcohol. *J. Exper. Med. 100: 615-627.*

33. Lieber, C.S., and DeCarli, L. (1974) An Experimental Model of Alcohol Feeding and Liver Injury in the Baboon. *J. Med. Primatol. 3: 153-163.*

34. Lieber, C.S., and DeCarli, L.M. (1975) Alcoholic Liver Injury: Experimental Models in Rats and Baboons. *Adv. Exper. Med. Biol. 59: 379-393.*

35. Lieber, C.S., and DeCarli, L.M. (1976) Animal Models of Ethanol Dependence and Liver Injury in Rats and Baboons. *Fed. Proc. 35: 1232-1236.*

36. Lieber, C.S., and Schmud, R. (1961) The Effect of Ethanol on Fatty Acid Metabolism: Stimulation of Fatty Acid Synthesis *in vitro*. *J. Clin. Invest. 40: 394-399.*

37. Lieber, C.S., Teschke, R., Hasumura, Y.C., DeCarli, L. (1975) Differences in Hepatic and Metabolic Changes After Acute and Chronic Alcohol Consumption. *Fed. Proc. 34: 2060-2074.*

38. Lindenbaum, J., and Lieber, C.S. (1976) Effects of Chronic Ethanol Administration on Intestinal Malabsorption in Man in the Absence of Nutritional Deficiency. *Ann. N.Y. Acad. Sci. 252: 228-234.*

39. Maribbio, A.T., Buchwald, H., Schwartz, M.A., and Varco, R. (1976) Hepatic Lesions of Central Pericellular Fibrosis in Morbid Obesity and After Jejunoileal Bypass. *Am. J. Clin. Path. 66: 684-691.*

40. Maxwell, M.H., and Kleeman, C.R. (eds.) (1975) *Clinical Disorders of Fluid & Electrolyte Metabolism*, Edition 2, McGraw Hill Book Co., N.Y.

41. Mendelson, J.H., and Mello, N.K., (1973) Studies of the Development of Alcohol Addiction in Infant Monkeys. *Ann. N.Y. Acad. Sci. 215: 145-161.*

42. Mezey, E. (1976) Intestinal Function in Chronic Alcoholism. *Ann. N.Y. Acad. Sci. 252: 215-227.*

43. Montgomery, R.L., Pick, J.R., Ellis, F.W., and Christian, E. (1978) Brain Lesions in Hippocampal Areas of Rhesus Monkeys Physically Dependent on Ethanol. 7th International Congress of Pharmacology, Paris.
44. Moxley, R.T. III, Pozefsky, T., and Lockwood, D.H. (1974) Protein Nutrition & Liver Disease After Jejunoileal Bypass for Morbid Obesity. *N. J. Eng. Med. 290: 921-926.*
45. Myers, R.D. (1978) Psychopharmacology of Alcohol. *Ann. Rev. Pharmacol. Toxicol. 18: 125-144.*
46. Nair, C.R., Nirankar, O.P., Chhuttani, P.N., and Datta, D.V. (1975) Impaired Hepatic Release of Triglycerides: A possible Cause of Acute Alcoholic Fatty Liver in the Rhesus Monkey. *Clin. Sci. Molec. Med. 48: 453-455.*
47. Nair, C.R., Saha, S., Chakravarti, R.N., Chhuttani, P.N., and Datta, D.V. (1973) Acute Alcoholic Fatty Liver in the Rhesus Monkey. *Clin. Sci. 44: 505-510.*
48. NAS-NRC (1978) Nutrient Requirements of Nonhuman Primates, NAS-NRC, Washington, D.C.
49. Nath, I., Sood, S.K., and Nayak, N.C. (1972) Experimental Siderosis and Liver Injury in The Rhesus Monkey. *J. Path. 106: 103-111.*
50. Ordonez, L.A. (1977) Control of the Availability to the Brain of Folic Acid, Vitamin B_{12} and Choline, in: *Nutrition and the Brain,* eds. R.J. Wurtman, J.J. Wurtman, Raven Press, N.Y. pp.205-248.
51. Ouellette, E.M., Rosett, H.M., Rosman, N.P., and Weiner, L. (1977) Adverse Effects on Offspring of Maternal Alcohol Abuse During Pregnancy. *N.E.J.Med. 297:*
52. Papara-Nicholson, D., and Telford, I. (1957) Effects of Alcohol on Reproduction and Fetal Development in the Guinea Pig. *Anat. Rec. 127: 438-439.*
53. Patek, A.J., Jr., Bowry, S., and Hayes, K.C. (1975) Cirrhosis of Choline Deficiency in the Rhesus Monkey. Possible Role of Dietary Cholesterol. *Proc. Soc. Exper. Biol. Med. 148: 370-374.*
54. Patek, A.J., Jr., Bowry, S.C., and Sabesin, S.M. (1976) Minimal hepatic changes in rats fed alcohol and a high casein diet. *Arch. Pathol. Lab. Med. 100: 19-24.*
55. Pieper, W.A. (1976) Great Apes and Rhesus Monkeys as Subjects for Psychopharmacological Studies of Stimulants and Depressants. *Fed. Proc. 35: 2254-2257.*
56. Pieper, W.A., and Skeen, M.V. (1975) Retention of Functional Tolerance to Ethanol in Rhesus Monkeys (*Macca Mulatta*). Pharmacol. Biochem. Behavior 3: 909-913.

57. Pieper, W.A., Skeen, M.J. (1977) Alcohol Withdrawal Reactions in Rhesus Monkeys, in: *Alcohol Intoxication and Withdrawal*, Vol. 3B, ed. M.M. Gross, Plenum, N.Y. pp.43-55.
58. Pieper, W.A., Skeen, M.J., McClure, H.M., and Bourne, P.G. (1972) The Chimpanzee as an Animal Model for Investigating Alcoholism. *Science 176: 71-73.*
59. Porta, E.A., Koch, O.R., Hartroft, W.S. (1970) Recent Advances in Molecular Pathology: a Review of the Effects of Alcohol on the Liver. *Exp. Molec. Path. 12: 104-132.*
60. Randall, C.L. (1977) Teratogenic Effects of *in utero* Ethanol Exposure in: *Alcohol and Opiates, Neurochemical and Behavioral Mechanisms,* ed. K. Blum, Academic Press, N.Y. pp. 91-107.
61. Randall, C.L., Taylor, J., and Walker, D.W. (1976) Ethanol-induced Malformations in Mice, Alcoholism: Clinical and Experimental Research. 1: 219-224.
62. Rebohcas, G., and Isselbacher, K.J. (1961) Studies on the Pathogenesis of Ethanol-induced Fatty Liver. *J. Clin. Invest.* 40: 1355-1362.
63. Regan, T.J., Ettinger, P.O., Haider, B., Ahmed, S.S., Oldewurtel, H.A., and Lyons, M.M. (1977) The Role of Ethanol in Cardiac Disease. *Ann. Rev. Med. 28: 393-409.*
64. Regan, T.H., Ettinger, P.O., Oldewurtel, H.A., and Haider, B. (1976) Heart Cell Responses to Ethanol. *Ann. N.Y. Acad. Sci. 252: 250-263.*
65. Riley, J.N., and Walker, D.W. (1978) Morphological Alterations in Hippocampus After Long-term Alcohol Consumption in Mice. *Science 201: 646-648*
66. Rogers, A.E., and Newberne, P.M. (1973) Alcoholic or Nutritional Fatty Liver and Cirrhosis. *Am. J. Path. 73: 817-820.*
67. Rubin, E., and Lieber, C. (1967) Early Fine Structural Changes in the Human Liver Induced by Alcohol. *Gastroenterology 52: 1-13.*
68. Rubin, E., and Lieber, C. (1968) Alcohol-Induced Hepatic Injury in non-Alcoholic Volunteers. *N.E.J.M. 278: 869-876.*
69. Rubin, E., and Lieber, C. (1974) Fatty Liver, Alcoholic Hepatitis and Cirrhosis Produced by Alcohol in Primates. *N.E.J.M. 290: 128-135.*
70. Rutherford, R.B., Boitnott, J.K., Donohoo, J.S., Marzolis, S., Sebor, J., and Zuidema, P.D. (1969) Production of Nutritional Cirrhosis in *Macacca Mulatta* Monkeys. *Arch. Surg. 98: 720-730.*

71. Sabesin, S.M., Hawkins, H.L., Bertram, P.D., Mann, J.A., and Peace, R.J. (1978) Alcoholic Hepatitis. *Gastroent. 74: 276-286.*
72. Sabesin, S.M., Hawkins, H.L., Kuken, L., and Ragland, J.B. (1977) Abnormal Plasma Lipoproteins and Lecithin-cholesterol Acyltransferase Deficiency in Alcoholic Liver Disease. *Gastroent. 72: 510-518.*
73. Schiff, L. (1975) *Diseases of the Liver*, ed. 4, J.B. Lippincott, Phila.
74. Seixas, F.A., and Eggleston, S. (1976) Work in Progress on Alcoholism. *Ann. N.Y. Acad. Sci. 273: 1-664.*
75. Seixas, F.A., Williams, K., and Eggleston, S. (1975) Medical Consequences of Alcoholism. *Ann. N.Y. Acad. Sci. 252: 1-399.*
76. Smuckler, E.A. (1975) Alcoholic Drink, It's Production and Effects. *Fed. Proc. 34: 2038-2044.*
77. Takada, A., Porta, E.A., and Hartroft, W.S. 1967 Regression of Dietary Cirrhosis in Rats Fed Alcohol and a "Super Diet". *Amer. J. Clin. Nutr. 20: 213-225.*
78. Takeuchi, J., Takada, A., Hasumura, Y., Matsuda, Y., and Ikegami, F. (1972) Acute Alcoholic Liver Injury and Choline Deficiency. *Meth. Achievment Exp. Path. 6: 81-110.*
79. Takeuchi, J., Takada, A., Kato, Y., Hasumura, Y., Ikegami, F., and Matsuda, Y. (1971) Hepatic Changes in Chronic Alcoholic Rats Following Periodic Acute Alcoholic Intoxications. *Amer. J. Clin. Nutr. 24: 628-636.*
80. Talbott, G.D. (1976) Primary Alcoholic Heart Disease. *Ann. N.Y. Acad. Sci. 252: 237-242.*
81. Tamura, T., Watson, J.E., Romero, J.J. and Halsted, C. H. (1978) Effect of Alcoholism on Hepatic Folate Metabolism. *Clin. Res. 26: 327A.*
82. Tuma, D.J., Barak, A.J., Schafer, D.F., and Sorrell, M.F. (1973) Possible Interrelationship of Ethanol Metabolism and Choline Oxidation in the Liver. *Canad. J. Biochem. 51: 117-120.*
83. VanNorren, D., and Padmos, P. (1977) Influence of Anesthetics, Ethyl Alcohol, and Freon on Dark Adaptation of Monkey Cone ERG. *Invest. Opthalmol. Visual Sci. 16: 80-83.*
84. VanWaes, L., and Lieber, C.S. (1977) Early Perivenular Siderosis in Alcoholic Fatty Liver: an Index of Progressive Liver Injury. *Gastroent. 73: 646-650.*

85. Vasdev, S.C., Chakravarti, R.N., Subrahmanyain, D., Jain, A.C., and Waki, P.N. (1975). Myocardial Lesions Induced by Prolonged Alcohol Feeding in Rhesus Monkeys. *Cardiovasc. Res. 9: 134-140.*

86. Wilgram, G.F. (1959) Experimental Laennec Type of Cirrhosis in Monkeys. *Ann. Int. Med. 51: 1134-1158.*

87. Wilgram, G.F., Lucas, C.C., and Best, C.H. (1958) Kwashiorkor Type of Fatty Liver in Primates. *J. Exper. Med. 103: 361-370.*

88. Williams K. (1976) The Cardiovascular System, Introduction. *Ann. N.Y. Acad. Sci. 252: 235-236.*

89. Wurtman, R.J., and Fernstrom, J.D. (1975) Control of Brain Monoamine Synthesis by Diet and Plasma Amino Acids. *Am. J. Clin. Nutr. 28: 638-647.*

INDUCTION OF OBESITY IN NONHUMAN PRIMATE MODELS OF HUMAN OBESITY

Barbara C. Hansen

Department of Physiology
University of Michigan
Ann Arbor, Michigan

I. INTRODUCTION

Monkeys were first used as models for the study of human obesity in the laboratory of S.W. Ranson in 1938 (30). The four succeeding decades have witnessed surprisingly few studies of obese nonhuman primates which have been of three general types: 1) descriptive reports of spontaneously obese monkeys, 2) short-term and long-term studies of monkeys made obese by hypothalamic lesions, and 3) examinations of the effects of overfeeding and increased adiposity on endocrine function and feeding behavior in monkeys.

Of primary importance to the use of any animal model is the degree to which it approximates the human condition. Table 1 presents a summary of selected anatomic, metabolic and behavioral characteristics which has been accumulated from multiple descriptions of spontaneous and experimental obesity and which attempts to assess the best "fit" of various models

ISBN 0-12-334250-3

TABLE I Summary of Several Characteristics of Spontaneous and Experimental Obesity

	1 Spontaneous Obesity Rat	2 Spontaneous Obesity Man	3 Experimental Obesity Man	4 Spontanoeus Obesity Monkey	5 Experimental Obesity Overfed Monkey	6 Experimental Obesity Hypothalamic Lesioned Monkey
Adipose tissue						
cell size	↑	↑	↑			
cell number	↑	↑ or Normal	Normal			
Glucose						
glucose tolerance (oral)		Normal or ↓	↓ slightly	Normal or ↓	Normal or ↓	
glucose tolerance (IV)		Normal or ↓	↓ slightly or N	Normal or ↓	Normal or ↓	↓
fasting	Normal	Normal or ↑	↑	Normal	Normal	Normal
Plasma insulin						
fasting	↑	↑	Normal or ↑	↑	Normal or ↑	↑
stimulation by glucose		↑	Normal or ↑	↓	↑	↓
Plasma glucagon						
fasting	↓	Normal or ↑				
stimulation by glucose		↓				

	1* Spontaneous Obesity Rat	2 Spontaneous Obesity Man	3 Experimental Obesity Man	4 Spontaneous Obesity Monkey	5 Experimental Obesity Overfed Monkey	6 Experimental Obesity Hypothalamic Lesioned Monkey
Plasma growth hormone						
fasting		↓ or Normal	Normal or ↓			
response to glucose		↓	Normal or ↓			↓
Adrenalcorticosteroids						
plasma cortisol (total)		↓ or Normal	Normal or ↓			
total 17 hydroxy-corticosteroids		↑	↑ but N/kg B.W.			
urinary free cortisol		↑ or Normal	↓			
Plasma lipids						
cholesterol	↑	↑ or Normal	Normal or ↑	↑		↑
triglycerides	↑	↑ or Normal	Normal or ↑	↑		↑
free fatty acids	↑	↑ or Normal	Normal or ↓	Normal or ↑	Normal or ↑	↑

* See Text for References.

of obesity. The major sources of data are indicated although the directions of the arrows generally represent a concensus of numerous researchers. Many other features or abnormalties of obesity have been identified, and are not included in this review, including reduced gonadotropins, altered water balance, changes in thyroid gland and pituitary functions, and adrenocortical abnormalities. Although many of the characteristics of obesity in man have been found in rodents and monkeys, a major question remains concerning which metabolic alterations are causal or primary and which are adaptive or secondary.

II. EXPERIMENTAL OBESITY IN MAN

Before considering the findings related to obesity in nonhuman primates, it is important to consider the classic series of studies on experimental obesity in man which were undertaken by Sims and his colleagues at the Univeristy of Vermont (first reported by Sims et al., (36), and including (34,35,37,38,31,32,26,23). Inmates of the Vermont State Prison volunteered for these, the only studies of long-term experimental overeating and obesity-induction in man. Most endocrine and metabolic changes of uncomplicated spontaneous human obesity were reproduced in direction, although not always in magnitude. Among the many conclusions of these studies, which have provided direction for subsequent studies of obesity in monkeys, were that all of these alterations were reversible, that the "luxuskonsumption" described by

---

*1. *Bray and York, 1971; Bryce, Johnson, Sullivan and Stern, 1977; Frohman, Goldman, Schnatz and Bernardis, 1971; Zucker and Antoniades, 1972.*
2. *Bagdade, et al., 1967; Berkowitz, 1964; Bray, 1976; Hirsch and Batchelor, 1976; Hirsch, et al., 1966; Horton, 1974; Opie and Walfish, 1963; Paulsen and Lawrence, 1968; Rabinowitz, 1970; Sjostrom and Bjorntorp, 1974.*
3. *Horton, 1974; O'Connell, et al., 1973; Salans, et al., 1971, 1974; Sims, 1976; Sims and Horton, 1968; Sims, et al., 1968, 1971, 1973.*
4. *Hamilton, et al., 1972; Walike, et al. 1977.*
5. *Current studies.*
6. *Anand, et al., 1955; Brooks, et al., 1942; Hamilton and Brobeck, 1963, 1964, 1965; Hamilton, 1972; Hamilton, et al., 1976; Ranson, et al., 1938.*

such German scientists as Neumann (25) may be a real adaptation to over or undernutrition, and that individual variations in response to over or undernutrition are great. Questions remain concerning the mechanisms of such adaptations to varying caloric intakes, the physiological nature of "set points", and the potential role of disturbances in such regulatory functions as etiologic agents in obesity.

III. SPONTANEOUS OBESITY IN MONKEYS

Because spontaneous human obesity is slow to develop, and difficult to reverse, and because of the potential for associated morbidity in experiments of obesity induction in man, the nonhuman primate as a model for obesity has received increasing attention. In evaluating the usefulness of this model, one of the first questions (although not necessarily most important) which has been raised is, "Do monkeys become spontaneously obese as do humans?" The answer is unequivocally--yes, at least under certain general environmental conditions.

Anthropologists who have examined the evolution of obesity in man and in monkeys, have identified both the occurence of obesity and the natural conditions which are closely associated with ponderosity (24). Militating against the development of obesity in both nonhuman primates and paleolithic man were the following factors: a) diet was largely herbivorous, b) high levels of exercise were necessary in order to obtain sufficient food, and c) significant adiposity leading to reduced agility and increased vulnerability to predators. Cultural or environmental changes have made possible the expression of obesity both in urbanized man and in zoo- or laboratory-contained monkeys and apes.

Genetic predispositions exist in man (2), although they are not as well defined as in rats and mice (8). In monkeys, such genetic predispositions have not been studied systematically, but were suggested by findings of Walike and colleagues (43) who identified an obese mother-daughter pair of pigtailed monkeys. The third generation is still too young to determine the presence of obesity.

Spontaneous obesity in *Macaca mulatta* has been described by Hamilton et al., (18) in seven "middle aged" male monkeys *(Macaca mulatta*, estimated to be 12-14 years old). These animals had been in the laboratory for long periods of time (up to 15 years) and maintained on Purina Monkey Chow (15%

protein) *ad libitum*. The classification of "obese" was made if body weight exceeded 15 kg. Monkeys with spontaneous obesity maintained normal fasting blood glucose levels, but had reduced insulin response to glucose loading. Glucose disappearance rates were significantly lower than those of lean monkeys. Free fatty acids, cholesterol, and triglyceride levels were all significantly elevated (Table 1).

Walike *et al.*(43) surveyed a primate colony containing 873 *Macaca nemistrina* in order to identify obese monkeys. Surprisingly, obesity was found only in females in this species. The average percent adipose tissue of total body weight for the five most obese females was 40.45 ± 7.26%, estimated by the method of tritiated water dilution for total body water. By contrast, the percent adiposity for 12 normal females was 12.72 ± 7.58%, and for 13 normal weight males, 8.66 ± 6.78% (Table II). Fasting glucose was slightly, although not significantly, higher in the obese, while glucose disappearance (K glucose) was lower in obese monkeys. Fasting free fatty acids did not differ; however, fasting insulin was markedly elevated in the obese group. Among the three skinfold measurements, abdominal, triceps and subscapular, only the abdominal measurement proved useful in discriminating obese monkeys, indicating the predominantly truncal distribution of adipose tissue in female *Macaca nemistrina*.

TABLE II Comparison of Normal and Obese Macaca Nemestrina

Criterion	*Normal* weight females N=12*	*Obese females N-5*
Abdominal skinfold, mm	*5.3 ± 1.8*	*13.2 ± 4.5*
Adiposity/body weight, %	*12.7 ± 7.6*	*40.4 ± 7.3*
Fasting glucose, mg/dl	*69.8 ± 7.3*	*76.4 ±18.8*
K glucose, %/min.	*3.1 ± 1.1*	*2.6 ± 0.8*
Fasting FFA, μE/ml	*1.1 ± 0.3*	*1.1 ± 0.4*
Fasting insulin, μU/ml	*59.2 ±38.3*	*186.4· ±67.2*

**Total body water greater than 63% and abdominal skinfolds less than 7.0 mm. Values are mean ± SD.*

Notable in this study was the low incidence of obesity in the colony as a whole and the complete absence of obese males. It was also observed that from 6 months of age males were heavier than females, and after 4 years of age the sex differences were pronounced (Figure 1).

Problems of Spontaneously Obese Monkeys as Models for Study of Human Obesity: Both the obese male *Macaca mulatta* of Hamilton et al. (18) and the obese female *Macaca nemistrina* of Walike et al. (43) provide excellent models of human obesity based on metabolic indications of decreased glucose tolerance and increased insulin resistance. The major difficulty with the spontaneously obese monkey as a model is its limited availability. Increased adiposity has not been identified in an animal weighing less than 7 kg and, thus, it is necessary to hold animals for long periods in order for obesity to be

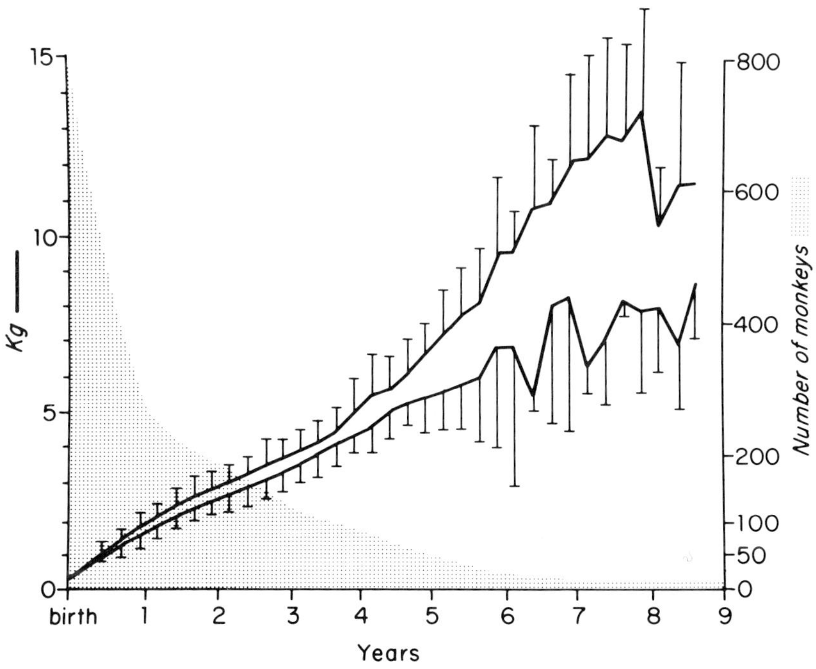

Fig. 1. Growth rate in a colony of 873 pig-tailed monkeys (Macaca nemistrina). (Reprinted by permission from Walike et al., J. Med. Primatol. 6: 151, 1977.

manifest. Among the more than 80 male *Macaca mulatta* studied in our laboratory over a seven-year period, one became spontaneously obese and glycosuric and was possibly prediabetic. A second obese monkey was received from another laboratory and had been in captivity for more than six years. A third had been overfed and returned voluntarily to its normal weight, then "spontaneously" developed obesity six months later by voluntarily overeating. Some monkeys have been chronically chair-restrained for the major portion of their time in the laboratory, sometimes exceeding 1 year; however, although showing normal intake and weight gain, obesity did not occur in association with the restricted activity of these monkeys. Thus, the likelihood of the spontaneous occurence of obesity in laboratory confined *Macaques* seems low, and often requires many years to become manifest.

IV. HYPERPHAGIA AND OBESITY FROM HYPOTHALAMIC LESIONS IN MONKEYS

Hyperphagia and obesity due to bilateral hypothalamic lesions was first described in one of fifty monkeys lesioned for this purpose by Ranson *et al.*, (30) perhaps testifying more to the inaccuracy of the method at that time than to the magnitude of the neural regulation of feeding behavior. This single monkey subsequently developed diabetes mellitus, a finding which has been supported by the more recent work of Hamilton and associates (13,14,15,16,17,18). Table III summarizes the results of a number of studies of "hypothalamic obesity" in monkeys. Notable is the low incidence of successful achievement of lesions resulting in obesity. It appears that this type of obesity cannot be as readily obtained in monkeys as in rats. In general the ventromedial hypothalamic-lesioned monkeys were hyperphagic, and utilized food more efficiently than controls. After initial large weight gains, food intake became relatively normal (17). Some monkeys also had glycosuria, and subsequent development of diabetes mellitus was not uncommon. Abnormal insulin production, hyperlipidemia, decreased K-glucose were also characteristic. The adipose tissue was found consistently to be primarily abdominal including subcutaneous and omental deposits. One animal has been studied by Hamilton (13) for more than 14 years following lesions of the ventromedial hypothalamus. Glucosuria was first noted seven months after surgery. Insulin therapy was required for the accompanying diabetes mellitus beginning 3 years after lesioning. The chief difficulty with

TABLE III Hypothalamic Obesity in Monkeys

Reference	No. of monkeys lesioned	No. of monkeys developing obesity	Starting wt. (kg) of obese monkeys	Ending wt. (kg) of obese monkeys	Distribution of adipose tissue	Comment
30	50	1	2.6	9	abdominal	developed diabetes
9	12	6	4-6	?	abdominal	low food intake no glycosuria
1	17	1	3.5-4.5	?	----	hyperphagic
15-18	?	9+	5.5	>10	abdominal	3 developed diabetes

the animal has been mastication, since its teeth have been worn off by the enormous amounts of lab chow ingested over this period.

Problems with the Use of Hypothalamic-lesioned Monkeys as Models of Human Obesity: The major advantage in the use of hypothalamic lesions for the production of obesity is the rapidity with which obesity and its metabolic alterations can be induced. Disadvantages, however, outweigh advantages. In addition to the difficulty of obtaining consistently effective location, size, and effect of hypothalamic lesions, several other factors suggest the value of seeking nonlesion models of obesity. Although hypothalamic lesions have been used for more than three decades, and more recently central feeding mechanisms, we still know little about the nature of the disturbance which results in obesity. "Knife-cut" studies in rats (33) have suggested the importance of fibers of passage as well as the nuclei themselves, and other researchers have implicated the limbic system, the reticular formation, and other extra-hypothalamic areas of the central nervous system in feeding regulation. The hypothalamus may be involved in the initiation of feeding, but it is clearly not the control center it was once thought to be.

Since the role of the hypothalamic area is not well understood, the obesity which results from its lesioning or stimulation (41) may be relevant only to the rare cases of human obesity which have, to date, been traced to hypothalamic disturbances (7). Clearly, electrolytic lesions destroy a part of the very system which one wishes to understand, and thus are less than optimal in the search for further understanding of feeding regulation.

Regardless of the role attributed to the hypothalamus and other areas in the central nervous system, one is still left with the question of the nature of the "input" to those cells. What is the "signal" which is sensed by the nervous system? Researchers have denervated nearly every source of nervous input to the brain without ascertaining that the loss of one afferent system prevents regulation of energy balance.

On the other hand, explorations of possible blood-borne factors, while incomplete, have also been unsuccessful in determining the nature of any humoral factor signaling feeding state or adiposity to the brain cells. One series of still unexplained studies in monkeys demonstrated that, during continuous cross circulation of blood between a normal weight, ad-lib fed monkey, and totally deprived monkey, there was no change in the food intake of the first monkey (45). Thus, although the blood from one monkey was delivering nutrients to two, it was not stimulated to double its intake

(Figure 2).

In summary, neither the role of various parts of the central nervous system, nor the signal(s) which serve the regulation of feeding behavior and body weight are as yet understood.

V. OVERFEEDING TO PRODUCE OBESITY IN MONKEYS

The infrequent occurrence of spontaneous obesity among monkeys, and the drawbacks to the use of hypothalamic-lesioned animals as models of human obesity, have led to the development of an overfeeding method to produce obese monkeys which includes the following protocol.

An intragastric silastic catheter (i.d. 0.062 in., o.d. 0.125 in.) is implanted so that one centimeter protrudes into the body of the stomach. A teflon cuff attached to the catheter is used to secure the cannula, and the distal tip is led out through the abdominal wall in the vicinity of the umbillicus. Simultaneously, a smaller catheter (i.d. 0.040 in., o.d. 0.085 in.) is implanted in the jugular vein to be used for blood sampling. The monkey is then placed in a specially designed nylon fabric dress, the skirt of which is secured to the middle tier of a 3-tiered primate restraining chair. Catheters exit between the middle and lower tier and are, therefore, inaccessible to the monkey. Following recovery from surgery and return of normal feeding and rate of weight gain, intragastric infusion is begun, using the same complete liquid diet which the monkey is consuming orally. The liquid diet (Ensure[R], Ross Laboratories) was formulated for use in adult man, and appears to be completely adequate for the maintenance of healthy adolescent and adult rhesus monkeys (See page 59). The use of a liquid diet and a suction-activated automated feeder permit the precise calculation of food intake, together with the recorded pattern of that intake (44).

In a series of experiments using this model, the initial intragastric infusion was calculated to equal the average oral intake of the monkey for the preceding two-week period (hereafter considered the 100% level). This amount was delivered intragastrically and continuously over 8 hours, the same time period generally available for feeding. As seen in Figure 3 (Monkey M), some oral intake usually continued for as long as the 100% intragastric infusion was maintained (2 to 4 weeks). Two monkeys were allowed to remain at the 100% IG infusion level for over one month, and in both, oral intake continued approximately 15-25% to the caloric total. Figure 4 shows the comparison between a normal weight monkey and an

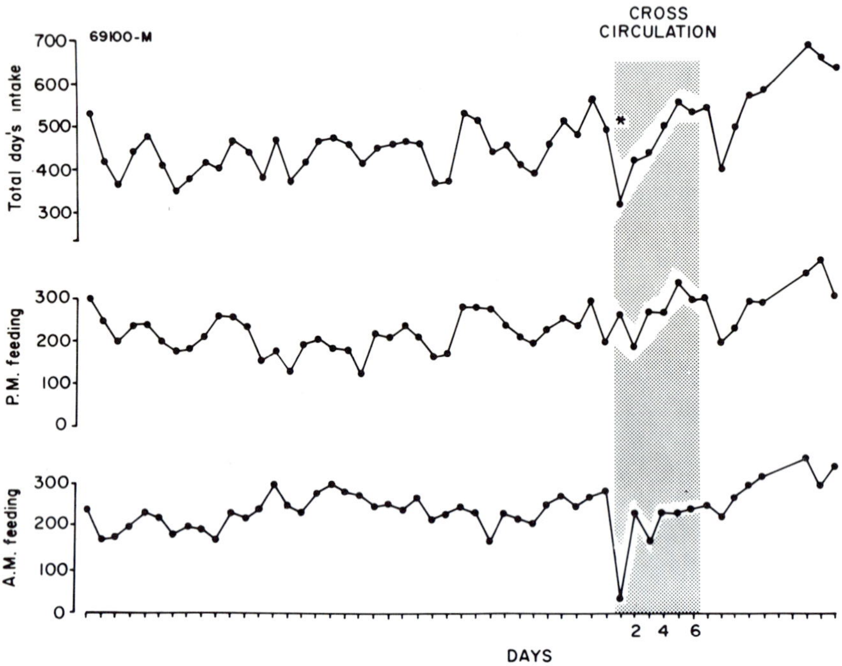

Fig. 2. Food intake of monkey M before, during and following 6 days of continuous cross circulation of this monkey with a totally deprived monkey. Because cross circulation was initiated immediately prior to the AM feeding, the intake of the first meal (and therefore, the total intake for day one) was depressed. The star indicates the projected total intake for the first day if the first meal had been similar to the mean of previous and succeeding days. (Redrawn from Walike et al., J. Comp. Physiol. Psych. 80: 372-381, 1972.)

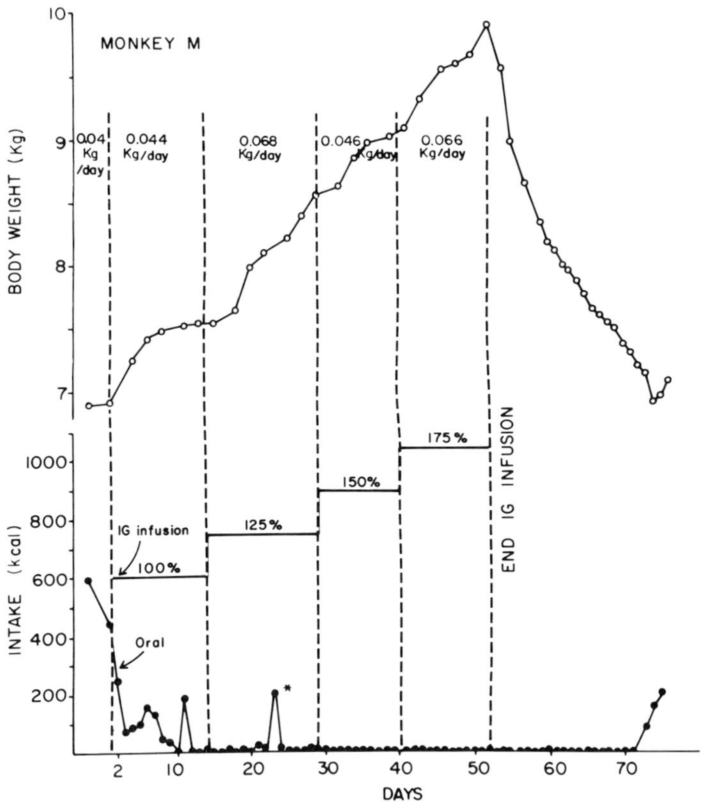

Fig. 3. Oral, intragastric and total intake of monkey M during progressive increases in overfeeding, weight gain, and spontaneous remission. The star indicates the amount of food taken orally when IG feeding pump was not turned on until PM that day.

overfed obese monkey which had a similar original body weight as the normal one before the overfeeding schedule started.

The level of intragastric infusion was increased in all monkeys over the subsequent 2 week period to 125%, 150% and finally 175%, or until evidence of diarrhea or vomiting indicated lack of tolerance. Most animals tolerated an infusion

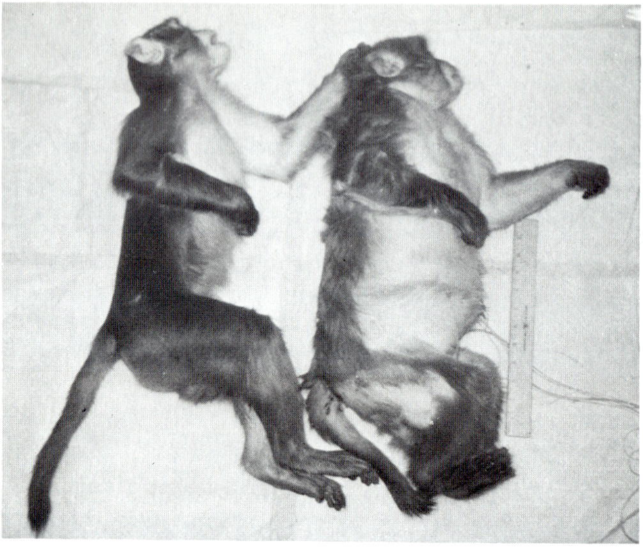

Fig. 4. Comparison between a normal weight monkey and an overfed obese monkey, the leg of which is shaved for adipose biopsy.

of 165 to 175% of baseline intake. Following cessation of overfeeding, oral intake generally remained suppressed until normal body weight was approached (Figure 3).

Although total intake remained fairly constant at each level above 125% (with oral intake suppressed), weight gain for each level of infusion was not linear. Figure 5 traces the body weight of six monkeys who were overfed for varying lengths of time, and then permitted to spontaneously return to normal body weight. Duration of overfeeding was adjusted depending upon the percent overweight desired. Monkeys gained 20 to 67% above their own initial body weight in these experiments. The rate of weight gain varied directly with the caloric load given to each monkey, as shown in Figure 6. Hyperinsulinemia usually developed parallel to the increase in body weight, while fasting glucose changed only slightly (Figure 7). Table IV indicates the initial insulin level, the level at maximal overfeeding, and variability seen between individual monkeys. All values are based on 16 samples drawn after a 16 hour fast. Analysis of mean insulin values for 17 monkeys indicates that fasting insulin is not correlated with body weight per se. On the other hand, analysis of baseline insulin values compared to insulin levels during overfeeding

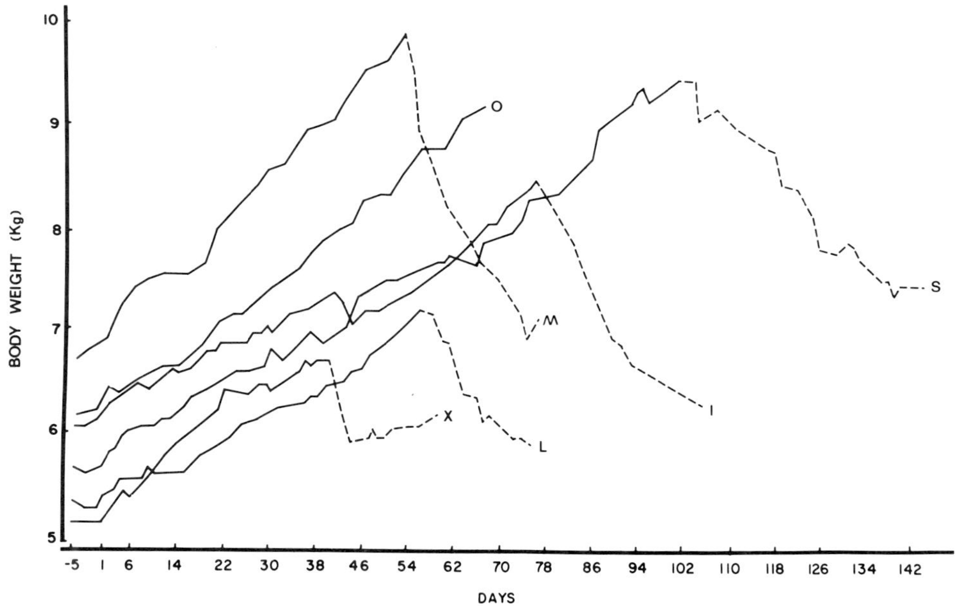

Fig. 5. Body weights of 5 monkeys during overfeeding (____) and remission (----) of obesity.

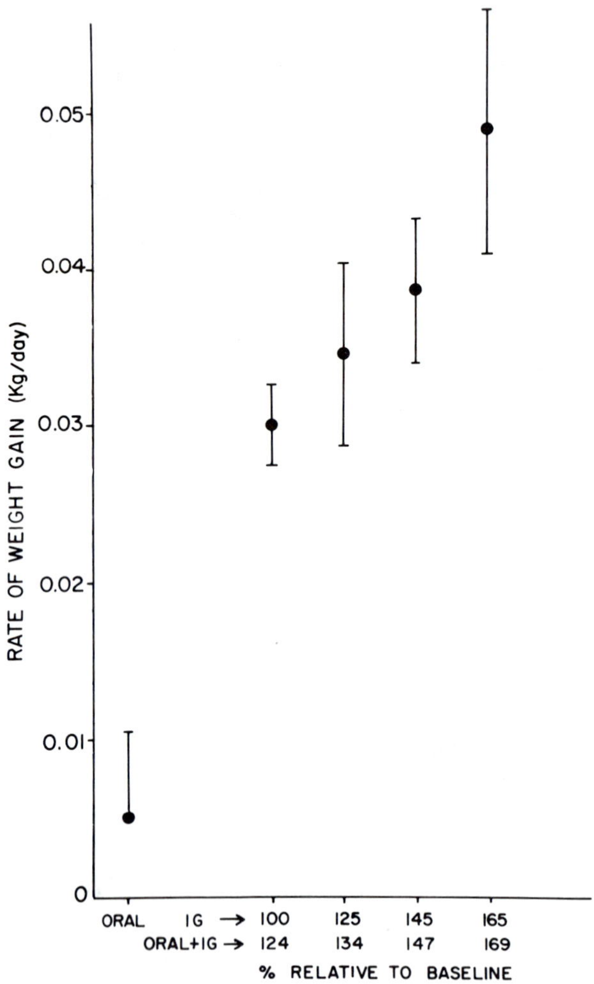

Fig. 6. Rate of weight gain compared to level of intragastric caloric load relative to baseline intake.

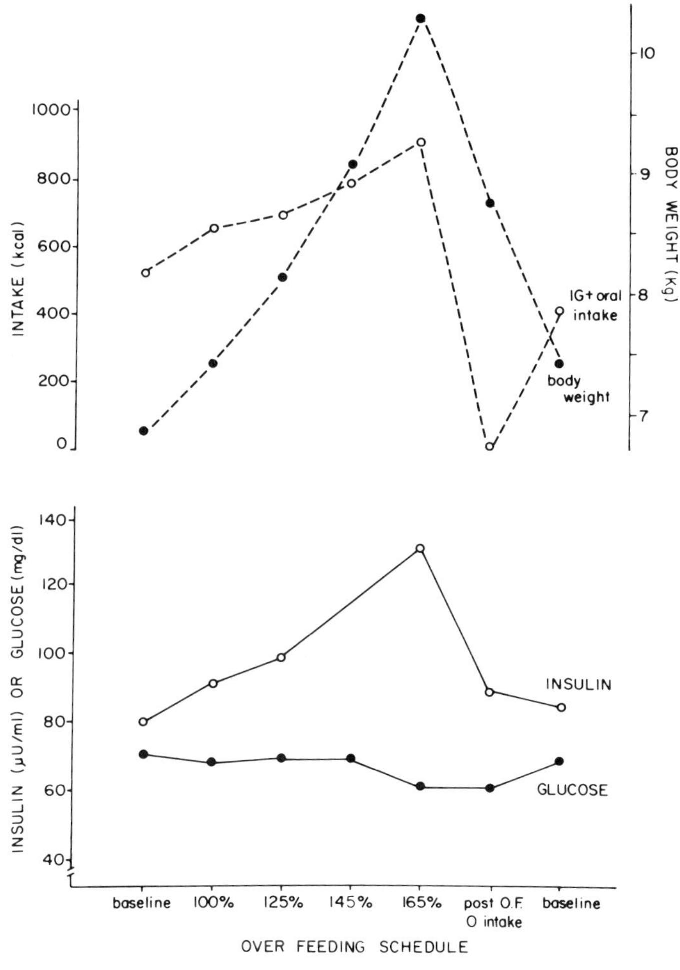

Fig. 7. Insulin, glucose, body weight and total intake during intragastric overfeeding of one monkey.

TABLE IV. Effect of Overfeeding on Mean Fasting Insulin

Monkey	*Insulin (μU/ml)*	
	Baseline	*At highest body weight*
C	53	100
F	69	197
I	117	92
L	56	67
S	62	90
X	127	136
Mean ± SE	81 ± 13	114 ± 19

of 6 monkeys indicate a significant correlation of basal insulin with adiposity or overfeeding (Table V). Oral and IV glucose tolerance tests were either normal or slightly impaired.

In another series of experiments in which rapid sampling of blood immediately preceded the initiation of feeding of monkeys, oscillations of insulin, glucose, glucagon, C-peptide, and pancreatic polypeptide were observed at 8-10 minute cycles (12). Examination of the effects of feeding and obesity on the frequency and amplitude of these oscillations in plasma levels has been initiated. Deprivation of food for 40 to 88 hours clearly damps the amplitude of these oscillations. Obesity and (or) overfeeding, by contrast, appear to increase the amplitude of oscillations of these pancreatic polypeptides; however, further studies are needed to verify this initial observation (19).

Critical Aspects in the Overfeeding Method for Modeling Human Obesity: Several aspects of this method require special attention if obesity is to be successfully achieved. Of primary importance is the use of a well-designed and carefully

TABLE V Correlation of Body Weight and Insulin Levels In Control and Overfed Monkeys

Criterion	Controls 17 monkeys	Baseline for overfeed 6 monkeys	During overfeeding of 6 monkeys
N (Expmts)	28	11	11
$\bar{x}$ Insulin	87.7	83.1	100.5
$\bar{x}$ BWT	15.5	15.9	17.0
r I x BW	.036	.061	669*

* $p < 0.05$

maintained restraint system. We have used a 3-tiered plexiglass chair with seat bars and foot bars. The monkey wears a nylon dress, the skirt of which is securely attached to the middle tier of the chair. Where a recent neck incision has been made, a collar is added to the dress, and the collar attached to the upper tier, thus preventing scratching of the fresh incision with its underlying jugular cannula. The plate at the belly level must be adjustable to accommodate the unusually large increase in abdominal girth with overfeeding. The other apertures of the chair must be carefully adjusted to match the particular monkey. Monkeys are inspected daily and chairs adjusted as needed in order to prevent the development of pressure sores. Stool and urine collections are accomplished using plastic bags cut to fit below the seat bars.

The use of a liquid diet, which is an excellent bacterial culture medium, increases the importance of daily cleaning of chairs. Cannula exit sites are cleaned as needed. A number of suitable commercially formulated liquid diets are available (See Page 59).

The intragastric overfeeding infusion procedure depends on the use of a large bore tube (i.d. 0.062 in.) and the regular flushing of the tube with water after each day's infusion to prevent inside coating of the tube with diet.

The key to any method of studying obesity is the adequate assessment and validation of the degree of obesity obtained. We have used four methods: body weight and weight change, abdominal skinfold measurement, tritiated water dilution to estimate lean body mass, and fat cell size determinations and estimation of fat cell number. Body weight alone is not sufficient to establish the presence of obesity, although this criteria has sometimes been used (18). In establishing the presence of obesity, the caliper measurement of abdominal skinfold can be used for gross determinations. We have validated general ranges of these values in obese and normal pigtailed monkeys (43). Norms based on large scale surveys, or other species, are not available. We routinely use the tritiated water dilution technique to estimate adiposity. The methodological error is estimated to be 4% (39). Fat cell size and number determinations, made on biopsies from multiple sites, have significant potential for indicating the nature of the induced obesity (21,42). No norms have yet been established for monkeys, and initial studies are now underway.

Problems with the Use of Intragastric Overfeeding as a Model for Human Obesity: While intragastric overfeeding probably produces obesity which is similar to, or identical with, human spontaneous obesity, there are several drawbacks to this model. The obesity can only be sustained by continued overfeeding, for, as in experimental obesity in humans, there is a strong tendency to return to normal body weight. Thus, the overfeeding procedure does not produce the "intractable" kind of obesity seen in most spontaneously obese humans.

The procedure itself has occasional attendant complications. Among those we have observed are diarrhea, septicemia (arising presumably from contamination of a venous catheter), and edematous scrotum and lower extremities (seen only once in an animal that was septicemic). Potential problems, which we have successfully avoided, include the development of "chair sores" from improper fitting of restraint, and the inadvertent dislodging of the gastric cannula (a likely consequence if the monkey's foot can reach the abdominal area). Our experience with this procedure has shown it to be unusually benign, with behavior, food intake, and weight gains following normal expectations for caged animals.

As in all studies of feeding and obesity, there remain the difficulties of assessing energy expenditure. We have made no attempt at this, but have assumed a relatively constant energy expenditure level as a consequence of chronic chair restraint and environmental temperature control.

V. FUTURE DIRECTIONS FOR RESEARCH ON OBESITY USING NONHUMAN PRIMATES

The potential studies which could be optimally and appropriately carried out on experimental and/or spontaneous obesity in monkeys abound. Among those which appear to be particularly promising and which may utilize the chronically catheterized overfed monkey are studies evaluating body weight, adiposity, feeding behavior, and endocrine function during systematic variations in the type of dietary constituents and caloric intake of the intragastric infusion. Such investigations should lead to increased understanding of the mechanisms of adaptation of various levels of food intake and adiposity. Changes in the size and numbers of fat cells under these varied diet and weight conditions must be monitored to provide further data, not available in man, on the nature of changes in adipose tissue structure and function. Such a primate model, combined with the administration of streptozotocin to produce diabetes, may lead to increased knowledge of which primary metabolic aberrations lead to specific secondary complications in diabetes. The interrelations of obesity and diabetes, the so-called "odd couple" (5), are still to be defined in terms of mechanism.

The overfed monkey also provides a model for further examining the interactions of carbohydrate and lipid metabolism, and the interactive effects of intake on hormone and substrate levels, and vice versa. It will be important to define the nature of the signals which cause the monkey to resume oral intake as body weight returns from obesity toward normal. The identification of such physiologic changes could be an important key to understanding the regulation of appetite.

REFERENCES

1. Anand, B.K., Dua, S., and Schoenberg, K. (1955) Hypothalamic control of food intake in cats and monkeys. *J. Physiol. 127: 143-152.*
2. Astwood, E.B. (1962) The heritage of corpulence. *Endocrinology 71: 337-341.*
3. Bagdade, J.D., Bierman, E.L., and Porte, D., Jr. (1967) The significance of basal insulin levels in the evaluation of insulin response to glucose in diabetic and nondiabetic subjects. *J. Clin. Invest. 46: 1549-1157.*

4. Berkowitz, D. (1964) Metabolic changes associated with obesity before and after weight reduction. *JAMA 187: 399-403.*

5. Bierman, E.L., Bagdade, J.D., and Porte, D., Jr. (1968) Obesity and diabetes: the odd couple. *Am. J. Clin. Nutr. 21: 1434-1437.*

6. Bray, G.A. (1976) The obese patient. Philadelphia: W. B. Sanders.

7. Bray, G.A., Gallagher, T.F., Jr. (1975) Manifestations of hypothalamic obesity in man: A comprehensive investigation of eight patients and a review of the literature. *Medicine 54: 301-330.*

8. Bray, G.A., and York, D.A. (1971) Genetically transmitted obesity in rodents. *Physiol. Rev. 51: 598-646.*

9. Brooks, C.M., Lambert, F.F., and Bard, P. (1942) Experimental production of obesity in the monkey *(Macaca mulatta). Fed. Proc. 1: 11.*

10. Bryce, G.F., Johnson, P.R., Sullivan, A.C., and Stern, J.S. (1977) Insulin and glucagon: plasma levels and pancreatic release in the genetically obese Zucker rat. *Horm. Metab. Res. 9: 366-370.*

11. Frohman, L.A., Goldman, J.R., Schnatz, S.D., and Bernardis, L.L. (1971) Hypothalamic obesity in the weanling rat: Effect of diet upon hormonal and metabolic alterations. *Metabolism 20: 501-512.*

12. Goodner, C.J., Walike, B.C., Koerker, D.J., Chideckel, E., Ensinck, J.W., Brown, A.C., Palmer, J., and Kalnasy, L.W. (1977) Insulin, glucagon and glucose exhibit synchronous, sustained oscillations in fasting monkeys *(Macaca mulatta). Science 195: 177-179.*

13. Hamilton, C.L. (1972) An observation of long-term obesity and diabetes mellitus in the monkey. *J. Med. Primatol. 1: 247-255.*

14. Hamilton, C.L., and Brobeck, J.R. (1963) Diabetes mellitus in hyperphagic monkeys. *Endocrinology 73: 512-515.*

15. Hamilton, C.L., and Brobeck, J.R. (1964) Hypothalamic hyperphagia in the monkey. *J. Comp. Physiol. Psych. 57: 271-278.*

16. Hamilton, C.L., and Brobeck, J.R. (1965) Control of food intake in normal and obese monkeys. *N.Y. Aca. Sci. 131: 583-592.*

17. Hamilton, C.L., Ciaccia, P.J., and Lewis, D.O. (1976) Feeding behavior in monkeys with and without lesions of the hypothalamus. *Am. J. Physiol. 230: 818-830.*

18. Hamilton, C.L., Kuo, P.T., and Feng, L.Y. (1972) Experimental production of syndrome of obesity, hyperinsulinemia and hyperlipidemia in monkeys. *Proc. Soc. Exp. Biol. Med. 140: 1005-1008.*
19. Hansen, B.W., Goodner, C.J., Koerker, D.J., Kalnasy, L. W., Brown, A.C., Ensinck, J.W., and Rubenstein, A.H. (1977) Factors influencing sustained oscillations of glucose, insulin, C-peptide, and glucagon in rhesus monkeys. *The Physiologist 20: 41.*
20. Hirsch, J., and Batchelor, B. (1976) Adipose tissue cellularity in human obesity. *Clin. Endoc. Metab. 5: 299-311.*
21. Hirsch, J., and Gallian, E. (1968) Methods for the determination of adipose cell size in man and animals. *J. Lipid Res. 9: 110-119.*
22. Hirsch, J., Knittle, J.L., and Salans, L.B. (1966) Cell lipid content and cell number in obese and non-obese human adipose tissue. *J. Clin. Invest. 45: 1023.*
23. Horton, E.S. (1974) Epidemiologic aspects of obesity. in: *Understanding and Successfully Managing Obesity.* L.P. Rubin, ed. Postgrade Med. Sympos. 51-61.
24. Montagu, A. (1966) Obesity and the evolution of man. *JAMA 195: 149-151.*
25. Neumann, R.O. (1902) Experimentalle Beigrage zur Lehr von dem taglichen Nahrunsbedarf des Menschen unter besonderer Berucksichtigung der nonwendigan Eiweissmenge. *Arch. Hyg. 45: 1.*
26. O'Connell, M., Danforth, E., Jr., Horton, E.S., Salans, L.P., and Sims, E.A.H. (1973) Experimental obesity in man. III. Adrenocortical Function. *J. Clin. Endoc. Metab. 36: 323-329.*
27. Opie, L.H., and Walfish, P.G. (1963) Plasma free fatty acid concentrations in obesity. *N.E.J. Med. 268: 757-760.*
28. Paulsen, E.P., and Lawrence, A.M. (1968) Glucagon hypersecretion in obese children. *Lancet 2: 110.*
29. Rabinowitz, D. (1970) Some endocrine and metabolic aspects of obesity. *Ann. Rev. Med. 21: 241-258.*
30. Ranson, S.W., Fischer, C., and Ingram, W.R. (1938) Adiposity and diabetes mellitus in a monkey with hypothalamic lesions. *Endocrinology 23: 175-181.*
31. Salans, L.B., Bray, G.A., Cushman, S.W., Danforth, E., Jr., Glennon, J.A., Horton, E.S., and Sims, E.A.H. (1974) Glucose metabolism and the response to insulin by humans in spontaneous and experimental obesity: Effects of dietary composition and adipose cell size. *J. Clin. Invest. 53: 848-856.*

32. Salans, L.B., Horton, E.S., and Sims, E.A.H. (1971) Experimental obesity in man: Cellular character of the adipose tissue. *J. Clin. Invest. 50: 1005-1011.*
33. Scalfani, A. (1971) Neural pathways involved in the ventromedial hypothalamic lesion syndrome in the rat. *J. Comp. Physiol. Psych. 77: 70-96.*
34. Sims, E.A.H. (1976) Experimental obesity, dietary-induced thermogenesis, and their clinical implications. *Clin. Endocrino. and Math. 5: 377-395.*
35. Sims, E.A.H., Danforth, E., Jr., Horton, E.S., Bray, G.A., Glennon, J.A., and Salans, L.B. (1973) Endocrine and metabolic effects of experimental obesity in man. *Recent Prog. Horm. Res. 29: 457-487.*
36. Sims, E.A.H., Goldman, R.F., Gluck, C.M., Horton, E.S. Kelleher, P.C., and Rowe, D.W. (1968) Experimental obesity in man. *Trans. Assoc. Am. Phys. 81: 153-170.*
37. Sims, E.A.H., and Horton, E.S. (1968) Endocrine and metabolic adaptation to obesity and starvation. *Am. J. Clin. Nutr. 21: 1455-1470.*
38. Sims, E.A.H., Horton, E.S., and Salans, L.B. (1971) Inducible metabolic abnormalities during development of obesity. *Ann. Rev. Med. 22: 235-250.*
39. Siri, E.A.: in Brozek, J., and Henschel, A. (1961) Conference on techniques for measuring body composition, National Academy of Science, National Research Council, Washington, D.C.
40. Sjostrom, L., and Bjorntorp, P. (1974) Body composition and adipose tissue cellularity in human obesity. *Acta. Med. Scand. 195: 201-211.*
41. Steffens, A.B. (1975) Influence of reversible obesity on eating behavior, blood glucose and insulin in the rat. *Am. J. Physiol. 228: 1738-1744.*
42. Stern, J.S., and Greenwood, M.R.C. (1974) A review of development of adipose cellularity in man and animals. *Fed. Proc. 33: 1952-1955.*
43. Walike, B.C., Goodner, C.J., Koerker, D.J., Chideckel, E.W., and Kalnasy, L.W. (1977) Assessment of obesity in pigtailed monkeys. *J. Med. Primatol. 6: 151-162.*
44. Walike, B.C., Campbell, D.J., and Hillman, R.A. (1971) A liquid diet feeder for monkeys. *J. Appl. Physiol. 31: 946-947.*
45. Walike, B.C., and Smith, O.A. (1972) Regulation of food intake during intermittent and continuous cross circulation in monkeys *(Macaca mulatta)*. *J. Comp. Physiol. Psych. 80: 372-381.*
46. Zucker, L.M., and Antoniades, H.N. (1972) Insulin and obesity in the Zucker genetically obese rat "fatty." *Endocrinology 90: 1320-1330.*

NUTRITION AND INFECTION INTERRELATIONSHIPS IN MONKEYS

Robert W. Wannemacher, Jr.
Clayton L. Hadick, Jr.
William R. Beisel

United States Army Medical Research
Institute of Infectious Diseases
Fort Detrick, Frederick, Maryland

INTRODUCTION

The nonhuman primate has been used extensively at USAMRIID as a model to study evolvement of infectious diseases due to organisms which are pathogenic for man (Table 1, Ref. 1-33). Approximately 25 different bacterial, rickettsial or viral diseases have been studied in these primates at USAMRIID. While rhesus *(Macaca mulatta)* and cynomolgus *(Macaca fascicularis)* monkeys have been used for a large number of model infections, African green *(Cercopithecus aethiops)*, squirrel *(Saimiri sciureus)* and silver-leaf *(Presbytis crystatus)* monkeys and chimpanzees have also proven to be valuable nonhuman primate species (Table I). Marked differences were observed in family and species susceptibility to the various infectious microorganisms. While *Rickettsia tsutsugamushi* (scrub typhus) produced a lethal illness in the silver-leaf

ISBN 0-12-334250-3

TABLE I Susceptibility of Various Species of Nonhuman Primates (Monkeys and Chimpanzees) to Infectious Microorganisms, Emphasizing those Studied at USAMRIID

Microorganism or Produce	Response[a]						Ref.
	Rhesus	Cyno	Afr. green	Squirrel	Silver-leaf	Chimp.	
Bacteria							
Streptococcus pneumoniae	++ (IV) ± (AR, IT)	++	ND[b]	± (IT)	ND	ND	1-7
Salmonella typhi	ND	ND	ND	ND	ND	++ (oral, AR)	8,9
Endotoxin	ND	ND	ND	ND	ND	+	10
Salmonella typhimurium	+ (IV)	++ (IV)	ND	ND	ND	ND	2,3,7
Endotoxin	+ (IV)	ND	ND	ND	ND	ND	11
Francisella tularensis SCHU S4	++ (AR, SC)	ND	ND	ND	ND	ND	12-14
Live, vaccine	± (AR) + SC	±	ND	ND	ND	ND	12,13
Pasteurella pestis	++ (SC) ND	ND	ND	ND	ND	ND	15
Klebsiella pneumoniae	ND	++	ND	ND	ND	ND	16
Bacillus anthracis	++ (AR, SC)	ND	ND	ND	ND	++ (SC)	17-19
Staphylococcal enterotoxin B	++ (IV)	ND	ND	ND	ND	ND	20,21
Rickettsia tsutsugamushi	ND	ND	ND	ND	++ (ID)	ND	22

Microorganism or Product	Response[a]						Ref.
	Rhesus	Cyno.	Afr. green	Squirrel	Silver-leaf	Chimp.	
Bacteria							
Rickettsia rickettsii	++ (SC,AR)	++ (SC)	ND	ND	ND	ND	23
Rickettsia conorii	ND	+ (SC)	ND	+ (SC)	ND	ND	24
Rickettsia sibirica	ND	+ (SC)	ND	+ (SC)	ND	ND	24
Viruses							
Influenza	± (AR,IT,SC)	ND	ND	++	ND	ND	4,5, 25,26
Yellow fever Asibi	++ (IP)	± (IP)	ND	ND	ND	ND	27,28
17D	± (IP)	ND	ND	ND	ND	ND	28
Venezuelan equine encephalomyelitis							
Trinidad	+ (IP)	ND	ND	ND	ND	ND	3,29
TC-83 vaccine	± (IP)	± (IP)	ND	ND	ND	ND	13
Japanese B encephalitis	++ (IN)	± (SC,IP,IV)	ND	ND	ND	ND	30
Bolivian hemorrhagic fever	++ (SC)	++ (SC)	++ (SC)	ND	ND	ND	31,32
Korean hemorrhagic fever	± (SC)	± (SC)	ND	ND	ND	ND	33

[a]++ = severe illness; + = no clinical illness, but neurological conversion. [b]Not done
IV = intravenous; IT = intratracheal; AR = aerosol; SC = subcutaneous; IP = intraperitoneal; IN = intranasal.

monkey, it produced no clinical illness in rhesus or cynomolgus monkeys (23). Similarly, *Salmonella typhi* (typhoid fever) was lethal in chimpanzees, but had little effect on rhesus or cynomolgus monkeys (8,9). Not only were differences noted between species and families of nonhuman primates, but also marked variations were observed between various members of the same genus. Yellow fever virus is extremely lethal for rhesus monkeys, but produces no clinical illness in the cynomolgus, even though both are members of the *Macaca* genus (3,27,28). Similarly, *Salmonella typhimurium* produces only a mild illness in rhesus, but an often lethal infection in cynomolgus monkeys (2,3,7), thus emphasizing the importance of screening a number of species of primates in order to find a model which simulates the progression of an infectious illness as seen in man.

The route of administration of the infectious organism can also influence the pattern of infectious disease produced in nonhuman primates. Japanese B encephalitis (JBE) virus produces a lethal infection in rhesus monkeys if it is administered intranasally (IN) but no clinical illness if injected subcutaneously (SC), intraperitoneally (IP) or intravenously (IV) (30). Similarly, the live, vaccine strain of *Francisella tularensis* produces clinical illness if injected SC, but has no effect if given by aerosol (AR) (12,13). Therefore, when developing infectious disease models, one must consider both species and route of administration in order to obtain the desired clinical illness.

Most nonhuman primate models have been utilized to study the pathogenesis of immunologic and serologic response to infectious disease, including therapy and the prophylactic value of various live and killed vaccines. Although severe anorexia is characteristic of most infectious diseases in nonhuman primates (3), only a limited number of studies have been concerned with the interaction of nutrition and infectious disease. However, several models have been developed to study the following: (a) the effects of chronic diet-induced hypercholesterolemia on the humoral and cellular immune functions of the rhesus monkey (34); (b) effects of gram-positive or gram-negative septicemia or endotoxemia on the ability of the rhesus monkey to handle a lipid load (35,36); and more recently (c) the value of parenteral and enteral (gavage) nutrition for assessing the interactions between nutrition and infectious disease. These three models will be described in more detail to demonstrate the use of nonhuman primates for study of these interactions.

II. HYPERCHOLESTEROLEMIA AND ALTERED IMMUNITY

Nutritional deficits are generally believed to reduce the ability of an individual to resist infection. In contrast, it has also been concluded by investigators who studied either bacterial or viral infections in different species of laboratory animals (37-39), that host resistance in experimentally induced infections may be diminished by the presence of obesity or long-term ingestion of a diet containing abnormally high quantities of fat. In an attempt to define the mechanisms by which lipid excess is detrimental to the host, eight rhesus monkeys were fed a diet containing excess fat and cholesterol, as described by Armstrong *et al.* (40), to produce chronic hypercholesterolemia; this diet contained both crystalline and egg yolk cholesterol. As a control, 8 monkeys were fed a normal biscuit diet low in fat. These diets were fed for 3 months, at which time the high fat, cholesterol group had a serum cholesterol value of 412 ± 38 mg/dl (mean ± SEM), while the controls had 123 ± 4 mg/dl (34). At this time all monkeys were inoculated SC with 15 mg ovalbumin in complete Freund's adjuvant; 3 weeks later they were given a SC booster of 15 mg of albumin in normal saline. As illustrated in Table 2, the appearance of hemagglutinin antibodies against albumin was similar in both dietary groups. The hypercholesterolemic monkeys exhibited an altered progression of precipitin response. Following the primary inoculation of albumin, 4 of 8 controls developed high titered precipitin responses to this antigen, while 7 of 8 demonstrated a precipitin responses to the booster and generally showed precipitin bands against more than one component of albumin (Table 2). In contrast, none of the hypercholesterolemic monkeys developed precipitins to the primary inoculation, 4 of 8 showed a precipitin response to the booster, and none had multiple band responses (34).

Two months after the ovalbumin booster, all monkeys were skintested with 5 μg of ovalbumin. A delayed positive reaction, with an area greater than 10 mm diameter at 24 and 48 h, occurred in equal proportions of monkeys in both groups (34). Four controls, however, displayed combined Arthus reactions during a 4-h period after skin testing, in contrast to none in the hypercholesterolemic group (Table II).

Tuberculin skin tests with 0.1 ml intermediate strength PPD were negative in all monkeys prior to initiation of the diet, but 5 hypercholesterolemic monkeys converted to positive two months after immunization with ovalbumin in complete Freund's adjuvant. Skin tests remained positive thereafter (34), whereas they remained negative in control monkeys fed

TABLE II *Immunological Response of Rhesus Monkeys Fed a Normal or Hypercholesterolemic Diet*

Immune Mechanisms	*Cholesterolemia*		
	Normal	*Hyper*	*P*
	Number responding per group		
Hemagglutinin response to ovalbumin	7/8	6/8	NS[a]
Primary precipitin response to ovalbumin	4/8	0/8	<0.05
Multiple precipitin bands to ovalbumin	5/8	0/8	<0.01
Anamnestic response to ovalbumin booster	8/8	4/8	<0.05
Arthus (immediate) skin test reaction to ovalbumin	4/6	0/6	<0.02
Positive TBC skin test	0/8	5/7	<0.01

[a]*Not significant.*

a normal diet (Table II). Since the possibility of active tuberculosis was not detected by extensive clinical, radiological and histological examinations or by culture, the tuberculin reactions seemed to suggest increased sensitivity to the mycobacterial components incorporated into the adjuvant.

Seven months after starting the diets, an acute infection was initiated in both groups of monkeys by IV inoculation of 10^8 virulent *Streptococcus pneumoniae*, type I, strain A-5. Within 24 h, all monkeys of both dietary groups developed fever, pneumococcal bacteremia, and a 2- to 4-fold increase in total blood neutrophils. During the brief period of infection, clinical responses of both groups were equivalent.

When tested 24 h after inoculation, leukocyte preparations from control monkeys showed an average incidence of 13% neutrophils that could reduce nitroblue tetrazolium (NBT) dye. The average value was significantly higher in the hypercholesterolemic monkeys, with 24% of their cells reducing NBT. Greater than 20% dye reduction was noted in 5 of 6 monkeys in the high cholesterol group. Another group of uninfected monkeys on a normal diet for 7 months averaged only 3 to 4% of peripheral blood neutrophils with the capability of reducing NBT.

The clearance rate of colloidal carbon from the blood was measured in monkeys 24 h after inoculation with pneumococci. During infection, the rate of clearance by the reticuloendothelial system in both normal and high cholesterolemic monkeys was significantly increased above that of uninfected controls fed the biscuit diet; however, infected hypercholesterolemic monkeys had a significantly smaller increase in clearance time than the average increase measured in infected control monkeys.

These findings combine to indicate that experimentally induced hypercholesterolemia in rhesus monkeys was associated with atypical humoral and cellular immunodefense responses. When studied 27 h after infection, there were no differences in gross pathology between groups, and none of the hypercholesterolemic monkeys had atheromatous lesions in their arteries. Since all the monkeys in these studies were killed at a single time, no information was obtained concerning the overall effect of hypercholesterolemia on survival following a pneumococcal infection. Accordingly, the observed differences in immune responsiveness could not be related in this study to changes in overall host resistance or to survival.

Therefore, a second study was initiated in which rhesus monkeys were fed either the normal biscuit diet or the high-fat diet of Armstrong *et al.* (40), with or without added cholesterol. AFter 3 months on these diets, the monkeys were challenged with a series of infectious organisms or toxins as outlined in Table III. At the end of 3 months, the monkeys fed the high-fat diet with added cholesterol had increased serum cholesterol concentrations similar to those observed in the first study, while monkeys fed the high-fat diet without cholesterol had normal concentrations with slightly elevated triglycerides. Survival after a normally nonfatal Venezuelan equine encephalomyelitis virus infection (3) was similar in all three dietary groups. All developed viremia and neutralizing antibodies of comparable titers.

Prior to assigning monkeys to given dietary groups, half had been immunized with staphylococcal enterotoxin B (SEB) to determine whether high fat and/or cholesterol would influence the immunological response to challenge with SEB. The first dose (1 μg/kg) was injected IV on day 0, the second (10 μg/kg), IV on day 7, and the third, intradermally (ID) 1 to 4 months later. All monkeys developed high hemagglutinin and precipitin titers. After 4 to 5 months on their assigned dietary regimen, both immune and nonimmune monkeys were injected IV with 10 μg/kg of SEB (Table III). Following challenge, survival was similar in immunized and nonimmunized monkeys from all three dietary groups. In monkeys fed the high-fat diet with or without cholesterol, bactericidal activity was not increased after SEB challenge in the nonimmune monkeys; leukopenia was absent following challenge in the immune group. Also the anamnestic hemagglutinin response was delayed after challenge in the hypercholesterolemic, SEB-immune monkeys.

Following tularemia immunization with a live vaccine strain of F. tularensis (Table III), all monkeys developed specific antibody titers, but the magnitude of the response was slightly less in those fed the high-fat diet with or without added cholesterol. All monkeys had positive tularemia skin tests with few differences among the three dietary groups. All monkeys then survived challenge with a normally lethal dose of virulent SCHU S4 strain of *F. tularensis*, but both high-fat groups had persistent neutrophilia and an antibody response which was delayed and decreased in magnitude when compared to monkeys fed the control diet.

Both studies indicate that experimentally induced hypercholesterolemia associated with a high-fat diet in rhesus monkeys results in altered humoral and cell-mediated defense responses. The hypercholesterolemic groups of monkeys clearly differed from controls in the various responses to different antigens and to infections themselves. However, this degree of altered immune function did not produce detectable differences in host resistance to infectious disease or depress protective antibody formation following immunization.

TABLE III Immunization Treatment Schedule for Monkeys Fed a Control, High-fat or High-fat Plus Cholesterol Diet

Time (months)	*Treatment*[a]
0	*Dietary regimens initiated*
3	*Venezuelan equine encephalomyelitis (VEE) (412 PFU, Trinidad strain, SC)*
4-5	*Staphylococcal enterotoxin B (SEB) (10 μg SEB/kg body weight, IV)*
10	*Tularemia immunization (10^6 live vaccine strain, ID)*
13	*Tularemia skin test - Foshay vaccine (0.1 ml, 1:1 - 1:100 dilution, ID)*
14	*F. tularensis challenge (4 x 10^4 SCHU S4 strain, SC)*

[a]*SC = subcutaneous*
IV = intravenous
ID = intradermal

These studies were both performed before development of currently available tests to measure various aspects of cellular immunity, such as T- and B-cell responses. Further, the monkeys did not overconsume the high-fat diet and develop obesity which has been observed in some dog models (38,39). Thus, the induced alterations in lipid metabolism in monkeys may not have been sufficiently long or severe to have affected their resistance to infectious disease. The studies do demonstrate, however, that nonhuman primates can be utilized to evaluate the effects of prior nutrition on susceptibility and immune responsiveness to infectious organisms.

III. LIPID METABOLISM DURING SEPSIS

In attempting to understand the mechanisms responsible for the elevated plasma lipids during certain infectious disease in man and experimental animals (37), rates of lipid disposal following either oral or IV lipid loading were determined during gram-positive or gram-negative sepsis in normally nourished rhesus monkeys (35,36). These studies were performed on the monkeys prior to and 48 h after IV inoculation with either 10^9 *S. typhimurium* or 10^8 *S. pneumoniae*. In the oral loading test, disposal of triglycerides was measured following intragastric administration of 4 g-kg of a corn oil emulsion (Lipomul, Upjohn Co., Kalamazoo, Michigan). The IV lipid tolerance test measured disappearance of triglycerides following the infusion of 0.5 g/kg of a soybean oil emulsion (Intralipid, Cutter Laboratories, Berkeley, California) over 30 minutes.

In the fasted, noninfected monkey given the oral lipid load, serum triglycerides increased from 20 to 40 mg/dl by 24 h. In contrast, serum triglyceride values increased to 150 mg/dl and 500 mg/dl 24 h after such loading of monkeys septic with *S. pneumoniae* or *S. typhimurium*, respectively. Following the IV infusion of lipid, serum triglycerides increased to approximately 900 mg/dl and decreased rapidly in the noninfected control monkeys to 50 mg/dl by 4 h after stopping the infusion. Despite a similar maximum elevation in serum triglycerides, the rate of disappearance was slightly slower in the monkeys infected with *S. pneumoniae*, while in the the monkeys septic with *S. typhimurium*, the rate of disappearance was reduced so that their 4-h post-infusion values had only fallen to approximately 600 mg/dl. The reduced rate of lipid disposal in infected monkeys correlated with a 25% or 75% decrease in post-heparin lipolytic activity in the monkeys inoculated, respectively, with *S. pneumoniae* or *S. typhimurium*.

A similar decrease in the rate of disappearance of an infused lipid load and depression in plasma post-heparin lipolytic activity were observed in monkeys 4 h after injection of 7 mg/kg body weight of *S. typhimurium* endotoxin. These observations suggest that the marked impairment in lipid disposal mechanism during gram-negative sepsis in rhesus monkeys was in part associated with endotoxin contained in the walls of this gram-negative microorganism.

The association of elevated fasting triglyceride values, intolerance to exogenous lipid loading, and decreased post-heparin lipolytic activity suggest that lipid disposal is impaired during bacterial sepsis. Further, the defective lipid disposal observed in these studies either produced or contributed to the elevated triglyceride values observed. Thus, the subhuman primate model has proven to be a useful means of measuring the effects of infection on the ability to handle a nutrient lipid load.

IV. PARENTERAL NUTRITIONAL MODEL IN NORMAL MONKEYS

In the older literature, it was suggested that increasing the caloric intake during certain bacterial infections in man could reduce wasting of body proteins (41). However, anorexia is a common symptom of most infectious diseases (42,43) which complicates the evaluation of nutritional support in septic man and experimental animals. To help determine whether various substrates could prevent wasting of body proteins during sepsis, techniques were adapted from clinical studies in man in order to develop parenteral and enteral nutrition models in monkeys.

Rhesus or cynomolgus monkeys, weighing 3 to 5 kg, were adapted to restraint in metabolic chairs for 3 days. Chairs were designed to allow quantitative collection of urine and feces. During the 3-day adaptation, the monkeys were fed *ad libitum* and had free access to water. Only those monkeys seen to be eating normally, maintaining body weight, having normal white blood cell counts and clinical chemistries, and showing no tendencies for excessive hyperactivity were accepted into the study.

After this adaptation, the monkeys were anesthetized with 10 mg/kg Ketamine (Parke-Davis Co., Detroit, Michigan) and prepared for surgery. In the parenteral nutrition model, an incision was made on the right side of the neck and the right carotid artery and internal jugular vein were dissected free of connecting fascia and skeletal muscle. Both vessels were ligated; sterilized silastic catheters (0.085 in, outside diameter, Dow Corning Corp., Midland, Michigan) were implanted under aseptic surgical conditions. The jugular catheter was inserted to the approximate point where the jugular vein joins the vena cava, and was utilized for the infusion of nutrient solution. The carotid catheter was inserted to reach the level of the aortic arch and was utilized to obtain arterial blood samples. The catheters were tunneled SC over the scapula and exited from the body at the base of the skull.

In the enteral nutrition model, a neck catheter was implanted only in the right carotid artery. A mercury-tipped, 7.3 French nasogastric tube (Hedco, Palo Alto, California) was inserted through the nares and esophagus into the lower stomach and utilized for the infusion of nutrient solutions.

In both models an incision was made in the upper right thigh and the femoral vein freed from surrounding fascia and skeletal muscle. The vein was ligated and a PE 190, polyethylene catheter (Clay Adams, Parsippany, New Jersey) was inserted up to the caudal vena cava just proximal to the bifurcation of the iliac veins. This was utilized to obtain venous blood samples from the lower hind-quarters.

After surgery, monkeys were returned to their metabolic chairs and the catheters were then connected under sterile conditions to 3-way stopcocks and placed in a sterile plastic "glove" box containing gauze saturated with Prepodyne (Poloxamer iodine complex). All catheters were kept patent by infusion with 0.5% saline containing 0.5 U/ml heparin at a flow rate of 0.5 ml/h. On the day after surgery a test solution was infused via the jugular vein in the parenteral model and by the nasogastric tube in the enteral model. The basic parenteral nutrition solution for monkeys is shown in Table 4; it supplied 0.55 g/kg/day of amino acid nitrogen (FreAmine II, McGraw, Irvine, California,) 85 cal/kg/day of dextrose) (McGraw), electrolytes, trace elements and vitamins (10 ml/liter of M.V.I., USV Pharmaceutical Corp., Tuckahoe, New York). The basic enteral nutrition solution is also presented in Table IV, with the major differences being that the enteral preparation contained nonprotein calories from complex carbohydrates (Sumacal, Hospital Diet Products, Corp., Buena Park, California) and a soybean oil emulsion (Intralipid). In addition, the enteral solution contained approximately 10 times the concentration of trace elements. Infusion rates were monitored by a small IV infusion pump (IVAC Corp., San Diego, California). Each pump was calibrated at the beginning and end of the experiment to insure accuracy of the infusion rate.

Immediately after starting the infusion, each monkey was administered a red stool marker via an intragastric bolus of 3 ml of an oral suspension of pyrvinium pamoate (Parke-Davis). Twentyfour-hour urine and red feces collections were made on each monkey for the next 6 days; 24-h urine volumes were recorded; each specimen was analyzed for pH, glucose, protein and ketones by dip-stick (Multi-Stix, Ames, Co., Elkhart, Indiana); samples were frozen at -20°C for future analysis. Fecal samples were placed in plastic 50-ml graduated tubes, digested and homogenized as described previously (6). Samples of urine were analyzed for total nitrogen, urea, amonia, creatinine, glucose and phosphorus by automated procedures (6). Samples of feces and nutrient solution were analyzed for total nitrogen.

Fasting blood samples were obtained on the day the monkeys were placed in metabolic chairs, immediately before surgery, before starting the nutrient infusion, and on days 1 to 3 and 6 of nutritional support. Red and white blood cell counts, hemoglobin, hematocrit and differential counts and platelets were measured. Plasma was separated from the blood and analyzed for glucose, urea, nitrogen, sodium, potassium, chloride, total proteins and albumin by automated procedures (6).

TABLE IV Composition of Basic Parenteral[a] and Enteral[b] Nutrition solutions for Monkeys

Variable	Parenteral	Enteral
Infusion rate (ml/kg/day)	100	125
Nutrients		
Amino acid N (g) (FreAmine II)	0.55	0.55
Dextrose (cal/kg/day)	85	0
Carbohydrate (Sumacal)	0	55
Lipid (Intralipid)	0	45
Electrolytes (mEq/liter)		
Sodium	12.8	7.18
Potassium	3.94	4.05
Chloride	12.3	6.44
Phosphorus	3.75	3.83
Calcium	0.68	0.51
Magnesium	1.69	1.57
Trace elements (μg/kg/day)		
Zinc	150	1000
Copper	30	350
Iron	100	1000
Vitamins		
Oil-soluble (IU/kg/day)		
A	940	940
D	94	94
E	0.5	0.5
Water-soluble (mg/kg/day)		
Ascorbic acid	47	47
Thiamine HCl	4.7	4.7
Pyridoxine HCl	1.4	1.4
Riboflavin	0.9	0.9
Niacinamide	9.4	9.4
Dexpanthenol	2.3	2.3

[a]The solution is prepared under a laminar-flow hood (with UV light) from sterile stock solution. Solutions are periodically checked for sterility after preparation. No positive cultures have been detected.

[b]645 mOsM solution.

Approximately 96 monkeys have been utilized in the parenteral and enteral nutrition studies; none have had an elevated body temperature which was not associated with experimentally induced sepsis. One monkey showed a positive blood culture for *Staphylococcus aureus* and 8 were positive for *Streptococcus epidermidis* on one of the 7 days studied. This latter organism is nonpathogenic and is often found on skin surfaces. Therefore, its presence may have represented contamination during collection of a sample. None of the catheters removed at the end of the experiments had positive cultures for microorganisms. This would suggest a low incidence of catheter sepsis, and that the model could be used for studies of experimentally induced infections.

Four rhesus control monkeys were utilized to evaluate the ability of the basic parenteral nutrition solution to maintain nutritional status in a noninfected host. As shown in Figure 1, monkeys were in positive nitrogen balance throughout the 6-day experimental period. Fecal excretion rates were very low and were absent in all monkeys on 4 of the 6 days of study. Body weights were similar at the time monkeys were placed in metabolic chairs, and when the experiment was

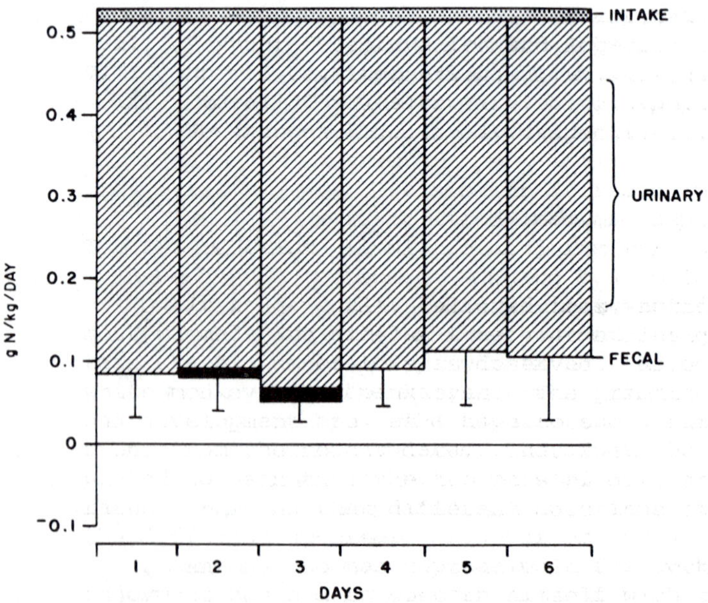

Fig. 1. Nitrogen intake (dots), urinary nitrogen (slanted lines), fecal nitrogen (black bars) and nitrogen balance (mean ± SEM) for 4 monkeys indicates that intake exceeded urinary and fecal excretion during the 6-day parenteral feeding trial.

terminated 10 days later (Table 5). Also body temperature was unchanged throughout the experiment; there was no indication of bacterial contamination.

No significant change was observed in the rate of urea nitrogen excretion; ammonia nitrogen was decreased compared to day 1 of the study, suggesting that the infused amino acids were being efficiently utilized by the monkey. Urinary creatinine excretion was similar for the 6 days of study (Table V), indicating that muscle mass was being maintained by the nutritional solution. Urinary glucose excretion was high on the day the infusions were started and then progressively decreased (Table V). Urinary phosphorus excretion also tended to decrease compared to day 1.

The stress of chairing the monkeys resulted in significant decreases in hemoglobin, hematocrit and lymphocyte concentration (Table 6, compare day -4 with day -1). Slight but nonsignificant decreases were also noted in white blood cell, with a slight increase in neutrophils. The stress of surgery caused an even more marked neutrophilia. No significant changes in hematology were associated with the infusion of the basic parenteral nutrition solution (Table VI).

Plasma glucose was slightly but significantly increased on days 1 and 3 at the beginning of the parenteral solution (Table VII). This increase in the plasma glucose concentration should not have exceeded tubular maximum for the kidney, but significant amounts of glucose were seen in the urine, especially on day 1, at the start of the infusions (Table V). The mechanism by which the glucose spilled into the urine during infusion of hyperosmolar dextrose solutions will require further investigation.

Blood urea nitrogen tended to decrease following initiation of the parenteral nutrition solution, indicating that positive nitrogen balance was not associated with the accumulation of nitrogen in extracellular fluids. No changes in plasma sodium, potassium or chloride were observed as a result of chairing or due to the parenteral nutrition solution (Table VII).

Total plasma protein, and more specifically plasma albumin tended to decrease as a result of the infusion of the basic parenteral nutrition solution. A similar decrease in plasma albumin concentrations has been observed in septic and nonseptic monkeys infused with the amino acid and hyperosmolar dextrose solution (6). A possible explanation is that the 10-fold increase in peripheral insulin values observed in monkeys infused with the amino acid and dextrose solution stimulated sequestration of amino acids into skeletal muscle (6).

TABLE V Sequential Changes in Body Weight, Body Temperature, Bacteremia[a] and Excretion of Urea, Ammonia, Creatinine, Glucose and Phosphorus

Parameter	Value ± SEM (n = 4) by days							
	-4	0	1	2	3	4	5	6
Body weight	3.7	ND[b]	ND	ND	ND	ND	ND	3.8
(kg)	± 0.2							± 0.2
Body temperature	ND	37.5	37.6	37.8	37.7	37.4	37.4	37.1
(C^{O})		± 0.2	± 0.1	± 0.1	± 0.2	± 0.1	± 0.1	± 0.1
Urea N	ND	ND	418	306	277	292	315	318
(mg/kg/day)			± 54	± 63	± 16	± 17	± 21	± 37
Ammonia N	ND	ND	41	33*	27*	27*	32*	31*
(mg/kg/day)			± 4	± 5	± 4	± 4	± 4	± 4
Creatinine	ND	ND	39.7	38.4	37.5	36.8	36.2	39.3
(mg/kg/day)			± 2.9	± 0.9	± 2.4	± 1.8	± 1.8	± 1.5
Glucose	ND	ND	857	130*	127*	124*	53*	107*
(mg/kg/day)			±150	± 32	± 59	± 40	± 10	± 45
(dip-stick)			(+++)	(+)	(±)	(±)	(-)	(-)
Phosphorus	ND	ND	38.5	28.5	17.5*	19.1*	17.0*	15.7*
(mg/kg/day)			± 3.0	± 3.1	± 3.2	± 3.9	± 1.6	± 3.6

[a]Bacteremia negative on all days.

[b]Not determined.

*$P < 0.05$ vs. day 1.

TABLE VI Sequential Changes in Blood Hematology

Parameter	Mean value ± SEM (n = 4) by days						
	-4	-1	0	1	2	3	6
Red blood count	5.30	4.75	4.43*	4.82	4.37*	3.98*	4.22*
(No./mm^3 x 10^{-6})	± 0.30	± 0.30	± 0.37	± 0.42	± 0.39	± 0.23	± 0.36
Hemoglobin	12.9	10.7*	10.3*	10.5*	10.0	9.5*	9.5*
(g/dl)	± 0.6	± 0.6	± 0.8	± 0.5	± 0.5	± 0.2	± 0.3
Hematocrit	38.8	33.7*	31.3*	34.0*	31.3*	29.1*	31.0*
(%)	± 1.1	± 1.7	± 2.3	± 2.4	± 2.0	± 1.1	± 2.2
White blood count	5.29	5.55	8.89	7.78	7.72	6.08	6.83
(No./mm^3 x 10^{-3})	± 1.09	± 1.12	± 1.90	± 0.74	± 1.32	± 1.27	± 1.47
Neutrophils	1.91	3.53	6.51*†	4.90*	5.22*	3.38*	4.35*
(No./mm^3 x 10^{-3})	± 0.38	± 1.13	1.70	± 0.39	± 1.28	± 0.84	1.17
Lymphocytes	3.11	2.39	2.00*	2.67	2.08	2.10*	2.16*
(No./mm^3 x 10^{-3})	± 0.47	± 0.56	0.27	± 0.08	0.31	± 0.30	0.30
Eosinophils	0.11	0.17	0.18	0.16	0.15	0.34	0.20
(No./mm^3 x 10^{-3})	± 0.05	± 0.06	0.09	± 0.08	± 0.08	± 0.06	± 0.07
Monocytes	0.05	0.06	0.09	0.08	0.23	0.30*	0.07
(No./mm^3 x 10^{-3})	± 0.05	± 0.06	± 0.09	± 0.08	± 0.15	± 0.06	± 0.07
Platelets	3.54	3.61	1.94	3.58	1.85	2.36	2.85
(No./mm^3 x 10^{-5})	± 0.43	± 0.38	± 0.36	± 0.66	± 0.54	± 0.71	± 0.36

*$P < 0.05$ vs. day-4.
†$P < 0.05$ vs. day-1.

TABLE VII Sequential Changes in Blood Chemistry

Parameter	Mean value ± SEM (n - 4) by days						
	-4	-1	0	1	2	3	6
Glucose	69	76	80	107*	87	106*	94
(mg/dl)	± 3	± 10	± 8	± 12	± 7	± 9	± 6
Blood urea nitrogen	14	16	12	10	11*	10*	9*
(mg/dl)	± 1	± 1	± 2	± 1	± 1	± 1	± 1
Sodium	155	151	147	148	144	148	149
(mEq/liter)	± 7	± 2	± 2	± 1	± 1	± 1	± 1
Potassium	3.8	3.2	3.2	3.2	3.1	4.1	4.1
(mEq/liter)	± 0.1	± 0.1	± 0.1	± 0.2	± 0.1	± 0.5	± 0.2
Chloride	104	104	107	108	109	106	109
(mEq/liter)	± 2	± 2	± 1	± 1	± 1	± 2	± 2
Total plasma protein	6.53	6.53	5.88	5.70*	5.30*	5.30*	6.08
(g/dl)	± 0.49	± 0.49	± 0.21	± 0.35	± 0.29	± 0.14	± 0.31
Plasma albumin	4.48	4.43	4.52	3.88*	3.40*†	3.33*†	3.40*†
(g/dl)	± 0.19	± 0.14	± 0.40	± 0.29	± 0.11	± 0.15	± 0.08

*$P < 0.05$ vs. day -4.
†$P < 0.05$ vs. day 0.

Data from these studies suggest that the basic parenteral nutrition solution supplying 0.55 g of nitrogen and 100 kcal/kg/day will maintain positive nitrogen balance, body weight and muscle mass. Very low incidences of bacterial infection (or contamination) were observed in this nonhuman primate model. Some changes were observed in blood hematology as a result of chairing monkeys and surgical implantation of catheters, and the parenteral nutrition solution tended to increase plasma glucose slightly and to decrease albumin concentrations. Despite these minor changes, the parenterally nourished rhesus monkey appears to be a useful model for studying the interaction of nutrition and infection on the metabolic and pathogenic response of the host.

Studies with the enteral nutrition model in nonhuman primates are preliminary but they do suggest that the solution described in Table 4 will maintain positive nitrogen balances and body weights in noninfected monkeys.

V. PARENTERAL NUTRITION MODEL DURING INFECTION

Recently the parenteral nutrition model in rhesus monkeys was utilized to determine whether substrate therapy could prevent wasting of body protein during sepsis (6). On the day after starting the infusion, the monkeys were injected via the femoral veins with live (test) or heat-killed (controls) *S. pneumoniae*. By day 2 the monkeys injected with live organisms became ill; both septic and control monkeys were started on antibiotic therapy. By day 3 after treatment was started, the septic monkeys had lysis of fever and negative blood cultures and they, along with controls, had their catheters removed and cultured.

During pneumococcal sepsis, monkeys were infused via a central venous catheter with a constant amount of amino acids with or without added isocaloric amounts of either dextrose or lipid emulsion. In monkeys infused with amino acids alone, pneumococcal sepsis resulted in a 4-fold increase in loss of body proteins, compared to calorie-restricted controls. Addition of 85 cal/kg/day of either dextrose or lipid markedly reduced body wasting associated with infectious disease. The calories from lipid were utilized by the septic host as a source of energy with a slightly reduced efficiency when compared to the isocaloric infusion with dextrose. Therefore, it could be concluded that septic monkeys seemed to be able to utilize fatty acids as an energy substrate. In addition, it appeared that the carbohydrate calories tended to favor synthesis of peripheral proteins (associated mainly with skeletal muscle), while lipid calories favored synthesis of visceral proteins, such as plasma albumin and acute-phase proteins.

Because of the presence of two cannulas from which blood samples could be drawn, this monkey model has proven to be a very effective technique for measuring the effect of infection and nutrition on the kinetics of glucose and alanine metabolism by isotope dilution methodology during continuous infusion (44). After obtaining a blood sample from the venous catheter, the monkeys were given a priming injection of 1 μCi/kg body weight each of 2-[^{3}H]glucose and U-[^{14}C]alanine (New England Nuclear, Boston, Massachusetts). Immediately after this priming dose, a continuous 3-h infusion of a solution that contained 3 μCi of each isotope was administered to the monkey at a rate of 1 ml/h. A 5-ml blood sample was drawn from the carotid catheter at 60 and 180 minutes. Plasma was promptly separated and deproteinized by the addition of sulfosalicylic acid (25 mg/ml). A sample of the filtrate was analyzed for alanine, [^{14}C]alanine, glucose and [^{3}H] glucose content. Data were then utilized to calculate glucose and alanine production and utilization, as well as the rate of gluconeogenesis. These studies have shown that both glucose production and utilization were significantly increased during pneumonococcal illness when compared to noninfected controls or monkeys studied after recovery from infection. Despite lysis of fever and disappearance of bacteremia, the newly convalescent monkeys do produce and utilize more glucose than control monkeys. During pneumococcal illness, there was a significant 2-fold increase in the rate of glucose production from alanine as compared to infected controls or to a period of early recovery in the same monkey. Further, the rates of alanine production and utilization were increased 2-fold during illness in monkeys infected with *S. pneumoniae* compared to noninfected controls or early after recovery of the infected group.

When percent of alanine utilization for glucose synthesis was calculated, it was observed that nearly 90% of the alanine utilized during pneumococcal sepsis was for the production of glucose compared to 60% in studies of control monkeys or the same monkeys early after recovery from infectious illness. These observations suggest that during calorie restriction, body proteins are broken down and amino acids are converted to alanine, which is subsequently utilized at an increased rate for glucose production.

Since the femoral vein catheter drained blood coming from the lower hind-quarter of the monkey and the carotid arterial catheter measured aortic blood, it was possible to measure arterial-venous differences mainly across skeletal muscle of the monkeys infused with the nutrient solutions. By this technique it was possible to determine that during sepsis in the calorie-restricted monkey, glucose and branched-chain amino acids were taken up at increased rates and lactate, alanine, and glutamine were released in elevated amounts from skeletal muscle (45). The results have been interpreted as indicating that during sepsis anaerobic glycolysis is elevated with the resultant release of lactate, where branched-chain amino acid oxidation is increased with the resultant synthesis of alanine and glutamine. These data suggest that during sepsis skeletal muscle will utilize branched-chain amino acids and glucose as a source of energy, resulting in an increased release of alanine, glutamine and lactate to be used as substrates by the liver and kidney for gluconeogenesis, ureagenesis and ammoniagenesis. This explains in part the marked catabolic effects of infectious disease.

Preliminary studies with the enteral nutritional model have suggested that during pneumococcal sepsis, stomach-emptying time and intestinal mobility may be decreased, with resulting build-up of infused fluid in the stomach and a tendency toward vomiting and nausea. This problem may be reduced by use of an enteral nutrition solution with a lower osmolar content.

VI. CONCLUSIONS

The nonhuman primate has been effectively utilized as a model to study infectious illness produced by microorganisms pathogenic for man. Consideration must be given to the species of nonhuman primate and the route of administration of microorganisms to produce an illness characteristic of a particular infectious disease.

Studies on interaction of infection and nutrition in non-human primates have been hampered by the marked anorexia which is common to most infectious diseases. However, monkey models have been effectively utilized to study the effect of prior nutrition on the immunological and survival response to infectious disease. These models have also been used to study the infected host's ability to handle a substrate load. More recently, parenteral and enteral models have been developed which allow for effective study of the interaction of nutrition and infection during various stages of disease. These models have been effectively utilized to study the protein-sparing effects on various nutrients as well as metabolic interactions of infectious disease.

Despite the problems of requiring chair-restraint and relatively small sample size, the monkey models utilized in these studies do provide useful data which can be applied to the problems of treatment of infectious disease of critically ill patients. The models have an advantage over clinical studies, in that the effects of infection can be evaluated in a subject not compromised by other problems such as injury, surgical complications, radiation, chemotherapy or organ failure. It is also possible to evaluate nutritional support attempts which may or may not be advantageous for the monkey's defense mechanism against infectious disease.

ACKNOWLEDGMENTS

In conducting the research described in this report, the investigators adhered to the "Guide for the Care and Use of Laboratory Animals," as promulgated by the Committee on the Revision of the Guide for Laboratory Animal Facilities and Care of the Institute of Laboratory Animal Resources, National Research Council. The facilities are fully accredited by the American Association for Accreditation of Laboratory Animal Care.

The views of the authors do not purport to reflect the positions of the Department of the Army or the Department of Defense.

REFERENCES

1. Woeber, K.A. (1971) Alterations in thyroid hormone economy during acute infection with *Diplococcus pneumoniae* in the rhesus monkey. *J. Clin. Invest. 50: 378-387.*
2. Fiser, R.H., Denniston, J.C., and Beisel, W.R. (1972) Infection with *Diplococcus pneumoniae* and *Salmonella typhimurium* in monkeys: changes in plasma lipids and lipoproteins. *J. Infect. Dis. 125: 54-60.*
3. Kastello, M.D., and Spertzel, R.O. (1973) The rhesus monkey as a model for the study of infectious disease. *Am. J. Phys. Anthropol. 38: 501-504.*
4. Berendt, R.F., McDonough, W.E., and Walker, J.S. (1974) Persistence of *Diplococcus pneumoniae* after influenza virus infection in *Macaca mulatta*. *Infect. Immun. 10: 369-374.*
5. Berendt, R.F., Long, G.G., and Walker, J.S. (1975) Influenza alone and in sequence with pneumonia due to *Streptococcus pneumoniae* in the squirrel monkey. *J. Infect. Dis. 132: 689-693.*

6. Wannemacher, R.W., Jr., Kaminski, M.V., Jr., Neufeld, H.A., Dinterman, R.E., Bostian, K.A., and Hadick, C.L. (1978) Protein sparing therapy during pneumococcal sepis in rhesus monkey. *J. Parenteral Enteral Nutr.* (in press).
7. Wannemacher, R.W., Jr. (1978) The cynomolgus monkey as a nonhuman primate model for *Streptococcus pneumoniae* infection. Unpublished data.
8. Edsall, G., Gaines, S., Landy, M., Tigertt, W.D., Sprinz, H., Trapani, R.J., Mandel, A.D., and Berenson, A.S. (1960) Studies on infection and immunity in experimental typhoid fever. I. Typhoid fever in chimpanzees orally infected with *Salmonella typhosa*. *J. Exp. Med. 112: 143-166.*
9. Tully, J.G., Gaines, S., and Tigertt, W.D. (1963) Studies on infection and immunity in experimental typhoid fever. V. Respiratory challenge of chimpanzees with *Salmonella typhosa*. *J. Infect. Dis. 113: 131-138.*
10. Tully, J.G., Gaines, S., and Tigertt, W.D. Studies on infection and immunity in experimental typhoid fever. VI. Response of chimpanzees to endotoxin and the effect of tolerance on resistance to organ challenge. *J. Infect. Dis. 115: 445-455.*
11. Fiser, R.H., Denniston, J.C., and Beisel, W.R. (1974) Endotoxemia in the rhesus monkey: alterations in host lipid and carbohydrate metabolism. *Pediatr. Res. 8: 13-17.*
12. Eigelsbach, H.T., Tulis, J.J., Overholt, E.L., and Griffith, W.R. (1961) Aerogenic immunization of monkey and guinea pig with live tularemia vaccine. Proc. Soc. Exp. Biol. Med. 108: 732-734.
13. Sawyer, W.D., Kuehne, R.W., and Gochenour, W.S., Jr. (1964) Stimultaneous aerosol immunization of monkeys with live tularemia and live Venezuela equine encephalomyelitis vaccines. *Milit. Med. 129: 1040-1043.*
14. Sawyer, W.D., Jemski, J.V., Hogge, A.L., Jr., Eigelsbach, H.T., Wolfe, E.K., Dangerfield, H.G., Gochenour, W.S., Jr., and Crozier, D. (1966) Effect of aerosol and age on the infectivity of airborn *Pasteurella tularensis* for *Macaca mulatta* and man. *J. Bacteriol. 91: 2180-2184.*
15. Marshall, J.D., Jr., Harrison, D.N., Rust, J.H., Jr., and Cavanaugh, D.C. (1971) Serological response of rhesus monkeys (*Macaca mulatta*) to immunization and infection with *Pasteurella pestis*. *Proc. Soc. Exp. Biol. Med. 138: 738-841.*

16. Knutsen, G.L. Effect of aerosol-disseminated respiratory infection in monkey models. In Annual Progress Report, FY 1977. U.S. Army Medical Research Institute of Infectious Diseases, Fort Detrick, Frederick, Md., 197-204.
17. Gochenour, W.S., Jr., Gleiser, C.A., and Tigertt, W.D. (1962) Observations on penicillin prophylaxis of experimental inhalation anthrax in the monkey. *J. Hyg. (Camb.) 60: 29-33.*
18. Berdjis, C.C., Gleiser, C.A., and Hartman, H.A. (1963) Experimental parenteral anthrax in *Macaca mulatta*. *Br. J. Exp. Pathol. 44: 101-115.*
19. Berdjis, C.C., and Gleiser, C.A. (1964) Experimental subcutaneous anthrax in chimpanzees. *Exp. Mol. Pathol. 3: 63-75.*
20. Crawley, G.J., Gray, I., LeBlang, W.A., and Blanchard, J.W. (1966) Blood binding, distribution and excretion of staphylococcal enterotoxin in monkeys. *J. Infect. Dis. 116: 48-56.*
21. Hodoval, L.F., Morris, E.L., Crawley, G.J., and Beisel, W.R. (1968) Pathogenesis of lethal shock after intravenous staphylococcal enterotoxin B in monkeys. *Appl. Microbiol. 16: 187-192.*
22. Walker, J.S., Cadigan, F.C., Vosdingh, R.A., and Chye, T. (1973) The Silvered leaf-monkey of Malasia, *Presbytis cristatus:* disease model for human scrub typhus. *J. Infect. Dis. 128: 223-226.*
23. Sammons, L.S., Kenyon, R.H., Hickman, R.L., and Pedersen, C.E., Jr. (1977) Susceptibility of laboratory animals to infection by spotted fever group rickettsiae. *Lab. Anim. Sci. 27: 229-234.*
24. Kenyon, R.H. Immunochemical studies with nonindigenous tick-borne rickettsiae. In Annual Progress Report, FY 1977, U.S. Army Medical Research Institute of Infectious Diseases, Fort Detrick, Frederick, Md., 45-57.
25. Berendt, R.F., and Hall, W.C. (1977) Reaction of squirrel monkeys to intratracheal inoculation with influenza/A/New Jersey/76 (swine) virus. *Infect. Immun. 16: 476-479.*
26. Stephen, E.L., Hilmas, D.E., Mangiafico, J.A., and Levy, H.B. (1977) Swine influenza virus vaccine: potentiation of antibody responses in rhesus monkeys. *Science 197: 1289-1290.*
27. Tigertt, W.D., Berge, T.O., Gochenour, W.S., Gleiser, C.A., Eveland, W.C., Vorder Bruegge, C., and Smetana, H. F. (1960) Experimental yellow fever. *Trans. N.Y. Acad. Sci., Series II, 22: 323-333.*

28. Stephen, E.L. Effectiveness of selected antiviral compounds against diseases. In Annual Progress Report FY 1977, U.S. Army Medical Research Institute of Infectious Diseases, Fort Detrick, Frederick, Md., 211-217.
29. Gleiser, C.A., Gochenour, W.S., Jr., Berge, T.O., and Tigertt, W.D. (1962) The comparative pathology of experimental Venezuelan equine encephalomyelitis infection in different animal hosts. *J. Infect. Dis. 110: 80-97.*
30. Harrington, D.G., Hilmas, D.E., Elwell, M.R., Whitmire, R.E., and Stephen, E.L. (1977) Intranasal infection of monkeys with Japanese encephalitis virus: clinical response and treatment with a nuclease-resistant derivative of poly (I)·poly (C). *Am. J. Trop. Med. Hyg. 26: 1191-1198.*
31. Kastello, M.D., Eddy, G.A., and Kuehne, R.W. (1976) A rhesus monkey model for the study of Bolivian Hemorrhagic fever. *J. Infect. Dis. 133: 57-62.*
32. Wagner, F.S., Eddy, G.A., Brand, O.M. (1977) The African green monkey as an alternate primate host for studying Machupo virus infection. *Am. J. Trop. Med. Hyg. 26: 159-162.*
33. French, G.R. Characterization and evaluation of selected hemorrhagic fever agents for vaccine development. In Annual Progress Report, FY 1977. U.S. Army Medical Research Institute of Infectious Diseases, Fort Detrick, Frederick, Md., 393-401.
34. Fiser, R.H., Jr., Denniston, J.C., McGann, V.G., Kaplan, J., Adler, W.H., III, Kastello, M.D., and Beisel, W.R. (1973) Altered immune functions in hypercholesterolemic monkeys. *Infect. Immun. 8: 105-109.*
35. Kaufmann, R.L., Matson, C.F., Rowberg, A.H., and Beisel, W.R. (1976) Defective lipid disposal mechanisms during bacterial infection in rhesus monkeys. *Metabolism 25: 615-624.*
36. Kaufmann, R.L., Matson, C.F., and Beisel, W.R. (1976) Hypertriglyceridemia produced by endotoxin: role of impaired triglyceride disposal mechanisms. *J. Infect. Dis. 133: 548-555.*
37. Beisel, W.R., Fiser, R.H., Jr. (1970) Lipid metabolism during infectious illness. *Am. J. Clin. Nutr. 23: 1069-1079.*
38. Newberne, P.M., Young, V.R., and Graviee, J.F. (1969) Effects of caloric intake and infection on some aspects of protein metabolism in dogs. *Br. J. Exp. Pathol. 50: 172-180.*

39. Fiser, R.H., Rollins, J.B., and Beisel, W.R. (1972) Decreased resistance against infectious canine hepatitis in dogs fed a high-fat ration. *Am. J. Vet. Res. 33: 713-719.*

40. Armstrong, M.L., Connor, W.E., and Warner, E.D. (1967) Xanthomatosis in rhesus monkeys fed a hypercholesterolemic diet. *Arch. Pathol. 84: 227-237.*

41. Lusk, G. The elements of the science of nutrition, 4th ed., Philadelphia, WB Saunders Co., 1928, 695-722.

42. Beisel, W.R. (1977) Magnitude of the host nutritional responses to infection. *Am. J. Clin. Nutr. 30: 1236-1247.*

43. Wannemacher, R.W., Jr., and Beisel, W.R. Metabolic response of the host to infectious disease. In: Richards JR, Kinney, J.M. (eds): Nutritional aspects of care in the critically ill, Edinburgh, Churchill-Livingstone, 1977: 135-159.

44. Wannemacher, R.W., Jr., Beall, F.A., Canonico, P.G., Dinterman, R.E., Hadick, C.L., and Neufeld, H.A. (1978) Glucose and alanine metabolism during bacterial infections in rats and rhesus monkey. *Metabolism* (in press).

45. Wannemacher, R.W., Jr., Dinterman, R.E., and Hadick, C. L. Metabolic fuel utilization by skeletal muscle (SM) of rhesus monkeys (RM) during pneumococcal sepsis. *Fed. Proc. 37: 849, 1978* (abstr.).

PRIMATES IN NUTRITIONAL AND DEVELOPMENTAL RESEARCH[1]

Arthur J. Riopelle

Department of Psychology
Louisiana State University
Baton Rouge, Louisiana

I. INTRODUCTION

It is generally believed that, nutritionally speaking, most mammals are similar once you get inside the skin. A certain skepticism of this is deserved. Rats are not little white monkeys with pink eyes. The primate data on nutrition and development are, as I hope to show, substantially different from the rodent data. That fact should not be interpreted to mean that the two sets of data are in contradiction, rather it means that different questions should be asked of the different species.

Anyone who has kept animals for work, for food, or for enjoyment has had to concern himself with the nutritional

[1]*Supported in part by a grant from the National Institute of Child Health and Human Development, NIH (HD-07470).*

ISBN 0-12-334250-3

requirements of his charges. The list of dietary constituents viewed as necessary for, or contributory, to good health is lengthy. Primates have requirements, too, and considerable is known about them. For me to summarize this research on primate requirements, element by element, would be to duplicate excellent reviews already published as well as the updated reviews of this symposium (1,2,3,4,5). The studies covered in these summaries, however, did not specifically address the problem of early development, a topic of considerable import.

II. NUTRITION AND DEVELOPMENT

Recent years have seen an expansion of research effort on early malnutrition. The expansion reflects both worldwide interest in the effects and costs of malnutrition and our hope that our efforts will produce a rationale for developing national and international programs to ameliorate the effects of early malnutrition and the elimination thereby of its potentially irreversible consequences.

If inadequate early nutrition merely resulted in a small body, it would hardly attract any attention despite the millions afflicted by its consequences (6). The promulgation of theories that warned of long-term, adverse consequences in the structural and functional integrity of the organism, including the possibility that intellectual functioning and personality development might be altered, heightened our awareness of this problem.

The expectation seems plausible in light of the well known fact that children born into families where the plane of nutrition is low are small at birth, have neurological and developmental anomalies, and have a high incidence of perinatal mortality, especially among those infants of low birthweight whose gestational period was longer than would be expected from their weight (7). Deviant behavior such as hyperkinesis, autism, involvement in childhood accidents, and difficulty in language development is frequently seen. Academic achievement, physical growth, and motor behavior are also adversely affected (8,9).

The major impetus to connect low intellectual achievement to early malnutrition came from the study by Stoch and Smythe (10). Subsequently, it was observed that monovular twins of widely disparate birthweights have intelligence quotients that reflect their birthweight: the lower the birthweight, the lower the I.Q. (11). Food deprivation in adults is a different matter altogether. The studies of Keys et al. (12) showed that although changes in behavior occurred during periods of extended starvation, there was no intellectual loss nor was there any permanent consequences of the starvation period.

A. Theoretical Background

Rationalization of these findings has been in terms of two dominant theories, both of which emphasize the importance of malnutrition during early development. Davison and Dobbing (13) have observed that morphological development occurs in four arbitrary stages. The first stage of brain development terminates when multiplication of neuroblasts ceases. Stage II lasts from then until the end of the period of rapid myelination and other growth. Stage III occurs throughout maturity, and Stage IV is the stage of senile regression. During the second stage there is a rapid increase in the size of the brain, a "growth spurt." Dobbing (14) noting teratological studies, proposed that vulnerability to adversity is greatest during the period of fastest growth. Insult from viruses, drugs, ionizing irradiation, and from inadequate nutrition at this stage produces gross anatomical deformity. Restricting the rate of growth during the growth spurt will retard development and possibly reduce its ultimate extent, resulting in a smaller adult with a smaller brain with fewer constituents, particularly myelin, and possibly some distortion of the normal pattern.

Winick's proposal (15) is based on Enesco and Leblond's principle (16) that except for a few organs which exhibit continuous cell division, growth begins with hyperplasia (increasing cell number) and follows with hypertrophy (increasing cell size). If cell proliferation cannot be completed within the genetically defined, restricted time allotments because of the unavailability of essential nutrients, permanently reduced cell number results despite later rehabilitation. The brain is no exception. While it is true that glial cells outnumber neurons 9:1, the effects should be the same on both components.

Prior to the advent of the research and the hypotheses just cited, it was thought that the birthweight of the fetus is not significantly influenced by dietary intake of the mother during pregnancy. Severe and prolonged malnutrition had to be experienced before conception if obstetrical performance was to be impaired. This belief held that the fetus is a perfect parasite, its tissues having a prior claim on the nutrients circulating in the maternal blood stream (17).

The Dobbing and the Winick hypotheses imply the contrary; that the mother rather is at best an imperfect filter, passing on to the fetus what remains after more or less adequately extracting her own requirements. The two animals thus compete for the nutrients, and the fetus might occasionally be the loser.

Both hypotheses are specified in terms of intake to the fetus or infant, and both have an important component of temporality. They assume that the mother has acquired no adequate reserves of the essential nutrients acquired in the distant past on which the fetus can draw, the labile stores, with the exception of fat, being limited to only a few days supply and they assert that rehabilitation will not be totally effected if it begins after the deadline for hyperplasia.

Studies of humans have supported the view that the incidence of early malnutrition is great and that the earlier the malnutrition occurs, the more deleterious is the effect on growth. The well known autopsy studies of Winick, Rosso, and Waterlow (18) of Chilean and Jamaican babies who died of malnutrition before two years of age showed that all the children had a subnormal number of brain cells. Parekh, Pherwani, Udani, and Mukherjee (19) found weight deficits in the brains of malnourished Indian children, and concluded that nutritional deficiency had its severest impact during the last three months of gestation or the first three months after birth. Naeye, Diener, and Dellinger (20) noted that stillborn children or those which died within 48 hours of birth and which came from economically poor families, i.e., those whose nutritional intake was inadequate, were smaller than those which came from non-poor environment.

Investigation of possible mechanisms by which the disability might arise have explored (a) the morphologic and physiologic integrity of the placenta as the proximate source of nutrients to the infant, (b) maternal protein-calorie deprivation during pregnancy, and (c) deprivation of the infant.

III. PLACENTAL SUFFICIENCY

A. Prenatal Infection

It is well known that prenatal infection, particularly of rubella, causes both prematurity and intrauterine growth retardation. During the viremia phase of maternal rubella, the virus reaches the placenta causing a villous placentitis and damage to the vascular endothelium. The persistence of infection with progressive damage to the placental vasculature may affect placental function. Viral infections of the embryo do not produce lesions characteristic for a specific virus, but instead cause developmental disorders in the infant (21).

B. Feto-placental Correlations

The placenta intervenes between mother and fetus *in utero*, and is the proximate source of fetal nutrition. Many have attempted to determine whether or not placental insufficiency is a cause of intrauterine growth retardation (IUGR). "Poor fetal growth could in theory result equally from: (i) conditions affecting the nutrient content of the maternal blood or its supply to the placenta, (ii) poor development, damage to, or specific abnormalities of the placental membrane affecting transport across the placenta, or (iii) disorders of the fetal placental circulation" (22).

Sedlers et al. (23) reported that in a group of infants in which half were Caucasians and half Negroes, those who survived the neonatal period had a mean placental weight of 446 g; for 272 infants who died neonatally the mean placental weight was 321 g; and for 234 stillborn the mean was 316 g. Racial differences, however, were only of the order of 24 g.

Although it is not certain whether placental size influences infant weight or vice versa, many have studied feto-placental weight ratios. Younoszai and Haworth (24) compared the birth weight: placental weight ratios of normal term, pre-term and IUGR infants. Normal fetuses weighed 7.9 times the placenta and IUGR infants 7.8. The ratio for pre-term infants was 6.2, significantly lower than for the other two groups. The exact values of the ratios have little meaning because different investigators discard different amounts in preparing the placenta for weighing. The similarity in ratios between the growth retarded (those whose birth weights were below the 10th percentile for their gestational age on the Lubchenco growth charts) and normals (between the 10th and 90th percentiles) is, as one might expect, whether either the fetus or the placenta was the controlling factor. The IUGR placentas weighed 311 g whereas the placentas for the normal infants weighed an average of 420 g. What is surprising in this study is the great range of ratios that are possible in either group. The lowest ratio of the normal term infant was approximately 4.6 and the highest was 11.7. Comparable values for the IUGR group were about 5.5 to 11.6. The range of ratios suggests that the materno-fetal coupling is very loose. Perhaps, because of this looseness we cannot expect to attribute birth deficiencies directly to placental defects.

The analysis of the role of the placental size to its functional capacity is complicated further by the possibility that the placentas of dysmature babies may be chemically different, but not smaller, than those of normal babies. Although the blood-freeweight of the placentas of dysmature babies has approximately the same relationship to the weight of the baby as do those of well-grown babies, dysmature placentas contain a greater proportion of insoluble and presumably metabolically inert protein (25).

C. Experimental Studies

RODENTS: Studies have been conducted to examine the effects of experimentally produced placental insufficiency on the growth and development of the fetus in rats, sheep, and monkeys. Wigglesworth (26) ligated the artery of one uterine horn of a rat during the latter part of pregnancy. The infants were stunted at birth and they showed organ changes characteristic of small malnourished babies. They had large well-developed brains and small livers, and the histological maturity of the organs was in accordance with gestational age. The ischemia produced defective fetuses in the neighborhood of the ligature, indicating that availability of oxygen and nutrients both to the placenta and to the fetus were compromised, adversely affecting the fetus.

PRIMATES: At 41 days the normal rhesus monkey fetus weighs about 1 g but the placenta weights between 4.5 and 6.5 g, more than five times the fetal weight. At the end of gestation, however, the fetus weighs four times the placenta (27). Simple feto-placental ratios are not entirely revealing because at 40 days gestation the fetus is about 92% water whereas at term it is about 75% water. The placenta contains 86% water at 40 days and 82% at 80 days. Thereafter the decline in water content is negligible.

Kerr et al. (28), interrupted pregnancies at 50, 75, 100, 150, and 175 days after conception. Although it is true that prematurely born human fetuses cannot be considered "normal" there is, nevertheless, remarkable similarity in the rates of growth of monkey and human fetal orangs and of the fetal body weight at varying percentages of pregnancy completed. Feto-placental weight ratios are quite similar in the two species.

To investigate the question of whether or not there is a decline in placental tissue with respect to the fetus during the last trimester as potential explanation of intrauterine growth retardation, I have converted the data of these two studies into the ratios of the dry fetus to the dry placenta tissue. Results are shown in Figure 1. The placenta does

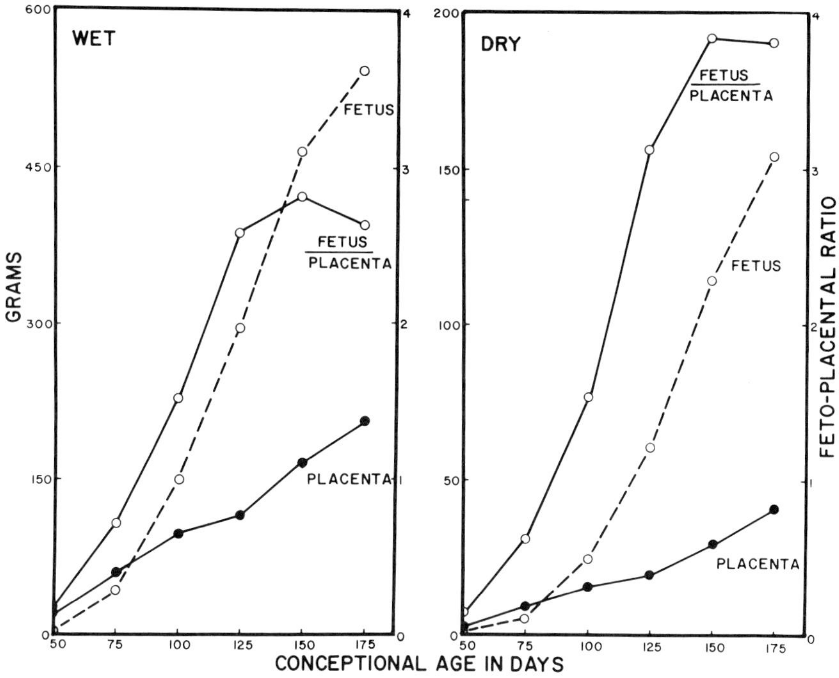

Fig. 1. Growth of fetus and of placenta during gestation based on both wet and dry weights of rhesus placenta.

not cease to grow before the fetus does; instead, the body weight of the rhesus monkey actually grows less rapidly than does the placenta during the last 50 days of pregnancy, opposite to what it was earlier in pregnancy. The notion that the infant outgrows the placenta seems less valid than earlier thought.

Myers et al. (29), ligated the fetal umbilical vessels leading to the secondary placental disc at the 100th gestational day in rhesus monkeys, isolating it from the fetus. Although there was high post-operative interruption of the pregnancy, seven of 13 animals reached 150-160 days of gestation and were delivered by cesarean section. The seven smallest of the 13 experimental animals weighed an average of 312 g at delivery in comparison with 475 g for the control animals that were sham operated and delivered at the same time. Thus, there was about a 1/3 reduction in body weight in half the animals. The brain, adrenals, kidney, and pituitary glands of these animals were the least affected and the lungs, pancreas, liver, and spleen were the most affected. The chemical compositions of the brain, liver, muscle, and carcass of these animals were examined by Hill et al. (30) who found that total DNA in the cerebellum of the IUGR group was low. Total amounts of RNA, protein, and water were low, while the concentrations were unchanged in both the cerebrum and cerebellum. The protein-DNA ratio was normal. Total amounts of DNA, RNA, protein and glycogen in the liver were low in the IUGR group.

Portman et al. (31) repeated the experiment involving ligation of the bridging vessels between the two discs of the placenta at day 90. All of their nine experimental animals survived to day 155, when they were delivered by cesarean section, but three of the nine died within the first two weeks. The others were kept for 40 days, after which they were sacrificed for analysis. Additionally, the infants from the mothers with the placental ligations were fed a low-protein formula (3.6% of energy as protein). In the placental ligation group there were significant reductions in body weight, total brain weight and the weights of brain stems and cerebellum but no significant reductions in total content of DNA, protein, or any of the lipid constituents of the cerebellum or cerebrum. They had significantly reduced quantities of glycolipids in the brain stem. We cannot ascribe all of the reduction of glycolipids to the prenatal treatment because the infants were malnourished for the first 40 days of postnatal life. This study does emphasize the hardiness of DNA and the possible functional reserve of the placenta in primates.

IV. PRENATAL DEPRIVATION

Interspecific comparisons of prenatally or of postnatally deprived animals cannot be made without running some risk of illegitimacy of comparison. The illegitimacy arises because the act of birth occurs at different points in fetal development in different species. For purposes of the present discussion, our primary interest is in man and the usefulness of other animals in providing information that will help us understand human development. We must defer questions of organ sensitivity at different times. The question immediately before us is, does limitation of the food intake of the pregnant human have adverse consequences for the birth status and development of the infant, not, what stage is the infant in when deprivation occurs?

Ideally, we would like to study the effects of single deficiencies, especially the most likely or the most important ones. A complication arises in studying the consequences of a single deficiency like protein, because we assume that administration of the experimental diets to the animal subject will result in no other deficiencies. It appears that species differ in regard to the response to a single-deficit diet; that is to say, it appears that rats, when fed a diet that contains less than about 10% protein, decrease their food intake so that other deficiencies accompany the protein lack. Some investigators have attempted to pair-feed in attempting to analyze only the protein deficiency, and although this tactic may be partly successful, the protein deficiency is merely added to the caloric deficiency.

RODENTS: Zamenhof, Van Marthens, and Margolis (32) maintained rats on an 8%- or a 27%- protein diet. The brain of the offspring of the low-protein mothers contained significantly less DNA and protein compared to the progeny of the well-fed females. There were fewer cells and less protein in each cell. It is generally accepted that rat brain cells at birth are predominantly neurons, so that the reduced DNA indicates that the dietary restriction resulted in permanent brain-neuron deficiency.

Nehrich and Stuart (33), however, reported that purified low-protein maternal diets caused a reduction in total weight and protein content but not in DNA content (cell number) of the cerebrum of newborn mice.

Considerable additional evidence has been reported to show that prenatal protein deprivation results in offspring that are undersized at birth (34,35,36,37). Naismith (17) is skeptical, however, that the birth weight of the fetus is substantially smaller when the mother consumes a low-protein diet during pregnancy. "...on a low plane of protein the fetus lives as a parasite, the tissues of the fetus have a prior claim on the nutrients circulating in the maternal blood stream" (p.25). "A reduction of 54% in the dietary protein value brought about a fall in mean fetal weight of no more than 11%. There was no reduction in litter size or in the total weight of the litter."

There is evidence from rat studies that reduction in maternal caloric intake has a similarly adverse effect on offspring size (38,39,40). Either caloric restriction or severe protein restriction during gestation can reduce the number of brain cells at birth (41). This is because protein deficiency results when caloric intake is inadequate, for then the protein ingested is consumed as energy rather than for tissue development. A 20% drop in birth weight and persistent stunting was reported for male offspring of mothers fed a diet containing 50% of the calories, but no reduction in, protein (42).

Cortical dimensions (thickness at several positions and within an area) were significantly reduced in offspring of rats that were maintained on calorie deficient diets. The reduction in cortical thickness was approximately double that expected from the reduction in cerebral weight, which suggests that the cerebral cortex itself is more affected by such malnutrition than the cerebrum as a whole. These deficiencies were most pronounced at birth and seemed to be somewhat recoverable with adequate nutrition (43).

Cowley and Griesel (44) studied intergenerational effects. They fed albino rats a low-protein diet from 21 days of age throughout their first pregnancy. Their offspring were also maintained on a low-protein diet. Retardation of growth was observed in both parent and first-generation experimental groups. Mortality rates were high in the offspring of the low-protein group. No differences were seen in exploratory behavior during the testing between the two parental groups. The offspring of the low-protein group, however, scored lower on the Hebb-Williams test of "intelligence."

Although the data are not entirely confirmatory, most reviewers of the literature concerning rodents would agree that on many measures of physical or behavioral integrity the offspring of rats fed diets during pregnancy that are deficient in proteins, calories, or both, are themselves deficient in a number of respects. Diets containing 10 to 15% protein have sometimes produced deficits in infant rats. It would appear that it is indeed possible to deprive the offspring by depriving the mother rat of protein or of calories.

PRIMATES: The picture in primates is significantly different. Only a few studies have been conducted of prenatal deprivation, and those that have were concerned primarily with protein deprivation. The bulk of the published data on this topic have come from ouw our laboratory. Our investigations began with an analysis of the protein utilization of nonpregnant adult female monkeys (45). Isocaloric diets containing 13.4%, 6.7%, 3.35%, and 1.68% were fed. Significant weight losses occurred in monkeys fed the three lowest protein concentrations despite continued high caloric intake. Upon return to the high-protein diet, weight recovery was rapid. When the experiment was repeated in pregnant monkeys with the 13.4%, 6.7%, and 3.35% protein diets, the animals fed the diet highest in protein ate more than the animals fed the intermediate- and the lowest-protein diets. Weight gains during pregnancy paralleled the differential food intakes so that when food consumption was analyzed relative to body weight it was noted that the animals on the highest-protein diet consumed slightly less than the other two groups (46). Even the animals fed the low-protein diets gained weight during pregnancy on diets which caused non-pregnant animals to lose weight. Furthermore, when all animals were placed on a high protein diet after delivery, the animals fed the low-protein diet during pregnancy did not gain weight as did the non-pregnant animals when returned to the high-protein diet (47). Fetal wastage was slightly, but not significantly, increased in the low-protein group (48), but the infants of the three groups weighed the same at birth and were of the same skeletal size and maturity (49). Subsequent growth (50) and behavior of the infants (51,52) were equal among the groups. It was concluded that within the range of protein concentrations used (3.35%, equivalent to about 1 g protein/kg of body weight per day, to 13.4%), efficiency of the pregnant mothers' metabolism increases sufficiently to enable the infant to grow normally without evident cost to the mother. Obviously, if the infants were not themselves deprived, then no behavioral, physiological, or anatomical test will differentiate among them.

That the mother gains rather than loses during pregnancy and that she does not deplete her tissues to supply the fetus (53) is a balancing factor against either the parasitism of the fetus or of the mother's jealously guarding her labile reserves at the expense of fetal growth. Hytten and Thomson (54), though they question the validity of balance studies on which Slemons' conclusions (53) was based, agree that under ordinary circumstances it appears valid for humans.

Although only a few studies with primates have been conducted on protein deprivation during pregnancy, and they differ in detail, there is a general concensus that dietary protein concentration below that found even in most preindustrial economies leaves the infant with little or no impairment. Cheek et al. (55) found that infants of mothers fed diets affording 1.2 g of protein/kg-day (about equal to that of Riopelle's lowest group) weighed an average of about 4% less than the control group infants when delivered by cesarean section at 157 days of gestation. Analyses of the data revealed no satistically significant changes in protein, DNA, RNA, cholesterol, phospholipid, water or cloride space of either tissue. The diet contained approximately 5% of the calories as protein.

Portman and his coworkers (31) conducted a similar experiment but in addition fed the infants a low-protein diet for 5 weeks before sacrifice. At time of cesarean delivery at 160 days gestational age, the mean weight of control infants was 485 g and of infants from protein-deficient mothers was 408 g. In this diet, protein supplied 3.6% of the calories. Portman et al. estimate that the protein intake for the deficient group was approximately 1.2 g/kg/day. At time of sacrifice at 195 days gestational age the infants of the protein-deficient mothers were smaller in body weight, brain weight, cerebrum, cerebellum, and in the brainstem weights. Parellaling the large reduction in the weight of the brain stems of the protein deficient monkeys, the protein and lipid contents were significantly reduced in that part of the brain.

Somewhat more drastic effects were shown by Kohrs' group (56) for pregnant rhesus monkeys fed a standard infant formula or an experimental diet containing only 25% as much protein but made isocaloric with supplemental lactose. The deficient animals received about 0.5 g protein/kg/body weight/day but the experimental animals reduced their food intake about 20% during the latter part of pregnancy. Infants were delivered at the 165th day of pregnancy by cesarean section. The birth weight of the infants from mothers fed the low-protein diet was depressed by about 15%. Maternal and fetal mortality were greater in the low-protein group.

It should be noted that with the exception of those born in our laboratories, the infants have been delivered by cesarean section at some specified number of days after conception. The point is worth noting because small differences in birth weight occurred for these animals. Gestation length is sensitive to maternal protein intake, and monkeys fed a high-protein diet have a considerably shorter gestation period than do those fed low-protein diets (57). Thus, it is not unlikely that some of the differences in birth weight would have been ameliorated had the mothers been allowed to carry their infants to spontaneous delivery.

Nevertheless, it appears likely that maternal intakes less than .5 g of high-quality protein/kg/day, or about 5% of the calories, especially if accompanied by a reduction in caloric intake, will exceed the adaptive capacities of the mother. Diets richer in nutrients than that level, are well within the adaptive capacity of the primate and cause no demonstrable deficit. Diets containing fewer than 5% of the calories as protein are quite atypical of human diets (58).

It is worth considering at this point why low-protein

intakes and probably low-calorie intakes (55) can be withstood by the primate and not by the rodent, the most common laboratory animal. Cheek (59) was the first to doubt whether prenatal protein-calorie restriction in primates would produce serious deleterious effects on the offspring. He believed that the syndrome described for rats may be peculiar to small animals. Primate growth is different from non-primate growth. Payne and Wheeler (60) showed that the metabolic stress of pregnancy for primates, represented by the ratio of total birth weight to the gestation time, is significantly below that of non-primates. The rate of accretion of protein and water for the rat fetus is 25 times higher than the monkey's and 60 times higher than that of the human.

The data for primates as a group fell clearly below the data for non-primates, prominent among which were species of the suborder of hystricomorphs, i.e., rodents comprising the porcupines, cavies, agouties, chinchillas, and their allies. The primates as a group could be distinguished from the other mammals by this analysis. The metabolic stress of pregnancy imposed on the primate mother is much less than for a non-primate of similar weight. Payne and Wheeler (60) suggest that the reduced stress of pregnancy in primates may be a consequence of their adaptation to arboreal life.

That may be a factor, but more important is an exploitation of the long period of fetal development. Consider, for example, the rat. Within 21 days it must produce a litter of perhaps ten pups weighing, with the other products of conception, 25% of the mother's weight. The rhesus monkey has 5.5 months to produce a single infant and other conceptional weight.[2] Several consequences follow: because the rodent produces a large litter, the mother cannot adjust the food supplies to the requirements of each of the individual fetuses, hence some may be undernourished. The rhesus mother with her single infant can engage in a metabolic dialogue with her fetus, adjusting the conditions of the pregnancy to keep the development of the fetus within acceptable bounds. One principal way in which the mother can adjust is to alter the length of her gestation period. Our studies (57) indicate that gestation length can be modified by the sex of the infant, the protein intake of the mother, her conception weight, and the season of the year of the pregnancy. The rate of growth of the infant is also a factor. Its effect is over and above that due to sex.

[2] *New World monkeys have higher fetal:maternal weight ratios. This is probably a reflection of their evolutionary specialization.*

The long gestation period of the monotoccus primate and the short gestation period of the polytoccus rodent represent quite different evolutionary adaptations. Obviously the ideal reproductive strategy of the rodent is to defer mating until the optimal conditions are present. Because of the short gestation period, delivery will likely occur in the same season and in the same nutritional conditions as conception. The primate, however, passes through 5.5 months between conception and delivery, a period spanning two seasons, during at least one of which the food supplied to the mother will be reduced either in quality or in quantity. Most monkeys deliver their babies just before or just after the beginning of the rainy season or spring. The Japanese macaque conceives in September or October and delivers in spring, so that gestation is carried through the winter months. In general, the timing of conceptions in the annual cycle of many monkey populations show some association with decreasing day length, favorable diet, and high levels of precipitation (61). To have evolved such a long gestation period primates must also have evolved ample means to accommodate periods of suboptimal nutrition during pregnancy.

Another relevant factor is the significance of the single versus multiple birth. In general, monotocus animals must produce infants that are physiologically competent at time of delivery to withstand most of the environmental insults they will encounter or the infants will die and the race will terminate. Only few of the infants in a litter of rats need survive in order for the species to continue. We can expect, therefore, that a higher percentage of rodents will be adversely affected by prenatal environmental insults than will be the case for primates. Accordingly, we are not surprised that rats born of protein deprived mothers have poor tooth development and are susceptible to caries (62). The principle holds not only for nutritional deficiency but for other environmental aberrations as well. The selection of the appropriate species for assessment of environmental pollutants, carcinogens, and other toxic agents is one of the crucial problems relating to the extrapolation of the results to human experience.

HUMANS: If our conclusions about the robustness of primates are correct, they ought to hold for human populations as well. There is an inevitable inconclusiveness about comparing feral animals with industrialized populations, which with canned goods and frozen foods can eliminate the seasonal variations in diet. The finding that protein levels as low as 3% during pregnancy may be adequate is paralleled by the report that protein was not limiting for birth weight in the diets of pregnant Guatemalan women (63).

Further nutritional parallels between the monkey and the human come from quite a different source. Wilmsen (64) reviewed the seasonal dietary intakes of Kalahari San in Botswana. These people, like rhesus monkeys in their natural habitat, experience periodic feast-famine conditions, but they are not malnourished. They time conception so that the modal number of births occurs during the months of March and April. The peak months of meat consumption are, however, in April, May, and June, at or after the peak birth season. Eaten most during the latter part of pregnancy are cucumbers and berries, foods which are low in protein. The San, like the monkeys, are pregnant during the period when protein intake is low. Since there is evidence that these people cover a wide variety of ecological niches, one cannot hold any particular environment responsible for the adaptive propensity to deliver the young after a protein-lean season. The point I wish to make is that seasonal fluctuations in diet are not outside of human experience, and such fluctuations in nutriture must be adapted to for survival. No doubt our progenitors encountered the same seasonal fluctuations as did the monkeys and they probably used similar mechanisms in adapting to them.

Finally, there is the often reported but controversial precocity of African infants (65). They are reported to be more vigorous, larger and better coordinated than Europeans. If people living in preindustrial societies are not retarded in development at birth, perhaps the nutritional intake, inadequate by our accepted industrial-society standards, is not in fact deficient.

V. POSTNATAL DEPRIVATION

More than 50 genera of primates are presently alive. Several have been used in studies of specific vitamin and mineral deficiencies but only rhesus monkeys have been used for studies in protein-calorie deficiencies in pregnancy. The situation is a little different for studies of postnatal deprivation, although rhesus monkeys have also been deprived postnatally (66). Boelkins and Hegsted (67) deprived infant squirrel monkeys *(Saimiri sciureus)* of protein. Either reduced calories, protein or both, prevented the squirrel monkeys from gaining weight over the 6 week deprivation period, but by the time the repleted animals were one year old their weights had returned to normal.

Cebus monkeys (mostly *Cebus albifrons*) in the same laboratory were fed a diet low in protein, low in calories, or low in both, from 8 to 28 weeks of age, after which they were returned to the control diet. Weight gain of the calorie-restricted animals was modest and that of the protein-restricted and of the combined protein-and calorie-restricted animals was negligible during the period. The control animals' weight, meanwhile, doubled. The erythrocyte volume of the protein-deficient animals decreased, but their number increased, so that a constant hematocrit was maintained. As expected, total plasma protein concentration dropped. Plasma cholesterol was elevated within the first few weeks of restriction. Oddly enough, calorie-restricted animals showed none of the above changes. Instead, they had elevated albumin and globulin concentrations but the cholesterol concentration was reduced (68). A puzzling finding was the absence of clinical evidence of protein deficiency in the group with the combined protein and calorie restriction (69).

Kerr et al., studied early malnutrition in rhesus monkeys (70). Infant monkeys beginning at one month of age were fed either the control diet Similac at the recommended, or at a 1:1, or at a 1:3 dilution. A fourth group was fed Similac that contained 50% of the normal protein content and the fifth group was fed Similac containing only 25% of the normal protein content. The last two diets were made isocaloric to the others by the addition of lactose. Supplements of sodium, calcium, chloride, potassium, magnesium, phosphorus, iron, and vitamins were all provided to make these formulas comparable to the control diet. Beginning at 7 months of age the animals were again fed the control diet to define any "catch-up" growth.

Animals fed the 1:1 dilution of the control diet doubled the volume of food intake during the experimental period, maintaining a constant inflow of nutrients. Their body weights remained within the normal range. Dilution of the diet to 1/3 resulted in a tripling of intake. The increase in food intake is truly remarkable when one considers that the animals ingested about 120% of their body volume each day. Reducing the dietary protein by 1/2 had little effect on intake or growth, but reducing it to 1/4 caused a slight increase in intake but a drastic reduction in growth, whether measured by body weight, head circumference, or body length. When the low-protein monkeys were refed the control diet their growth rate increased sharply, even though their intake was no greater than that of the control animals. Mortality was significantly increased in the two extreme diet groups.

After analyzing the biochemical and cytochemical composition of major organs, Kerr, Helmuth, Campbell and El Lozy (71) reported that the animals fed dilute and the low-protein diets suffered growth failure in cerebral hemispheres, lung, liver, kidney and muscle and, with few exceptions, in their total-organ content of water, protein, lipid, glycogen, DNA, and RNA. The reduced organ size and DNA content indicates slowing of normal meiotic processes. Only the animals fed the low-protein diet showed a reduction from controls in total lipid content of the cerebral hemisphere. The cerebral lipid content of animals fed the dilute diet varied between 7 and 10% below the mean value for the control animals whereas that in brains of the animals fed the low protein diet varied from 13 to 28% the control value. The most significant decrease occurred in the cerebrosides, the lipids characteristic of myelin (72). These data show that monkeys, not surprisingly, are susceptible to cellular and lipid defects if the nutritional intake is deficient in protein or is greatly diluted.

VI PRELIMINARY APPRAISAL OF THE CELLULAR HYPOTHESIS

The initial motivation, for the conduct of the studies cited above came from the notion that malnutrition during the period of cell replication or of maximum growth rate would lead to permanent damage in the infant because cell proliferation would take place. Unless there is considerable distortion in the literature due to the failure of investigators who find an absence of cellular loss to submit or to have their work accepted for publication, the published data imply partial broad substantiation of the first half of the Winick hypothesis. A number of rat studies have shown that by depriving the mother during pregnancy the offspring have a decreased DNA content in the brain and other organs. Although the same does not occur with moderate protein deficiency in primates, it appears that ligating the placental vessels may produce the same result in monkeys as well.

Because investigators chose to elaborate this theoretical position rather than to challenge it with critical experiments, we do not yet have enough data to demonstrate that the cell replication period is in fact determined solely by genetic constraint, that inanition does not prolong the period of potential cell proliferation, or that subsequent timely rehabilitation will not stimulate development of the missing cells. On the one hand, Winick and Noble (73), Guthrie and Brown (74), and Culley and Lineberger (75) showed that early malnutrition resulted in permanent decrease in brain size and DNA content despite rehabilitation. On the other hand, Rajalakshmi, Ali, and Ramakrishman (76) found that undernutrition from birth until 28 days of age did not affect rat brain size or its DNA protein content, and Benton et al. (77) noted that the effects of undernutrition on rat brain size, cerebral size, and its phospholipid and cholesterol content could be overcome if *ad libitum* feeding was given from the third to the sixth weeks of age. Moreover the finding that the increase in brain DNA for male rats requires less time than it does for female rats suggests that hormones may play a role in determining the length of the period during which cell proliferation can take place. Balazs et al. (78) assert that postnatal cell formation in the brain is influenced by hormones: myotonic activity is inhibited by cortical steroids; active cell replication is prematurely terminated in animals treated with relatively large doses of thyroid hormones; and in thyroid deficiency the rate of cell acquisition is selectively retarded in the cerebellum. Final cell numbers, however, are similar to normal because of the prolongation of the period of intensive cell proliferation. With respect to muscle, Cheek et al. (79), has summarized the information relating to the DNA unit which indicates considerable DNA recovery potential. Christensen and Crampton (80), for example, showed that one-year-old rats forced to exercise had more muscle DNA than controls.

It would appear, then, that although considerable data are available to show that DNA in various organs is reduced following early malnutrition, we cannot say yet with certainty that the quantity of DNA in an organ after early malnutrition must stay smaller than normal. It would seem appropriate that in view of the critical theoretical, as well as practical, importance of a finite limit on the period of cell proliferation that considerably more research should be directed towards evaluating this point and to ascertain under what condition, if any, cell proliferation can be accelerated or that cell death can be retarded. Obviously, small laboratory rodents are the experimental animals of choice for these studies.

VII. POSTNATAL DEPRIVATION AND BEHAVIOR

Zimmermann et al. (81) reported on monkeys that were protein deprived after the age of 90 days. From an extensive series of behavior studies conducted while these animals were undergoing nutritional deprivation, Zimmermann concluded that if they had any deficiency, it was in motivation and in attention. The deprived animals could learn many skills as rapidly as normal animals, and the behavioral differences that were seen could just as easily have arisen from the fact that the animals were being tested while deprived as from any permanent deficit in cerebral organization.

A. Behavior After Rehabilitation

For an appraisal of permanent effects of early malnutrition uncomplicated by possible concurrent malnutrition, interest must be directed to the animal's status after repletion. Elias and Samonds (68) rehabilitated nutritionally deprived cebus monkeys by placing them on an adequate diet. At the end of 22 weeks of rehabilitation, weights and long-bone lengths approached normal. When the deprivation period was terminated, physical size was more retarded than behavior was. After rehabilitation, the protein-calorie restricted group had not recovered fully in exploratory behavior although a protein-restricted and a calorie-restricted group had done so.

B. Synergism of Diet and Environment

When Levitsky and Barnes (82) reported that the behavioral deficits usually found in malnourished rats were absent in groups reared in highly stimulating environments, the suggestion was raised that malnutrition, like social isolation, acted to attenuate the assimilation of environmental information which could only be overcome by supplemental stimulation. In a human family situation, then, the malnutrition may reduce the amount of behavioral interaction among the members, further reducing the sensory stimulation of the victim (83).

Studies of the combined nutritional and environmental deprivation or enrichment have been conducted with primates (67,68,84). Typically, such studies involve an "enriched" group and an "isolated group," the enrichment and isolation indicating the amount of interaction that is permitted with other animals and with stimuli in the laboratory. Visual attention to a novel stimulus habituated, i.e., declined, more slowly in squirrel monkeys which were reared in various nutritionally and intellectually non-optimal conditions during infancy than in their control counterparts.

Both dietary and rearing restriction produced stereotyped behaviors such as rocking and self clutching in cebus monkeys (68). The effects of combined nutritional and environmental deprivation were rather complex, for the animals which were deprived of both protein and calories behaved more like the control animals than did either the only-protein-restricted group. Under enriched rearing all groups were moderately active, but among the groups raised in isolation, the most active were the control and the protein-restricted groups.

No simple interpretation emerges from these studies; it is doubtful that one will, considering the experimental complexities. In the first place, both experimental enrichment and isolation are abnormal for the monkey; equally, neither condition represents anything like a human condition, so that extrapolation to the human situation is risky. Second, the isolation conditions are difficult to define and to translate from experiment to experiment and from species to species. Third, and perhaps the most important, complicated experimental designs require many animals; in very few laboratories are they available simultaneously. Consequently, three or four years may be required to complete the groups required in the experimental design; in the meantime the conditions in the laboratory have changed, seasons have come and gone and so have graduate students and caretakers. Despite the best efforts to maintain standardization of conditions, variations in procedures can occur. The usual effect will be to inflate some component of one of the interaction terms in the analysis of the results; that one term may be hailed, erroneously, as evidence for interaction of nutrition and social experience. Finally, the complex factorial designs that involve multiple diet conditions, produce numerous interaction terms in the analysis of variance. Eventhough one such term may achieve statistical significance, it might not correspond to the particular degrees of freedom associated with the most meaningful interpretation. Moreover the most important comparisons may not involve interaction, but simple addition. For example, the combination of an adverse nutritional plane with a disadvantageous rearing condition alone, and each of the two single effects, should be worse than the control condition. But what I have just described is an additive, not an interactive effect. Accordingly, significant interactions should be viewed with caution if not alarm. They have been reported in the literature, but upon examination they differ considerably from experiment to experiment, and most are impossible of straight-forward interpretation.

VIII. NORMAL DEVELOPMENT

The literature on nutrition and development is large. The list of nutrients required in the diet of the young is extensive, and in the long run, each item of the diet can be considered as a potential limiting nutrient. Rather than considering further malnutrition, however, I should like to turn now to the role of nutrition in helping us understand normal development and its underlying mechanism. One particular result has been observed in several species of primates at several laboratories but it has not gotten the attention that it deserves.

A. The Regulation of Infantile Food Intake and Its Consequences

When a human infant goes home from the hospital, a few days after delivery, the zealous mother records the daily intake of the bottlefed infant as a guide for future planning. After a few days she is somewhat more casual about recording it, being occupied with the many new duties that confront her, and she soon forgets to record the amount eaten. Complete records, when available, come from hospitals to which infants are carried when management problems arise. Accordingly, the data from hospitals may best be applicable to abnormal growth.

Rats, in contrast, are usually reared in litters, so individual intakes are unknown. Miller (85), though he hand reared and fed baby rats, keeping adequate records in the process, did so for only five days. Monkeys, as we have seen, are reared individually, and several investigators have kept records of the early growth and food intake of their animals. In 1970, Waisman and Kerr (86) published curves on the daily food intake relative to body weight for the developing normal rhesus monkey. In the same year, Ausman, Hayes, Lage, and Hegsted (87) published comparable curves for the cynomolgus monkey *(Macaca fascicularis)*, the *(Cebus)*, and the squirrel monkey *(Saimiri sciureus)*. Obvious differences between the species of monkeys were noted at birth. The macaques were physically and mentally more advanced and alert at birth and responded to and expected more individual attention. They were independent and explored their environment readily. The cebus and the squirrel monkeys were much slower in their motor development and were relatively unimpressed by the environment around them. Despite these obvious differences in maturity, detailed by Elias (88), a maturational pattern of food intake similar for all the species became evident. Figure 2 shows the number of calories ingested each day for each kilogram of body weight. These curves are redrawn from the reports just cited and from our own data. The amount of food ingested per unit of body weight is highest around 20

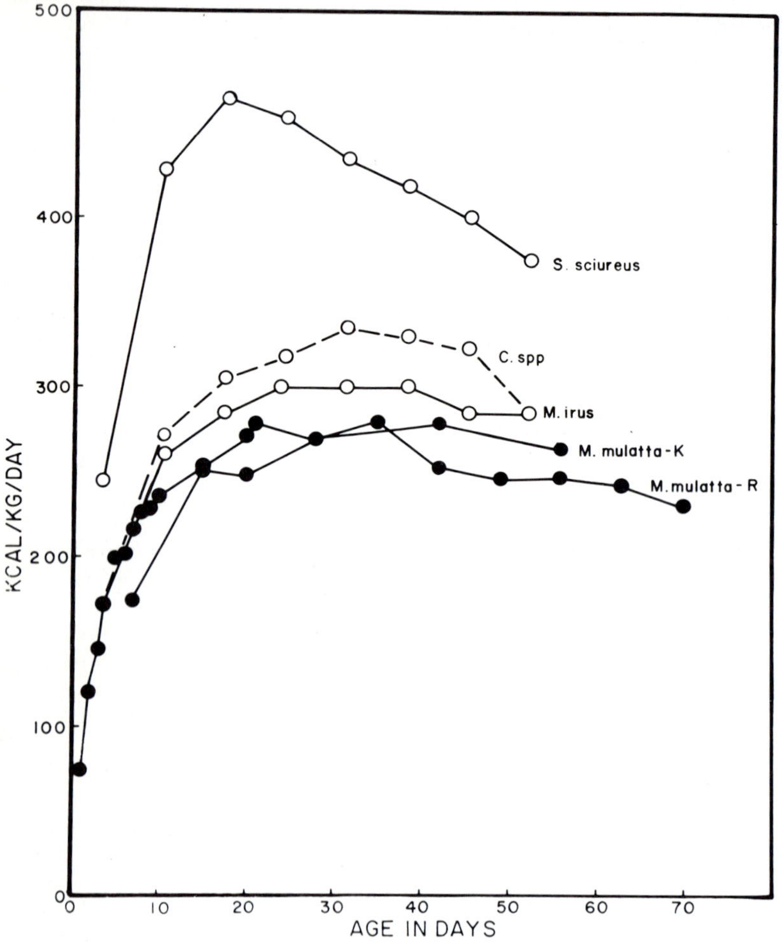

Fig. 2. Food intake of infant primates in relation to body weight.

days of age in the squirrel monkey and between 25 and 40 days in the other species. The peaks of the curves for individual animals probably are much sharper than shown here for grouped data because individual peaks might differ by several days. The phenomenon seems to be rather general among primates, and it is unfortunate that Miller stopped feeding his rats after five days while his animals were still on the left (ascending) side of the curve. He did not have a chance to observe if there was a subsequent decline in the intake relative to body weight.

The peak may simply be an artifact that indicates that the stomach enlarges rapidly for the first few weeks of life and then stays relatively steady, the rest of the body, meanwhile, increasing in size so that the ratio of intake to body size declines. If that were the only relevant fact, the phenomenon would be of very little interest. However, Figure 3, for rhesus monkeys only, which shows weight gain in relation to kilocalories ingested, reveals a similar though somewhat earlier maximum. At that time the animals are gaining most rapidly relative to the amount consumed. In other words, the food efficiency seems to be highest during the second and third weeks of life for rhesus monkeys. The fact that these two phenomena appear in different species and in different laboratories and that the peaks come at a defined time of life, suggests, first that this is a highly regulated phenomenon associated, perhaps, with the emergence of the functional adequacy of some endocrine organ releasing a hormone or with the maturation development of the brain which controls and regulates metabolism and metabolic efficiency, and secondly that these measures may be useful for the assessment of normal and abnormal growth reflecting, perhaps, normal or abnormal utilization. Little is known of the factors controlling feeding efficiency.

IX. OVERALL CONCLUSION

The possibility that protein malnutrition, thought to afflict many in the world today, might be a factor directly affecting the functional integrity of the organism and in particular the central nervous system, has led to a series of studies involving rats and monkeys on the effects of different kinds of malnutrition alone and in combination with a variety of social circumstances. The effects described by Winick for rats, particularly the reduction in cell number as measured by quantity of DNA, has under certain circumstances

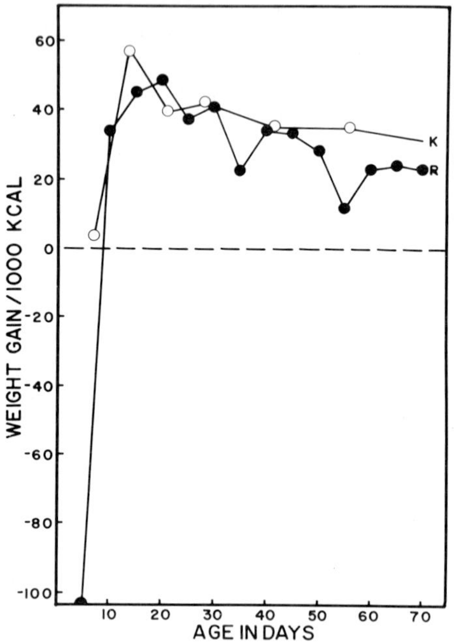

Fig. 3. Weight gain of rhesus monkeys per 1000 kcal of food intake. Data from Kerr (K) and Riopelle (R).

been corroborated by the data from the primate laboratory. The degree of nutritional deprivation required to produce these deficits, particularly during the prenatal period, suggests that the monkeys are considerably more robust than previously thought, and they have probably evolved ways of coping with a period of undernutrition during the long gestation period. The mechanisms involved in producing the damage to the offspring and the infant are probably the same in primates and in rodents, but the levels of deprivation required to produce the effects are sufficiently different between them to suggest that the magnitude of the human problem is not as great as originally anticipated.

Studies on postnatal deprivation, which at first would seem simpler and capable of producing a more readily understandable result, have in fact not done so. Attempts to conduct complex experiments with multiple groups using monkeys have their own built-in complications due to the relative unavailability of large numbers of monkeys at one time. This virtue should be exploited in the rat.

One caveat should be mentioned. None of the experimental studies duplicates the conditions of life-long impoverishment and malnutrition that is experienced by many humans. Whereas we might feel confident that primates, including man, may be more robust than rodents, we do not know the consequences of prolonged malnutrition on the viability or the capability of the adult or of its offspring living under similar circumstances. This review has emphasized nutritional effects; the role of hunger is yet unexplored.

REFERENCES

1. Portman, O.W. (1970) Nutrient requirements of monkeys. Washington, D.C.: National Academy of Sciences-National Research Council Publication 990.
2. Greenberg, L.D. (1970) Nutritional requirements of macaque monkeys. In Harris, R.S., Ed., Feeding and nutrition of non-human primates. New York: Academic Press, pp. 45-64.
3. Kerr, G.R. (1972) Nutritional requirements of subhuman primates. *Physiological Reviews, 52: 415-467.*
4. Day, P.L. (1944) The nutritional requirements of primates other than man. *Vitamin and Minerals, Vol. 2, 71-105.*
5. National Academy of Sciences - National Research Council. (1978) Nutrient requirements of nonhuman primates. No. 4. Washington, D.C.: National Academy of Sciences, 1-83.
6. F.A.O. Third World Food Survey (Freedom from Hunger Campaign Basic Study II), Rowe, 1963, p. 51.
7. Gruenwald, P. (1964) Infants of low birthweight among 5,000 delivered. *Pediatr., 34: 157-162.*
8. Drillien, C.M. (1968) Studies of mental handicap, II: Some obstetric factors of possible aetiological significance, *Arch. Dis. Childh., 43: 283-294.*
9. Caputo, D.V., and Mandell, W. (1970) Consequences of low birthweight, *Devel. Psychol. 3: 363-383.*
10. Stoch, M.B., and Smythe, P.M. (1963) Does undernutrition during infancy inhibit brain growth and subsequent intellectual development? *Arch. Dis. Childh., 38: 546-552.*
11. Willerman, L., and Churchill, J.A. (1967) Intelligence and birth weight in identical twins. *Child Dev. 38: 623-629.*

12. Keys, A., Brozek, J., Henschel, A., Mickelsen, O., and Taylor, H.L. (1950) The Biology of Human Starvation, Vol. 1, Vol. 2. Minneapolis: The University of Minnesota Press, 1-1385.
13. Davison, A.N., and Dobbing, J. (1968) The developing brain. In Davison and Dobbing; Applied Neurochemistry, Oxford: Blackwell, 253-286.
14. Dobbing, J. (1968) Vulnerable period in developing brain. In A.M. Davison, and J. Dobbing, Applied Neurochemistry, Oxford: Blackwell, 287-316.
15. Winick, N. (1968) Nutrition and cell growth. *Nutr. Rev. 26: 195-197.*
16. Enesco, M., and LeBlond, C.P. (1962) Increase in cell number as a factor in the growth of the organs and the tissues of the young male rat. *J. Embrol. Exp. Morphol. 10: 530-561.*
17. Naismith, D.J. (1969) The fetus as a parasite. *Proc. Nutr. Soc. 28: 25-31.*
18. Winick, N., Rosso, P., Waterlow, J. (1970) Cellular growth of cerebrum, cerebellum, and brain stem in normal and marasmic children. *Exper. Neurol. 26: 393-400.*
19. Parekh, U.C., Pherwani, A., Udani, P.M., Mukherjee, S. (1970) Brain weight and head circumference in fetus, infant, and children of different nutritional and socio-economic groups, *Indian Ped., 7: 347-358.*
20. Naeye, R.L., Diener, N.M., and Dellinger, W.S. (1969) Urban poverty: Effects on Prenatal nutrition. *Science 166: 1206.*
21. Flamm, H. (1966) Some considerations concerning the pathogenesis of prenatal infections. In H.F. Eichewall, Ed. The prevention of retardation through control of infectious diseases. Washington: U.S. Dept. of Health, Education, & Welfare, pp. 79-87, cited in W.T. Hughes, Infection and Intrauterine Growth Retardation, *Pediatr. Clin. N.A. 1970, 17: 119-124.*
22. Wigglesworth, J.S. (1966) Fetal growth retardation. *Brit. Med. Bull., 22: 13-15.*
23. Sedlers, A., Berends, H., Kim, H.S., Stone, D.F., Weiss, W., Deutschberger, J., and Jackson, E. (1967) The placental weight-birth weight relationship. *Devel. Med. Child Neurol., 9: 160-171.*
24. Younoszani, N.K., Haworth, J.C. (1969) Placental dimensions and relations in pre-term, term, and growth-retarded infants. *Am. J. Obstet. & Gynec. 103: 265-271.*

25. Garrow, J.S., and Hawes, S.F. (1971) The relationship of the size and composition of the human placenta to its functional capacity. *J. Obstetr. Gynaecol. Brit. Commonwealth., 78: 22-28.*
26. Wigglesworth, J.S. (1964) Experimental growth retardation in the fetal rat. *J. Path. Bact., 88: 1-13.*
27. Behrman, R.E., Seeds, A.E., Jr., Battaglia, F.C., Hellegers, A.E., and Bruns, P.D. (1964) The normal changes in mass and water content in fetal monkey and placenta throughout gestation, *J. Pediatr., 65: 38-44.*
28. Kerr, G.R., Kennan, A.L., Waisman, H.A., and Allen, J.R. (1969) Growth and development of the fetal rhesus monkey, I. Physical growth. *Growth 33: 201-213.*
29. Myers, R.E., Hill, D.E., Holt, A.B., Scott, R.E., Mellits, E.D., and Cheek, D.B. (1971) Fetal growth retardation produced by experimental placenta insufficiency in the rhesus monkey, I, Body weight organ size. *Biol. Neonate., 18: 379-394.*
30. Hill, E.E., Myers, R.E., Holt, E.B. Scott, R.E., and Cheek, D.B. (1971) Fetal growth retardation produced by experimental placental insufficiency in the rhesus monkey, II. Chemical composition of the brain, liver, muscle, and carcass. *Biol. Neonate., 19: 68-82.*
31. Portman, O.W., Alexander, M., Neuringer, M., Novy, M., Illingsworth, R., and Uno, H. (1977) The effects of perinatal malnutrition on lipid composition of neural tissues from rhesus monkeys. *J. Nutr. 107: 2228-2235.*
32. Zamenhof, S., VanMarthens, E., and Margolis, F.L. (1968) DNA (cell number) and protein in neonatal brain: Alteration by maternal dietary protein restriction. *Science 160: 322-323.*
33. Nehrich, H., and Stuart, J.A. (1978) The effect of prenatal protein restriction on the developing mouse cerebrum. *J. Nutr. 108: 368-372.*
34. Zeman, F.J., and Stanbrough, E.C. (1969) Effect of maternal deficiency on cellular development in the fetal rat. *J. Nutr. 99: 372-382.*
35. Zeman, F.J., Shrader, R.E., and Allen, L.N. (1973) Persistent effects of maternal deficiency in postnatal rats. *Nutr. Repts. Int. 7: 421-436.*
36. Hawrylewicz, E.J., Kissane, J., Blair, W.H., and Hepner, C.A. (1973) Effect of maternal protein nutrition on neonatal lung development and minor chondral function. *Nutr. Repts. Inc. 7: 253-270.*
37. Rosso, P. (1977). Maternal-fetal exchange during protein malnutrition in the rat. Placental transfer of glucose and a non-metabolizable glucose analog. *J. Nutr. 107: 2006-2010.*

38. Chow, B.F., and Lee, C.J. (1964) Effect of dietary restriction of pregnant rats on body weight gain of the offspring. *J. Nutr. 82: 10.*
39. Zamenhof, S., VanMarthens, E., and Gravel, L. (1971) DNA (cell number) in neonatal brain: alteration by maternal dietary calorie restriction. *Nutr. Rept. Int. 4: 269-274.*
40. Smart, J.L., Adlard, B.P.F., and Dobbing, J. (1972) Effect of maternal undernutrition and other factors on birthweight in the rat. *Biol. Neonate, 20: 236-244.*
41. Winick, M. (1972) Neurological correlates of malnutrition in animals and man. In W.M. Moore, M.M. Silverburg, and M.S. Read, eds., Nutrition, growth, and development of North America Indian Children, D.H.E.W. Publication Number (NIH) 72-26, pp. 139-150.
42. Rider, A.A., and Simonson, M. (1973) Effect on rat offspring of maternal diet deficient in calories but not in protein. *Nutr. Repts. Int. 7: 361-370.*
43. Clark, G.M., Zamenhof, S., Van Marthens, E., Gravel, L., and Crugar, L. (1973) The effect of prenatal malnutrition on dimensions of cerebral cortex. *Brain Research 54: 397-402.*
44. Cowley, J.J., and Griesel, R.D. (1966) The effect of growth and behavior on rehabilitating first and second generation low protein rats. *An. Behav. 14: 506-514.*
45. Riopelle, A.J., Hill, C.W., Li, S-C., Wolf, R.H., Seibold, H.R., and Smith, J.L. (1974) Protein deprivation in primates, I. Non-pregnant adult rhesus monkeys. *Am. J. Clin. Nutr. 27: 13-21.*
46. Riopelle, A.J., Hill, C.W., Li, S-C., Wolf, R.H., Seibold, H.R., and Smith, J.L. (1975) Protein deficiency in primates, IV. Pregnant rhesus monkeys. *Am. J. Clin. Nutr. 28: 20-28.*
47. Riopelle, A.J., Shell, W.F. (1978) Protein deprivation in primates, XI. Determinants of weight change during and after pregnancy. *Am. J. Clin. Nutr. 31: 394-400.*
48. Riopelle, A.J., Hill, C.W., and Li, S-C. (1975) Protein deprivation in primates, V. Fetal wastage and neonatal status of infants born of deprived mothers. *Am. J. Clin. Nutr. 28: 983-993.*
49. Riopelle, A.J., Hale, P.A., and Watts, E.S. (1976) Protein deprivation in primates, VII. Determinants of size and skeletal maturity at birth in rhesus monkeys. *Hum. Biol., 48: 203-222.*

50. Riopelle, A.J., and Favret, R. (1977) Protein deprivation in primates, XIII. Growth of infants born of deprived mothers. *Hum. Biol. 43: 331-333.*

51. Riopelle, A.J., Hale, P.A., and Hill, C.W. (1976) Protein deprivation in primates, VIII. Early behavior of progeny. *Dev. Psychobiol. 9: 465-476.*

52. Hillman, N.M., Khalid, S.R., and Riopelle, A.J. (1978) Protein deprivation in primates, X. Test performance of juveniles born of deprived mothers. *Am. J. Clin. Nutr. 31: 388-393.*

53. Slemons, J.M. (1919) The nutrition of the fetus. *Am. J. J. Obstet. Dis. Women 80: 194-210.*

54. Hytten, F.E., and Thomson, A.M. (1970) Maternal Physiological adjustments. Committee on Maternal Nutrition/Food and Nutrition Board, Maternal Nutrition and the Course of Pregnancy. Washington, D.C.: National Academy of Sciences, 41-73.

55. Cheek, D.B., Holt, A.B., London, W.T., Ellenberg, J.A., Hill, D.E., and Sever, J.L. (1976) Nutritional studies in the pregnant rhesus monkey - the effect of protein-calorie or protein deprivation on growth of the fetal brain. *Am. J. Clin. Nutr. 29: 1149-1157.*

56. Kohrs, M.B., Harper, A.E., and Kerr, G.R. (1976) Effects of a low-protein diet during pregnancy of the rhesus monkey: I. Reproductive efficiency. *Am. J. Clin. Nutr. 29: 136-145.*

57. Riopelle, A.J., and Hale, P.A. (1975) Nutritional and environmental factors affecting gestation length in rhesus monkeys. *Am. J. Clin. Nutr. 28: 1170-1176.*

58. Hegsted, D.M. (1977) Safe allowance of protein (Letter). *Am. J. Clin. Nutr. 30: 465-466.*

59. Cheek, D.B. (1973) Brain growth and nucleic acids: the affect of nutritional deprivation. In F. Richardson, Ed., Brain and intelligence: the ecology of child development. Hyattsville, Md.: National Education Press, 237-256.

60. Payne, P.R., and Wheeler, E.F. (1968) Comparative nutrition in pregnancy and lactation. *Proc. Nutr. Sci. 27: 129-138.*

61. Lancaster, J.B., and Lee, R.B. The annual reproductive cycle in monkeys and apes. In DeVore, I. (Ed.) Primate behavior. Field studies of monkeys and apes. New York: Holt, Rinehart and Winston, 486-513.

62. Holloway, P.J., Shaw, J.H., and Sweeney, E.A. (1961) Effects of various sucrose: casein ratios in purified diets on the teeth and supporting structures of rats. *Arch. Orol. Biol. 3: 185-200.*

63. Habicht, J. P., Yarborough, C., and Klein, R.E. (1974) Assessing nutritional status in a field study of malnutrition and mental development: Specifically, sensitivity and congrueity of indices of nutritional status. In Cravioto, J., Hambraeus, L., and Vahlquist, B. Symposium of the Swedish Nutritional Foundation, Vol. 12, Early malnutrition and mental development. *Uppsala: Almqvist and Wiksell, 35-42.*

64. Wilmsen, E.N. (1978) Seasonal effects of dietary intake on Kalahari San. *Fed. Proc., 37: 65-72.*

65. Warren, N. (1972) Infant African precocity. *Psychol. Bull. 78: 353-367.*

66. Zimmermann, R.R., Geist, C.R., and Strobel, D.A. (1975) Behavioral deficiencies in protein-deprived monkeys. In G. Serban, Ed., Nutrition advances in behavioral biology. New York: *Plenum, 33-64.*

67. Boelkins, R.C., and Hegsted, D.M. (1977) Consequences of early protein and energy malnutrition in monkeys. Symposium, Behavioral effects of energy and protein deficits, Washington, D.C., in press.

68. Elias, M.F., and Samonds, K.W. (1977) Protein and calorie malnutrition in infant cebus monkeys: Growth and behavioral development during deprivation and rehabilitation. *Am. J. Clin. Nutr. 30: 355-366.*

69. Fleagle, J.K., Samonds, K.W., and Hegsted, D.M. (1975) Physical growth of cebus monkeys, *Cebus albifrons*, during protein or calorie deficiency. *Am. J. Clin. Nutr., 28: 246-253.*

70. Kerr, G.R., Allen, J.R., Scheffler, G., and Waisman, H.A. (1970) Malnutrition studies in the rhesus monkey. I. Effect on physical growth. *Am. J. Clin. Nutr. 23: 739-748.*

71. Kerr, G.R., Helmuth, A., Campbell, J.A., and El Lozy, M. (1976) Malnutrition studies in the *Macaca mulatta*. V. Effect on biochemical and cytochemical composition of major organs. *Am. J. Clin. Nutr. 29: 868-879.*

72. Kerr, G.R., and Helmuth, A.C. (1973) Malnutrition studies in *Macaca mulatta*. III. Effect on cerebral lipids. *Am. J. Clin. Nutr. 26: 1053-1059.*

73. Winick, M., and Noble, A. (1966) Cellular response in rats during malnutrition at various ages. *J. Nutr. 89: 195-197.*

74. Guthrie, H.A., and Brown, M.L. (1968) Effect of severe undernutrition in early life on growth, brain size and composition in adult rats. *J. Nutr. 94: 419-426.*

75. Culley, W.J., and Lineberger, R.O. (1968) Effect of undernutrition on the size and composition of the rat brain. *J. Nutr. 96: 375-381.*

76. Rajalakshmi, R., Ali, S.Z., and Ramakrishnan, C.V. (1976) Effect of inanition during the neonatal period on discrimination learning and brain biochemistry in the albino rat. *J. Neurochem. 14: 29-34.*

77. Benton, J.W., Moser, H.W., Dodge, T.R., and Carr, S. (1966) Modification of the schedule of myelination in the rat by early nutritional deprivation. *Pediatrics, 38: 801-807.*

78. Balazs, R., Patel, A.J., and Hajós, F. (1975) Factors affecting the biochemical maturation of the brain: Effects of hormones during early life. *Psychoneuroendocrinology 1: 25-36.*

79. Cheek, D.B., Holt, A.B., Hill, D.E., and Talbot, J.L. (1971) Skeletal muscle cell mass and growth: the concepts of the deoxyribonucleic acid unit. *Pediatric Research 5: 312-328.*

80. Christensen, D.A., and Crampton, E.W. (1965) Effects of exercise and diet on nitrogenous constituents in several tissues of adult rats. *J. Nutr. 86: 369-375.*

81. Zimmermann, R.R., Geist, C.R., Strobel, D.A., and Cleveland, T.J. (1974) Attention deficiencies in malnourished monkeys. In J. Cravioto, L. Hambraeus, and B. Vahlquist, Eds. Symposia of the Swedish Nutrition Foundation, XII. *Upsala: Almqvist and Wiksell, 205-209.*

82. Levitsky, D.A., and Barnes, R.H. (1972) Nutrition and environmental interaction in the behavioral development of the rat: Long term effects. *Science 176: 68.*

83. Frankova, S. Interaction between early malnutrition and stimulation in animals. In J. Cravioto, L. Hambraeus, and B. Vahlquist. Symposia of the Swedish Nutrition Foundation, XII. Early Malnutrition and Mental Development. *Uppsala: Almqvist and Wiksell, 205-209.*

84. Goldberger, L.I. (1977) Interaction of nutritional and environmental deprivation in infancy in effecting long term incidental long term deficits in squirrel monkeys. Symposium, Behavioral Effects of Energy and Protein Deficits, Washington, D.C. (in Press).

85. Miller, S.A. (1970) Nutrition in a neonatal development of protein metabolism. *Fed. Proc. 29: 1497-1502.*